Sustainable Carbon Materials
from Hydrothermal Processes

Sustainable Carbon Materials from Hydrothermal Processes

Edited by

MARIA-MAGDALENA TITIRICI

School of Engineering and Materials Science,
Queen Mary, University of London, UK

WILEY

This edition first published 2013
© 2013 John Wiley & Sons, Ltd

Registered office
John Wiley & Sons Ltd, The Atrium, Southern Gate, Chichester, West Sussex, PO19 8SQ, United Kingdom

For details of our global editorial offices, for customer services and for information about how to apply for permission to reuse the copyright material in this book please see our website at www.wiley.com.

Library of Congress Cataloging-in-Publication Data

Sustainable carbon materials from hydrothermal processes / edited by Maria-Magdalena Titirici.
    pages cm
    Summary: "The first book on hydrothermal carbonization (HTC) for the production of sustainable, versatile and functional carbonaceous materials"– Provided by publisher.
    Includes bibliographical references and index.
    ISBN 978-1-119-97539-7 (hardback)
    1. Hydrothermal carbonization.   2. Energy conversion.   I. Titirici, Maria-Magdalena.
    TP156.C3S87 2013
    621.042–dc23

                                                                                                    2013004413

A catalogue record for this book is available from the British Library.

ISBN: 9781119975397

Set in 10/12pt Times by Aptara Inc., New Delhi, India
Printed and bound in Malaysia by Vivar Printing Sdn Bhd

1   2013

Contents

List of Contributors

Niki Baccile, Laboratoire de Chimie de la Matière Condensée de Paris, Collège de France, France

Daniela Baris, Karlsruhe Institute of Technology (KIT), Institute for Catalysis Research and Technology, Germany

Nicole D. Berge, Department of Civil and Environmental Engineering, University of South Carolina, USA

Nicolas Brun, Max Planck Institute of Colloids and Interfaces, Germany

Diego Cazorla-Amorós, Materials Institute and Inorganic Chemistry Department, Universidad de Alicante, Spain

Rezan Demir-Cakan, Department of Chemical Engineering, Gebze Institute of Technology, Turkey

Camillo Falco, Institute for Advanced Sustainability Studies, Earth, Energy and Environment Cluster, Germany

Tim-Patrick Fellinger, Max Planck Institute of Colloids and Interfaces, Germany

Antonio B. Fuertes, National Council for Scientific Research (CSIC), Instituto Nacional del Carbon (INCAR), Spain

Bo Hu, Division of Nanomaterials & Chemistry, Hefei National Laboratory for Physical Sciences at Microscale, Department of Chemistry, University of Science and Technology of China, China

Claudia Kammann, Department of Plant Ecology, Justus-Liebig-University Giessen, Germany

Andrea Kruse, Karlsruhe Institute of Technology (KIT), Institute for Catalysis Research and Technology, Germany, and University Hohenheim, Institute of Agricultural Engineering, Conversion Technology and Life Cycle Assessment of Renewable Resources, Germany

Shiori Kubo, Absorption and Decomposition Technology Research Group, National Institute of Advanced Industrial Science and Technology, Japan

Judy Libra, Department of Technology Assessment and Substance Cycles, Leibniz Institute for Agricultural Engineering Potsdam-Bornim (ATB), Germany

Dolores Lozano-Castelló, Materials Institute and Inorganic Chemistry Department, Universidad de Alicante, Spain

Juan Pablo Marco-Lozar, Materials Institute and Inorganic Chemistry Department, Universidad de Alicante, Spain

Kyoung Ro, USDA-ARS Coastal Plains Soil, Water, and Plant Research Center, USA

Marta Sevilla, National Council for Scientific Research (CSIC), Instituto National del Carbon (INCAR), Spain

Maria-Magdalena Titirici, School of Engineering and Materials Science, Queen Mary, University of London, UK

Nicole Tröger, Karlsruhe Institute of Technology (KIT), Institute for Catalysis Research and Technology, Germany

Hiromitsu Urakami, Max Planck Institute of Colloids and Interfaces, Germany

Jens Weber, Max Planck Institute of Colloids and Interfaces, Colloid Chemistry, Germany

Robin J. White, Institute for Advanced Sustainability Studies, Earth, Energy and Environment Cluster, Germany

Peter Wieczorek, Artec Biotechnologie GmbH, Germany

Stephanie Wohlgemuth, Max Planck Institute of Colloids and Interfaces, Germany

Shu-Hong Yu, Division of Nanomaterials & Chemistry, Hefei National Laboratory for Physical Sciences at Microscale, Department of Chemistry, University of Science and Technology of China, China

Li Zhao, Chinese Academy of Sciences, National Center for Nanoscience and Technology, China

Hai-Zhou Zhu, Division of Nanomaterials & Chemistry, Hefei National Laboratory for Physical Sciences at Microscale, Department of Chemistry, University of Science and Technology of China, China

Preface

To alleviate our dependence on fossil fuels and consequently reduce the risks of completely destroying the planet, humanity seeks novel and sustainable technologies. Scientists have the duty to provide solutions and create new materials without using scarce elements, but to use those precursors generously provided by nature at no cost. Such materials should be able to perform important functions in our modern society.

With regard to applications, carbon has played and will continue to play a very important role. Carbon can take many different forms and, strangely enough, although known since ancient times as a natural product of biomass coalification, today is mostly manufactured using fossil-based precursors. This should no longer be the case as fossil fuels are diminishing at a rapid rate and they are generating huge amounts of CO_2 in the Earth's atmosphere, extinguishing our ecosystem.

New and sustainable carbon materials are therefore of upmost importance. This book presents a novel technology able to produce carbon materials from biomass in water at low temperatures, mimicking the natural process of coal formation (hundreds of millions of years) in the synthetic laboratory (in a few hours), called *hydrothermal carbonization (HTC)*.

The process of HTC was first reported by Bergius in 1913 (Nobel Prize winner) and recently rediscovered as an alternative aqueous solution to modern carbon materials by the scientists working at the Max Planck Institute of Colloids and Interfaces.

Dr. Titirici, the editor of this book, was the scientist in charge of the development of this technology into novel and exiting materials for twenty-first century applications. She was the leader of a group of young and highly motivated researchers (the authors of various chapters of this book) who during a period of only 5 years made hydrothermal carbon technology an important addition to carbon science. HTC is now a well-established and recognized technology with many different products and important applications. All these developments would have not been possible without the support of Professor Markus Antoinetti, the Director of the Max Planck Institute of Colloids and Interfaces, who sustained Dr. Titirici's group with research funding and important scientific discussions.

Chapter 1 offers an overview on the state-of-the-art of various green carbon materials from carbon nanotubes to graphene, activated carbons, Starbon® products, and ionic liquid-derived materials, together with a brief history of the HTC process.

Chapter 2 describes various possibilities for introducing porosity in such HTC-derived materials in a broader context of porous carbon materials in general. These include the use of structural-directing agents such as "soft" or "hard" templates as well as bioinspired approaches to generate porosity.

Chemically activated carbons are described in Chapter 3. The chapter offers a general overview on activated carbons produced from lignocellulosic biomass while the use of hydrothermal carbons as precursors for producing activated carbons is also discussed in detail.

Chapter 4 gives a nice and comprehensive overview on how HTC can be elegantly used to produce valuable carbon hybrid materials for practical application.

Functionalization is a challenging task in carbon science. However, this is not the case in HTC. The low temperatures utilized in preparing hydrothermal carbons allows easy functionalization either in one step or via postfunctionalization. This is described in Chapter 5.

Some clarifications related to the formation mechanism of such HTC-derived materials, their chemical structure, morphological features, and pore properties are provided in Chapter 6. In addition, a comprehensive introduction to the use of ^{13}C solid-state nuclear magnetic resonance applied to biomass-derived carbons as well as practical and theoretical examples on how gas adsorption can be applied to determine the porosity of various carbon materials are also provided.

Maybe one of the most impressive developments of HTC-derived materials is their wide range of applications, often outperforming other fossil-derived carbon nanomaterials. Therefore, Chapter 7 is the most extensive of this book. For most of these applications, a brief state-of-the-art is provided. Topics such as renewable energy (rechargeable batteries, supercapacitors), electrocatalysis (fuel cells), heterogeneous catalysis, photocatalysis, gas storage, water purification, sensors, and medical applications are discussed.

In Chapter 8, the efficiency of HTC to convert unconventional precursors such as municipal waste is discussed. The debate is then switched to other agricultural biomass resources while the perspective of using hydrothermal carbons for soil applications and its impact on water streams and environment are also considered.

Chapter 9 refers to the HTC process from an industrial perspective. The production of large amounts (tonnes) of hydrothermal carbons in a continuous fashion from biomass precursors and their potential utilization is discussed. This chapter also touches on aspects such as hydrothermal gasification and hydrothermal liquefaction of biomass.

The book is directed towards a broad readership, including advanced undergraduate- and graduate-level students in nanotechnology, applied chemistry, and chemical engineering, researchers in carbon science, nanotechnology, pollution control, gas separations, water treatment, and renewable energy, scientists working in the field of biomedical applications who might get inspiration for new potential materials as well as biomass investors looking for alternative technologies to convert biomass into useful products.

Maria-Magdalena Titirici
School of Materials Science and Engineering
Queen Mary University of London
February 2013

1

Green Carbon

Maria-Magdalena Titirici
School of Engineering and Materials Science, Queen Mary, University of London, UK

1.1 Introduction

In the early part of the twentieth century, many industrialized materials such as solvents, fuels, synthetic fibers, and chemical products were made from plant/crop-based resources (Figure 1.1) [1, 2]. Unfortunately, this is no longer the case and most of today's industrial materials, including fuels, polymers, chemicals, carbons, pharmaceuticals, packing, construction, and many others, are being manufactured from fossil-based resources. Humankind is still living mentally in a world where petroleum resources have absolute power. However, crude oil resources are rapidly diminishing. It is predicted that this will lead to serious conflicts in the world related to its distribution and control. What it is of even more concern is that essentially such fossil-fuel-derived products eventually end up as CO_2 in the Earth's atmosphere. Several important findings from climate research have been confirmed in recent decades and have finally been accepted as facts by the scientific community. These include a rapid increase in the CO_2 concentrations in the atmosphere during the last 150 years, from 228 ppm to the 2007 level of 383 ppm [3]. This increase is our own fault and is due to the burning of fossil fuels.

What will the world look like in 2050? It is believed that if we continue relying on fossil fuels, we may face an ecological collapse of unprecedented scale due to the degradation of natural capital and loss in ecosystem services. However, we have the capability to reverse this dark and warring perspective of an ecological fiasco, and shape a future where we can live in harmony with nature. For this to happen, scientists have the most important responsibility and joint efforts from multidisciplinary scientific fields are of upmost importance to achieve this goal.

Sustainable Carbon Materials from Hydrothermal Processes, First Edition. Edited by Maria-Magdalena Titirici.
© 2013 John Wiley & Sons, Ltd. Published 2013 by John Wiley & Sons, Ltd.

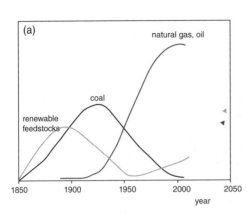

Figure 1.1 *(a) Raw materials basis of the chemical industry in an historical perspective. (Reprinted with permission from [2]. © 2004 Académie des sciences. Published by Elsevier Masson SAS. All rights reserved.) (b) View on sustainable materials for a sustainable future. (Reprinted with permission from [4]. © 2012 Materials Research Society.)*

One of the most important issues is the production of renewable energy to cure our addiction to oil. Solar and wind energy are expected to play the most important roles in the future. Available solar and wind energy depends strongly on geography and local climate, and varies greatly with season, time of day, and weather. This creates additional subsidiary challenges of cost-efficient energy storage and transportation. This requires high-performance materials in smart grids, batteries, fuel cells, solar cells, and gas storage or efficient catalysts to convert renewable resources in transportation fuels.

The paradigm shift from petroleum hydrocarbons to bio-based feedstock provides remarkable opportunities for the chemical processing industry and enables production of sustainable materials capable of performing the above-mentioned functions strongly linked with a sustainable future [4].

Nature offers an abundance of opportunities for shaping structural and functional materials in its wide variety of raw materials, including carbohydrates, nucleotides, and proteins. In this respect, Koopman *et al.* emphasized the importance of developing new starting materials from biomass from an industrial point of view [5]. Biomass is the most abundant renewable resource on Earth. An approximate estimation of terrestrial biomass growth amounts to 118 billion ton year^{-1}, dried [6]. About 14 billion ton year^{-1} are produced in agricultural cycles and out of this about 12 billion ton year^{-1} are essentially discharged as waste. Obviously, there is enough biomass available at almost no cost to be used in many different ways. Here, we will point out three of them, with the focus on the last one:

- The greatest potential for biomass utilization is the generation of biofuels as a sustainable alternative for transportation with no CO_2 emissions. This can be achieved either by fermentation [7], gasification [8], or catalytic liquefaction [9].
- One aspect of green chemistry refers to the use of biomass to provide alternative starting materials for the production of chemicals, vitamins, pharmaceuticals, colorants, polymers, and surfactants [10]. Industrial white biotechnology highlights the use of

microorganisms to provide the chemicals. It also includes the use of enzyme catalysis to yield pure products and consume less energy [11]. Examples using these techniques include composite materials such as polymeric foams and biodegradable elastomers generated from soybean oil and keratin fibers [12]. Plastics such as polylactic acid [13] along with biomass-based polyethers [14], polyamides [15], and polyurethanes [16, 17] have also been developed. The list of such biomass-derived products, commercially available or under development, is obviously much larger, but is beyond the scope of this book [18].

- Work on the conversion of biomass and municipal waste materials into carbon is still rare, but is a significantly growing research topic. This is not surprising given the enormous potential of carbon to solve many of the challenges associated with sustainable technologies presented in Figure 1.1b.

Carbon (derived from the Latin *carbo* for coal and charcoal) is one of the most widespread and versatile elements in nature, and is responsible for our existence today. Humans have been using carbon since the beginning of our civilization. Carbon exists in nature in different allotrope forms from diamond to graphite and amorphous carbon. With the development of modern technology and the need for better-performing materials, a larger number of new carbon materials with well-defined nanostructures have been synthesized by various physical and chemical processes, such as fullerenes, carbon nanotubes (CNTs), graphitic onions, carbon coils, carbon fibers, and others. Carbon materials have been recognized with major awards 3 times in the last 17 years: fullerenes (1996 Nobel Prize in Chemistry), CNTs (2008 Kavli Prize in Nanoscience), and graphene (2010 Nobel Prize in Physics). To date, it is probably fair to say that researchers on carbon materials are encountering the most rapid period of development, which we would like to call the "Back to Black" period.

Despite its widespread and natural occurrence on Earth, carbon has been mainly synthesized from fossil-based precursors with sophisticated and energy-consuming methodologies, having as a consequence the generation of toxic gases and chemicals. The pressures of an evolving sustainable society are encouraging and developing awareness amongst the materials science community for a need to introduce and develop novel porous media technology in the most benign, resource-efficient manner possible. In particular, the preparation of porous carbon materials from renewable resources is a quickly recognized area, not only in terms of application/economic advantages, but also with regard to a holistic sustainable approach to useful porous media synthesis. Carbon has been created from biomass from the very beginning, throughout the process of coal formation. Nature has mastered the production of carbon from biomass and we only need to translate it into a synthetic process. Therefore, we need now to reinvent the "Green Carbon" period.

Within this first chapter, I will first provide a short overview on the state-of-the-art concerning the production of green carbon materials and then a short history of the hydrothermal carbonization (HTC) technique, which is the main focus of this book.

1.2 Green Carbon Materials

By green carbon, I mean materials that are synthesized from renewable and highly abundant precursors consuming as little energy as possible (e.g., low temperatures), and avoiding the use and generation of toxic and polluting substances. In addition, they should perform important technological tasks. These prerequisites are not trivial to achieve. Below, I will provide some examples from the literature where the synthesis of such materials has been targeted.

1.2.1   CNTs and Graphitic Nanostructures

Many potential applications have been proposed for CNTs, including conductive and high-strength composites; energy storage and energy conversion devices; sensors; field emission displays and radiation sources; hydrogen storage media; and nanometer-sized semiconductor devices, probes, and interconnects [19]. Some of these applications are now realized in products. Others are demonstrated in early-to-advanced devices and one, hydrogen storage, is clouded by controversy. Nanotube cost, polydispersity in nanotube types, and limitations in processing and assembly methods are important barriers for some applications.

The demand for this raw material in the nanotechnology revolution is rising explosively. As this trend continues and nanomaterials become simple commodities, mundane production issues, such as the limitation of available resources, cost of production materials, and amount and cost of energy used in nanomaterial synthesis, will become the key cost drivers and bottlenecks. Many efforts have been made to find simple technologies for the mass production of CNTs at low cost. A review on this topic has been recently published by Dang Sheng Su [20]. For mass production, the catalyst is considered as the key factor for CNT growth. The transition metals (Fe, Co, Ni, V, Mo, La, Pt, and Y) are active for CNT synthesis [21]. Any effective production process that leads to a large reduction in costs will lead to a breakthrough of CNT applications. Investigations into new inexpensive feedstocks as well as more efficient catalyst/support combinations suitable for the mass production of CNTs are required.

In one example, Mount Etna lava was used as a catalyst and support for the synthesis of nanocarbon [22]. The main component is silicon (SiO_2, 48 wt%) and the total amount of iron as Fe_2O_3 is as high as 11 wt%, distributed among silicate phases and Fe–Ti oxides (Figure 1.2). The presence of iron oxide particles in the porous structure of Etna lava (Figure 1.2) makes these materials promising for the growth and immobilization of carbon nanofibers (CNFs). For chemical vapor deposition (CVD) growth (700 °C), the crushed powder was put into a horizontal quartz reactor and reduced with hydrogen prior to CVD treatment. Ethylene was used as a carbon source. A mixture of CNFs and CNTs grown on lava rock was obtained (Figure 1.2), with nanofibers dominating. Transmission electron microscopy (TEM) analysis revealed that the CNFs and CNTs obtained on lava exhibited a graphitic wall structure, but normally did not have a regular tubular or fibrous form. The diameter distribution of the obtained CNTs and CNFs was broad, ranging from a few nanometers to several micrometers.

Although the estimated volume of emitted lava was about $10–11 \times 10^6$ m^3, while the volume of tephra exceeded 20×10^6 m^3, there are still issues associated with the availability of such catalyst. The advantage of using lava, which avoids the wet chemical preparation of an iron catalyst, is challenged by issues such as collection, transportation, and purification that may consume additional energy in the whole process, while a similar amount of energy is also exhausted when alumina is produced on an industrial scale.

In another example, the same group used a special type of red soil from Croatia as a catalyst support for the synthesis of nanocarbons. The composition of soil was a mixture

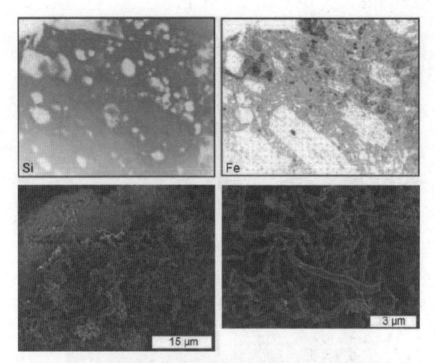

Figure 1.2 *Scanning electron microscopy (SEM) images showing the elemental distribution of silicon and iron in lava stone granulate (top) and CNFs grown on lava (bottom). (Reproduced with permission from [20]. © 2009 WILEY-VCH Verlag GmbH & Co. KGaA, Weinheim.)*

of aluminum, iron, silicon, calcium, and magnesium oxides. Ethylene was used as carbon source for the CNT growth through a CVD process. CNFs grown on the red soil were found to exhibit a broad diameter distribution. The quality of the CNFs was comparable to that produced using lava rock as the catalyst/support as reported above [20].

Endo and his group used garnet sand pulverized from natural garnet stones (Ube Sand Kogyo, US\$1.4 kg^{-1}) as a catalyst and support, and cheap urban household gas (US\$1.1 m^{-3}) as a carbon source for the CVD process [23]. After CVD, the 200-mm granulates of garnet powder (Figure 1.3a) were coarsened to about 400 mm (Figure 1.3b) and were covered with CNTs (Figure 1.3c and d). About 25–30% of the weight from the sand/CNT composite corresponded to the CNTs. The produced CNTs had diameters typically in the range of 20–50 nm and exhibited well-ordered structures with large-diameter hollow cores (Figure 1.3e and f). The graphitization degree of the walls was much higher than that of the CNFs prepared with lava and soil catalysts, and in addition the resulting CNTs could be very easily separated from the garnet sand by simply using an ultrasonic bath in a water suspension.

Other low-cost natural catalysts used in the production of CNTs were bentonite [24], natural minerals such as forsterite, disposide, quartz, magnesite, and brucite [25] or biomass-derived activated carbons previously modified with iron by an impregnation method [26, 27]. The later method resulted in hierarchically structured carbon, consisting of CNFs supported on activated carbon.

Figure 1.3 *(a) Photograph of the garnet sand used to produce CNTs (inset: SEM image showing the average diameter of the sand particles; average size around 200 mm). (b–d) SEM images of the CNTs grown on the surface of the garnet sand particles (in parts b and c, G and T indicate the garnet particle and CNT, respectively; part d corresponds to the CNTs only). (e and f) TEM images showing the central hollow core of a typical as-grown CNT (e), and the highly linear and crystalline lattice of the wall (f). (Reproduced with permission from [23]. © 2008 WILEY-VCH Verlag GmbH & Co. KGaA, Weinheim.)*

In another study, the intrinsic iron content of biomass-derived activated carbons (especially from palm kernel shell, coconut, and wheat straw) was directly used as a catalyst for CNF synthesis [28]. The step involving preparation of iron particles on the activated carbon was circumvented and the overall process was simplified.

So far, only examples of how low-cost and naturally abundant catalysts have been successfully integrated in the production of CNTs have been given. However, the precursors used were gases of fossil fuel origin. The natural materials originating from biomass, such as coal, natural gas, or biomass itself, can be used as a carbon source for nanocarbon synthesis.

The feasibility of producing CNTs and fullerenes from Chinese coals has been investigated [29]. When used as a carbon source, camphor ($C_{10}H_{16}O$; a botanical carbon material) was reported to be a highly efficient CNT precursor requiring an exceptionally low amount of catalyst in a CVD process [30]. CNTs can also be obtained by heating grass in the presence of a suitable amount of oxygen [31]. Fabrication of CNTs with carbohydrates could be expected when all the other possibilities have been tested. It is interesting that the well-known formation mechanism of CNTs (i.e., generating active carbon atomic species followed by assembling them into CNTs) cannot be applied here. Tubular cellulose in grass is directly converted into CNTs during the heat treatment.

With respect to developing different methods other than CVD for CNT production, hydrothermal treatment represents a "greener" solution [32], provided that the precursors also belong to the same category. Calderon Moreno *et al.* used the hydrothermal process to reorganize amorphous carbon at a moderate temperature of 600 °C and a pressure of 100 MPa into nanographitic structures such as nanotubes and nanofibers in the absence of catalysts [33, 34]. (Figure 1.4). High-resolution TEM observations and Raman characterization provided evidence that carbon atoms rearrange to form curved graphitic layers during

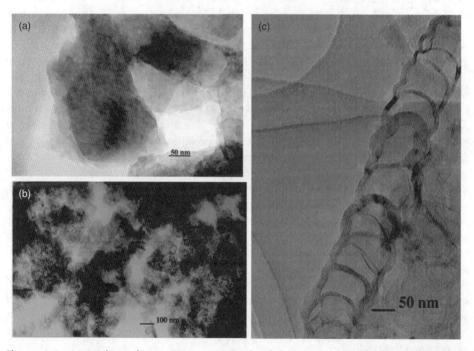

Figure 1.4 (a) High-resolution TEM micrograph of the amorphous carbon particles used as starting material. (b) High-resolution TEM micrograph of the bulk microstructure after hydrothermal treatment, showing the interconnected nanocells formed by curled graphitic walls. (c) A chain of connected cells illustrating how the graphitic carbon forms a single interconnected structure with multiple individual nanocells. (Reproduced with permission from [33]. © 2001 Elsevier.)

hydrothermal treatment. The growth of graphitic multiwall (MW) structures in hydrothermal conditions takes place by different mechanisms than in the gas phase. Hydrothermal conditions provide a catalytic effect caused by the reactivity of hot water that allows the graphitic sheets to growth, move, curl, and reorganize bonds at much lower temperatures than in the vapor phase in inert atmospheres. Such reorganization is induced by the physical tendency to reach a more stable structure with lower energy, by reducing the number of dangling bonds in the graphitic sheets. The mechanism by which amorphous carbon rearranges into curled graphitic cells in the hot hydrothermal fluid is complex and involves the debonding of graphitic clusters from the bulk carbon material in hydrothermal conditions. Closed graphitic lattices can be favored at increasing temperatures or more chemically reactive environments.

An interesting and low-cost approach to high-quality MWCNTs was reported by Pol *et al.* who described a solvent-free process that converts polymer wastes such as low-density and high-density polyethylene into MWCNTs via thermal dissociation in the presence of chemical catalysts (cobalt acetate) in a closed system under autogenic pressure [35]. The readily available used/waste high-density polyethylene is introduced for the fabrication of the MWCNTs. A digital image of such feedstock is shown Figure 1.5a. The grocery bags are extruded from a machine that works in the following manner: for the length of the bag, polyethylene molecules (Figure 1.5a, inset) are arranged in the long chain direction, allowing maximum lengthwise stretch and possessing greater strength. As shown in Figure 1.5b, the MWCNTs grew outwards forming bunches 2–3 mm in size. Each bunch was comprised of hundreds of MWCNTs growing outwards. Under the above-mentioned experimental conditions, polyolefins will reduce to carbon, further producing MWCNTs around the cobalt nanocatalyst obtained from the dissociation of cobalt acetate. The diameter of the MWCNTs was 80 nm and a length of more than 1 μm (Figure 1.5c) was observed within 2 h of the initial reaction time; thus, the growth of MWCNTs is a function of reaction time. A higher percentage of low-density polyethylene was used for making soft, transparent grocery sacks (Figure 1.5d), shrink/stretch films, pond liners, construction materials, and agriculture film. In low-density polyethylene, the molecules of polyethylene are randomly arranged (Figure 1.5d, inset). The as-formed MWCNTs obtained from the thermolysis of waste low-density polyethylene in the presence of cobalt acetate catalyst in a closed system are shown in Figure 1.5e. The MWCNTs were randomly grown during 2 h of reaction time, not analogous to high-density polyethylene. The dissociation of low-density polyethylene with cobalt acetate catalyst also created around 1000 psi pressure. In both cases, the grown MWCNTs were tipped with nanosized metallic cobalt particles. Transmission electron micrographs further confirmed the hollow tubular structures of MWCNTs. The energy dispersive spectroscopy (EDS) (Figure 1.5f) and X-ray diffraction (XRD) pattern (Figure 1.5f, inset) of MWCNTs prepared from the mixture of low-density polyethylene and cobalt acetate confirms that the MWCNTs are comprised of graphitic carbon and trapped cobalt. It needs to be mentioned that in the absence of the catalysts, micrometer-sized hard spheres are obtained instead [36].

Given that polyethylene-based plastics need hundreds of years to degrade in atmospheric conditions and innovative solutions are required for polymer waste, this technology represents a very environmentally friendly and low-cost method to produce CNTs.

Graphitic carbon nanostructures have been synthesized from cellulose by Sevilla via a simple methodology that essentially consists of two steps: (i) hydrothermal treatment of

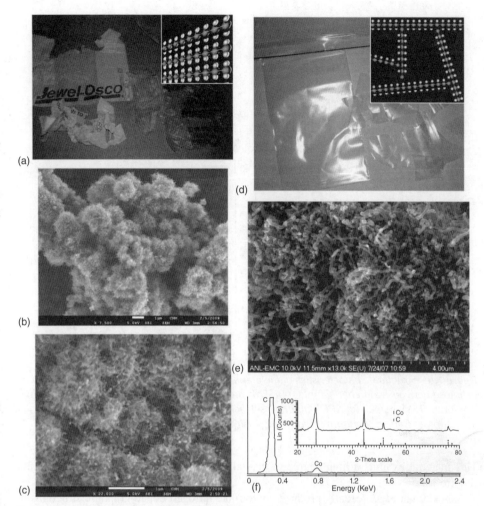

Figure 1.5 *(a) Digital image of high-density polyethylene (inset: arrangement of polyethylene groups) polymer wastes. (b) Field emission (FE)-SEM image. (c) High-resolution SEM image of as-prepared MWCNTs using a mixture of high-density polyethylene and cobalt acetate. (d) Digital image of low-density polyethylene (inset: arrangement of polyethylene groups) polymer wastes. (e) FE-SEM image of MWCNTs prepared from low-density polyethylene. (f) EDS measurements of as-prepared MWCNTs fabricated from low-density polyethylene with cobalt acetate catalyst (inset: powder XRD pattern). (Reproduced with permission from [35]. © 2009 Royal Society of Chemistry.)*

cellulose at 250 °C and (ii) impregnation of the carbonaceous product with a nickel salt followed by thermal treatment at 900 °C [37]. The formation of graphitic carbon nanostructures seems to occur by a dissolution/precipitation mechanism in which amorphous carbon is dissolved in the catalyst nanoparticles and then precipitated as graphitic carbon around the catalyst particles. The subsequent removal of the nickel nanoparticles and amorphous carbon by oxidative treatment leads to graphitic nanostructures with a coil morphology.

Figure 1.6 *Structural characteristics of the graphitic carbon nanostructures obtained from the cellulose-derived hydrochar sample. (a) SEM microphotograph, (b) TEM image (inset: High-resolution TEM image), (c) XRD pattern (inset: selected area electron diffraction pattern), and (d) first-order Raman spectrum. (Reproduced with permission from [37]. © 2010 Elsevier.)*

This material exhibits a high degree of crystallinity and a large, accessible surface area (Figure 1.6).

Ashokkumar *et al.* recently produced onion-like nitrogen-doped graphitic structures by simple high carbonization of collagen – a waste derivative from the leather industry (Figure 1.7). The leather industry generates voluminous amounts of protein wastes at a level of 600 kg ton^{-1} skins/hides processed, as leather processing is primarily associated with purification of a multicomponent skin to obtain a single protein, collagen. This synthetic route from biowaste raw material provides a cost-effective alternative to existing CVD methods for the synthesis of functional nanocarbon materials and presents a sustainable approach to tailor nanocarbons for various applications [38].

To summarize this subsection, some progress has been achieved in the synthesis of CNTs using either natural catalysts and/or natural precursors. Several studies have shown that natural materials can be used for the synthesis of nanomaterials, aimed at developing low-cost, environmentally benign, and resource-saving processes for large-scale production. Catalyst-free CNTs have also been successfully synthesized from amorphous carbon under hydrothermal conditions. The examples provided show promising potential and an interesting perspective on nanocarbon syntheses using these inexpensive resources. Unfortunately, when using such low-cost technologies, uniform diameters and homogeneous

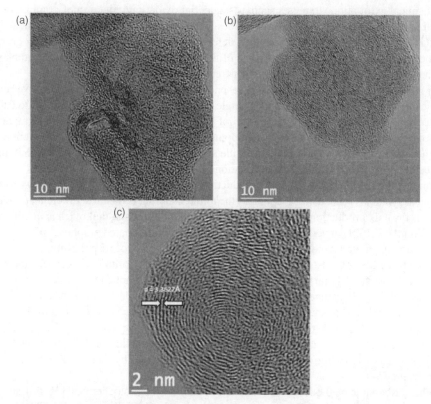

Figure 1.7 High-resolution TEM images of the carbon material derived from collagen waste by treating at 1000 °C for 8 h. (a and b) Polyhedral and spherical onion-like nanostructures showing the presence of graphitic layers with significant defects. (c) Spherical carbon nano-onion structure showing highly defective shells separated by 0.3363 nm. (Reproduced with permission from [38]. © 2012 Royal Society of Chemistry.)

structures are difficult to achieve. Although the investigations were performed to look for a cost-effective method for mass production of CNTs, studies regarding the sustainability of using such natural organic materials are still required.

1.2.2 Graphene, Graphene Oxide, and Highly Reduced Graphene Oxide

Since the award of the Nobel Prize in 2010, graphene has been the new star of carbon science. Graphene is not a new material and it is known to form graphite by parallel stacking, as well as fullerenes and CNTs by rolling into two-dimensional nanostructures. The delay in its discovery as an individual material can be partially attributed to the single-atom-thick nature of the graphene sheet, which was initially thought to be thermodynamically unstable [39]. However, graphene is not only stable, but also exhibits impressive electronic and mechanical properties (charge carrier mobility $= 250\,000\,\text{cm}^2\,\text{V}^{-1}\,\text{s}^{-1}$ at room temperature [40], thermal conductivity $= 5000\,\text{W}\,\text{m}^{-1}\,\text{K}^{-1}$ [41], and mechanical stiffness $= 1\,\text{TPa}$ [42]).

Chemical exfoliation strategies such as sequential oxidation/reduction of graphite often result in a class of graphene-like materials best described as highly reduced graphene oxide (HRG) [43, 44], with graphene domains, defects, and residual oxygen-containing groups on the surface of the sheets. Indeed, none of the currently available methods for graphene production yields bulk quantities of defect-free sheets.

In general, methods for producing graphene and HRG can be classified into five main classes: (i) mechanical exfoliation of a single sheet of graphene from bulk graphite using Scotch tape [45, 46], (ii) epitaxial growth of graphene films [47], (iii) CVD of graphene monolayers [48], (iv) longitudinal "unzipping" of CNTs [41, 49], and (v) reduction of graphene derivatives, such as graphene oxide and graphene fluoride [50, 51], which in turn can be obtained from the chemical exfoliation of graphite.

Among all these methods, chemical reduction of exfoliated graphite oxide (GO), a soft chemical synthesis route using graphite as the initial material, is the most efficient approach towards the bulk production of graphene-based sheets at low cost. Stankovich *et al.* [50, 52] and Wang *et al.* [53] carried out the chemical reduction of exfoliated graphene oxide sheets with hydrazine hydrate and hydroquinone as the reducing agents, respectively.

Since these first reports a significant effort has been made to find greener technologies to reduce exfoliated graphene oxide sheets to defect-free graphene. Xia *et al.* reported an electrochemical method as an effective tool to modify electronic states via adjusting the external power source to change the Fermi energy level of the electrode materials surface. This represents a facile and fast approach to the synthesis of high-quality graphene nanosheets on a large scale by electrochemical reduction of the exfoliated GO at a graphite electrode, while the reaction rate can be accelerated by increasing the reduction temperature [54].

Other sustainable methods for the reduction of GO involve photochemical [55, 56], sugars [57], L-ascorbic acid, iron [58], zinc powder [59], vitamin C [60], microwaves [61], baker's yeast [62], phenols from tea [63], bacteria [64, 65], gelatin [66], supercritical alcohols [67], and others.

Despite all these milder methods towards GO reduction and although it could become an industrially important method to produce graphene, until now the quality of this liquid exfoliated graphene is still lower than mechanically exfoliated graphene due to the destruction of the basal plane structure during the oxidation and incomplete removal of the functional groups. In addition, the oxidation of graphite is a tedious method involving very aggressive substances such as $KMnO_4$, $NaNO_3$, and H_2SO_4.

Recently, many research groups have published several CVD methods for growing large-sized graphene on wafers. For the growth of epitaxial graphene on single-crystal silicon carbide (SiC) [47], the cost of this graphene is high due to the price of the 4H-SiC substrate. Also, metals such as copper [68], nickel [48, 69], iron [70], cobalt [71], and platinum [72] have been used as catalytic substrates to grow mono-, bi-, or multilayer graphene. The CVD method is limited to gaseous carbon sources such as methane or acetylene.

The group of Tour has come up with a solution to the use of gas precursors and showed that large-area, high-quality graphene with controllable thickness can be grown from different solid carbon sources (e.g., polymer films or small molecules) deposited on a metal catalyst substrate at temperatures as low as 800 °C. Both pristine graphene and doped graphene were grown with this one-step process using the same experimental setup [73]. The same group expanded this concept to any solid precursor such as waste food and insects (e.g., cookies,

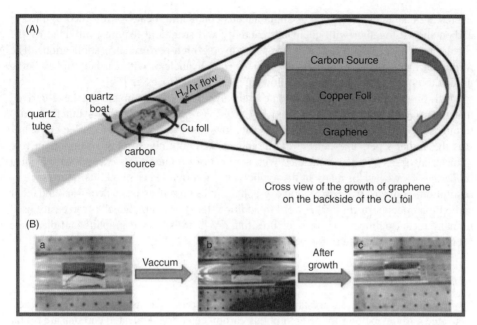

Figure 1.8 *(A) Diagram of the experimental apparatus for the growth of graphene from food, insects, or waste in a tube furnace. On the left, the copper foil with the carbon source contained in a quartz boat is placed at the hot zone of a tube furnace. The growth is performed at 1050 °C under low pressure with a H₂/Ar gas flow. On the right is a cross view that represents the formation of pristine graphene on the backside of the copper substrate. (B) Growth of graphene from a cockroach leg. (a) One roach leg on top of the copper foil. (b) Roach leg under vacuum. (c) Residue from the roach leg after annealing at 1050 °C for 15 min. The pristine graphene grew on the bottom side of the copper film (not shown). (Reproduced with permission from [74]. © 2011 American Chemical Society.)*

chocolate, grass, plastics, roaches, and dog feces) to grow graphene directly on the backside of a copper foil at 1050 °C under H_2/Ar flow (Figure 1.8) [74]. The nonvolatile pyrolyzed species were easily removed by etching away the frontside of the copper. Analysis by Raman spectroscopy, X-ray photoelectron spectroscopy (XPS), ultraviolet (UV)/Vis spectroscopy, and TEM indicates that the monolayer graphene derived from these carbon sources is of high quality. Using this method, low-valued foods and negative-valued solid wastes are successfully transformed into high-valued graphene, which brings new solutions for the recycling of carbon from impure sources.

Hermenegildo Garcia *et al.* showed that chitosan, a nitrogen-containing biopolymer, can form high-quality films on glass, quartz, metals, and other hydrophilic surfaces. Pyrolysis of chitosan films under argon at 800 °C and under an inert atmosphere gives rise to high-quality single-layer nitrogen-doped graphene films (over 99% transmittance) as evidenced by XPS, Raman spectroscopy, and TEM [75].

Ruiz-Hitzky *et al.* demonstrated the possibility of preparing graphene-like materials from natural resources such as sucrose (table sugar) and gelatin assembled to silica-based porous solids without any requirement for reducing agents. The resulting materials show interesting characteristics, such as simultaneous conducting behavior afforded by the sp^2 carbon sheets, together with chemical reactivity and structural features, provided by the silicate backbone, which are of interest for diverse high-performance applications. The formation mechanism of supported graphene is still unclear, with further studies being needed to optimize its preparation following these green processes [76].

Much progress has been made to date in the synthesis of sustainable graphene-derived materials. Given that the field is relatively new it is expected that new synthetic break-throughs are soon to come for the large-scale, low-cost synthesis of defect-free graphene. It is the author's personal believe that graphene will continue to play an important role in materials science when associated with applications related to its exceptional physical properties. However, for many of the applications described later in the literature, such as adsorption, catalysis, or energy storage, graphene in its pure form is not necessary and other carbon materials perform just as well. In addition, the word "graphene" is too easily used in many recent publications for structures that are just disordered graphite and that have been known in the literature for many years.

1.2.3   Activated Carbons

So far we have discussed crystalline forms of carbons such as CNTs and graphenes. Activated carbons belong to the amorphous carbon category. Activated carbons are by far the oldest and most numerous category of carbons prepared from renewable resources. A comprehensive description is behind the scope of this book chapter. Many reviews exist in the literature on this topic [77]. In addition, Chapter 3 provides a very solid introduction to activated carbon with a focus on activated carbons prepared from lignocellulosic materials. Activated carbons are prepared either by chemical or physical activation from biomass or waste precursors at temperatures between 600 and 900 °C. They are microporous and used mainly for adsorption processes (i.e., water purification), and recently in supercapacitors [78] and gas storage [79]. One main disadvantage of activated carbons is the impossibility to predict their resulting porosity and to control their pore properties. I will not say more here about activated carbons, but direct the interested reader to Chapter 3, which is dedicated to activated carbons from biomass and from hydrothermal carbons.

1.2.4   Starbons

The Starbon® technology was developed in the group of Professor James Clark at the University of York and it is based on the transformation of nanostructured forms of polysac-charide biomass into more stable porous carbonaceous forms for high-value applications [80]. This approach opens routes for the production of various differently structured porous materials and presents a green alternative to traditional materials based on templating meth-ods. The principle of this methodology relies on the generation of porous polysaccharide precursors that can then be carbonized to preserve the porous structure.

This material synthesis strategy was initially focused on the use of mesoporous forms of the composite polysaccharide starch (from where the name is derived), but evolved into a generic tunable polysaccharide-based route. The technology involves: (i) native polymer expansion via polysaccharide aqueous gel preparation, (ii) production of solid mesoporous polysaccharide, via solvent exchange/drying, and (iii) thermal carbonization/dehydration.

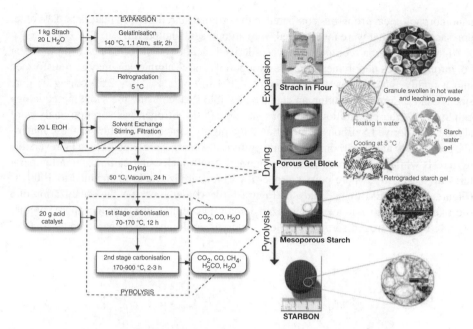

Figure 1.9 *Diagrammatic representation of the main processing steps in the production of starch-derived Starbon materials. (Reproduced with permission from [81]. © 2009 Royal Society of Chemistry.)*

The resulting carbon-based materials are highly porous and mechanically stable in the temperature preparation range from 150 to 1000 °C. At temperatures above 700 °C the carbonization process leads to the synthesis of robust mesoporous carbons with a wide range of technologically important applications, including heterogeneous catalysis, water purification, separation media, as well as potential future applications in energy generation and storage applications.

Starbon material production comprises of three main stages (Figure 1.9). Starch (typically from high amylose corn starch) is transformed into a gel by heating in water. The resulting viscous solution is cooled to 5 °C for typically 1–2 days to yield a porous gel. Water in the gel is then exchanged with the lower surface tension solvent ethanol. The resulting material is then filtered and may be oven-dried to yield a predominantly mesoporous starch with a surface area of typically 180–200 m^2 g^{-1} [82, 83]. In the final stage the mesoporous starch can be doped with a catalytic amount of an organic acid (e.g., *p*-toluenesulfonic acid).

The surface area of the as-prepared materials increased with increasing the carbonization temperature from 293 m^2 g^{-1} at 300 °C up to 600 m^2 g^{-1} to 800 °C. A nice aspect of this technology, similar to hydrothermal carbons, is the fact that the surface polarity and the porosity can be modulated with temperature.

Inspired from systematic studies on the starch system, the same authors investigated the use of other linear polysaccharides in the preparation of second-generation Starbon materials. It was anticipated that the utilization of differing polysaccharide structures and functionality may allow access to materials with differing textural properties and nanomorphological properties compared to the original starch-derived materials. Two other polysaccharides that were investigated were alginic acid and pectin.

Alginic acid is a complex seaweed-derived acidic polysaccharide with a linear polyuronide block copolymer structure. Nonporous native alginic acid may be transformed into a highly mesoporous aerogel ($S_{BET} \sim 320$ m^2 g^{-1}; $V_{meso} \sim 2.50$ cm^3 g^{-1}; pore size around 25 nm), presenting an acidic accessible surface using the same methodology employed for the preparation of porous starches [84]. Nitrogen gas sorption analysis of alginic acid-derived Starbons demonstrated the highly mesoporous nature of these materials, particularly at low carbonization temperatures. Isotherms presented a type IV reversible hysteresis while mesoporous volumes contracted with increasing carbonization temperatures up to 500 °C, where porous properties were stabilized and maintained to 1000 °C (Figure 1.10e). TEM image analysis (Figure 1.10a–d) demonstrates the organization of a rod-like morphology into mesoscale-sized domains, generating the large mesopore volumes observed from nitrogen sorption studies. Materials could also be prepared up to 1000 °C

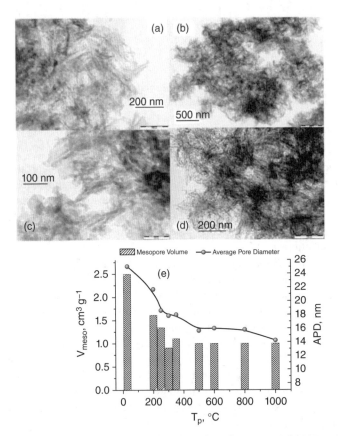

Figure 1.10   *TEM images of alginic acid (AS1)-derived Starbon materials at T_p = (a) 300, (b and c) 500, and (d) 1000°C. (e) Impact of increasing carbonization temperature on the mesoporous properties of alginic acid-derived Starbon materials. (Reproduced with permission from [84]. © 2010 WILEY-VCH Verlag GmbH & Co. KGaA, Weinheim.)*

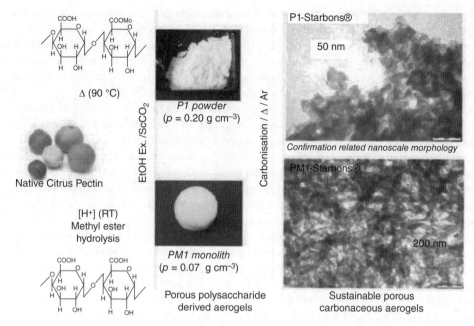

Figure 1.11 *Representation of routes to porous polysaccharide-derived materials from pectin and the corresponding TEM images of pectin-derived Starbon-type materials. (Reproduced with permission from [85]. © 2010 WILEY-VCH Verlag GmbH & Co. KGaA, Weinheim.)*

with no decrease in the quality of the textural properties or alteration in the structural morphology. This approach to the generation of second-generation Starbon materials uses no additive catalyst and simply relies on the decomposition of the acidic polysaccharide itself to initiate the carbonization process.

Pectin, an inexpensive, readily available, and multifunctional polyuronide, occurs as a major cell wall component in land plants. Common commercial sources include fruit skins – a major commercial waste product. Gelation of native citrus pectin in water and subsequent recrystallization yielded a semitransparent gel, which was converted to the porous aerogel via solvent exchange/drying outlined above [85]. Supercritical CO_2 extraction of ethanol was found to yield a low-density ($\rho = 0.20$ g cm^{-1}), highly porous powder aerogel (Figure 1.11, PI-powder). Addition of hydrochloric acid yielded a viscous solution that could be poured into any desired shaped vessel and cured at room temperature to yield a very dimensionally strong gel, which upon water removal (via solvent exchange/supercritical CO_2 drying) yielded extremely-low-density materials ($\rho = 0.07$ g cm^{-3}; Figure 1.11, PMI-monolith).

Direct heating of the pectin aerogels under an inert atmosphere allowed direct access to the carbonaceous materials. Promisingly, the resulting Starbon-type materials prepared from the two different pectin aerogel precursors presented significantly different textural and mesoscale morphologies compared to materials prepared from either alginic acid or acid-doped starch.

Pectin-derived Starbon materials demonstrate the flexibility of this material synthesis approach in terms of textural and morphological properties, and also further exemplify the impact of polysaccharide structure and the metastable gel state, necessary to generate the mesoporosity in these materials. The utilization of pectin provides two extra routes (excitingly here from the same polysaccharide) for the production of second-generation Starbon materials with not only differing porous properties, but also with remarkably variable mesoscale morphology, accessible via a simple change in the gelation route and the resulting chemical modification of polysaccharide structure (Figure 1.11).

Thus, using various polysaccharide precursors from plant material, Clark *et al.* successfully prepared well-defined porous carbon materials with tunable porosity, morphology, and surface groups. This of course opens the doors to many different applications.

Starbon materials have been intensively applied as heterogeneous catalysts. The ability to adjust the surface properties and hydrophobicity/hydrophilicity balance of mesoporous Starbons provides the possibility to achieve highly active, selective, and reusable water-tolerant solid acids. The esters of diacids derived from fermentation processes find use in the manufacture of polymers, fine chemicals, perfumes, plasticizers, and solvents. Thus, using Starbons as solid catalysis modified with sulfonic groups, the researchers from York successfully esterified various substrates (e.g., succinic, fumaric, levulinic, and itaconic acids) in aqueous ethanol, providing high conversions and selectivities to their respective esters [86]. The rates of esterification of diacids (succinic, fumaric, and itaconic) were found to be between 5 and 10 times higher for Starbon acids compared to those of commercial solid acids (e.g., zeolites, sulfated zirconias, acidic clays, and resins) or microporous commercial sulfonated carbons (DARCOs and NORITs).

Starbon acids were also found to have a temperature-dependent optimum of catalytic activity (that could be controlled by the preparation temperature of the parent Starbons and consequently by modification of their surface properties) as well as a substrate-dependent catalytic activity maximum. Starbon acid activities peaked at around 400, 450, and 550 °C for succinic, fumaric, and itaconic acids, respectively, with sharply reduced activities below or above this preparation temperature.

The same sulfonated carbons proved also to be efficient catalysts for preparation of aromatic amides via *N*-acylation of amines under microwave irradiation [87]. Quantitative conversions of starting material were typically achieved in 5–15 min with very high selectivities to the target product, applicable to a wide range of compounds (including aromatic and aliphatic amines), substituents, and acids. Starbons acids provided starkly improved activities compared to other acid catalysts including zeolites, Al-MCM-41, and acidic clays.

The resulting Starbon materials have also been successfully hybridized with various nanoparticles and applied in heterogeneous catalysis for various applications. A comprehensive review on this topic is found in White *et al.* [81].

Another very interesting and important application is the application of the alginic acid-derived materials prepared at 1000 °C as the stationary phase in liquid chromatography for the separation of a mixture of carbohydrates [84]. The separation potential was demonstrated for the representative highly polar neutral sugars glucose (mono-), sucrose (di-), raffinose (tri-), stachyose (tetra-), and verbascose (pentasaccharide). This allowed the generation of designer porous graphitic carbon-type stationary phases, whereby the surface polarity could be moderated by selecting the carbonization temperature employed to control the degree or extent of graphitic structure development.

Starbon technology provides a useful and sustainable route to highly mesoporous carbonaceous materials. Flexibility in terms of preparation temperature provides carbonaceous materials with tunable surface chemistry properties – arguably a material feature not accessible via conventional hard or soft template routes. By selection of gelation conditions, polysaccharide type, and carbonization temperature, a wide range of carbonaceous materials may be synthesized using inexpensive and readily available renewable sugar-based precursors. The drawbacks of this technology are the lack of well-defined pore size, with most of the materials exhibiting broad pore size distributions. In addition, the pore properties are unpredictable as in the case of activated carbons. In principle, this should be overcome by the use of either hard or soft templates (preferably also derived from biomass), which should lead to hierarchically porous materials.

Another disadvantage of this technology is that it is limited to the use of polysaccharides in their pure form, which requires additional isolation and purification from the derived biomass parent material.

1.2.5   Use of Ionic Liquids in the Synthesis of Carbon Materials

Recently, it has been shown that carbon materials can also be obtained by the direct carbonization of some particular ionic liquids. Thus, Paraknowitsch *et al.* designed a set of ionic liquids entirely composed of carbon, nitrogen, and hydrogen atoms using a combination of nitrogen-containing cations (i.e., 1-ethyl-3-methylimidazolium (EMIm) or 3-methyl-1-butylpyridine (3-MPB)) and anions (i.e., dicyanamide (dca)) [88]. Within the same context, Lee *et al.* designed ionic liquids composed of different cations that contained imidazolium groups (not only EMIm, but also 1-butyl-3-methylimidazolium (BMIm) and 1,3-bis(cyanomethyl)imidazolium (BCNIm)) and anions that contain nitrile groups (e.g., $[C(CN)_3]^-$) (Figure 1.12) [89].

The nitrogen-rich character of these ionic liquids allowed, by direct combustion, obtaining nitrogen-doped carbons with remarkable nitrogen contents of up to 18 at% [89]. Interestingly, the authors demonstrated that the carbonization yield depended on the nitrile character of the anions so that the resulting carbon network can be cross-linked via both cations and anions.

This approach based on the use of ionic liquids as carbonaceous precursors was also applied to the preparation of porous carbons with high surface area. Kuhn *et al.* first reported on an ionothermal polymerization method using a molten salt ($ZnCl_2$) and simple aromatic nitriles (e.g., 4,40-dicyanobiphenyl and 4-cyanobiphenyl), which provided carbonaceous polymer networks with well-defined bimodal micro- and (nonperiodic) mesoporosity [90]. Later on, Lee *et al.* used ionic liquids composed of nitrile-functionalized imidazolium-based cations (e.g., $[BCNIm]^+$ or 1-cyanomethyl-3-methylimidazolium $[MCNIm]^+$)) and non-nitrile-functionalized anions (e.g., $[Tf_2N]^-$ and $[beti]^-$), with surface areas of up to 780 m^2 g^{-1} [91]. Anions release was detrimental in terms of both carbonization yields and nitrogen contents.

Textural properties without compromising nitrogen contents can be also obtained with the aid of traditional structure-directing agents. The use of porous silica nanoparticles promoted more than a 10-fold increase in the surface area of the resulting carbons as compared to those obtained in the absence of templates [88] (around 1500 versus 70 m^2 g^{-1}, respectively).

More intriguing is the use of ionic liquids as solvents for the formation of a silica gel from either hydrolysis or condensation of regular orthosilicate precursors (e.g., TEOS) or

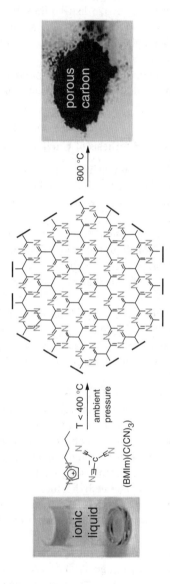

Figure 1.12 Reaction scheme of the trimerization of a nitrile-containing anion, leading to the formation of an extended framework. (Reproduced with permission from [89]. © 2010 WILEY-VCH Verlag GmbH & Co. KGaA, Weinheim.)

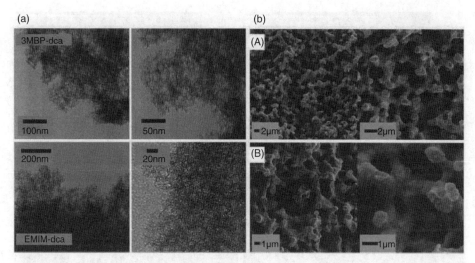

Figure 1.13 *(a) TEM images of LUDOX-templated ionic liquid-derived materials. (b) 3MBP-dca, (bottom) EMIM-dca. (b) SEM images of (A) silica monolith template and (B) nitrogen-doped carbon monolith. (Reproduced with permission from [93]. © 2010 Royal Society of Chemistry.)*

by coagulation of LUDOX®-silica nanoparticles. Upon carbonization of the ionic liquids and subsequent silica dissolution hierarchical porous (with pores at both the nano- and microscale) carbon monoliths could be obtained (Figure 1.13) [88, 92, 93]. The application of this approach to ionic liquids composed of dca (as anion) and a long-alkyl-chain pyridinium derivative (as cation) has recently resulted in the formation of composite (silica- and nitrogen-doped carbon) microparticles with a well-defined mesoporous structure [94].

However the "green" character of burning (carbonizing) ionic liquids is questionable. Even if such materials showed good properties for various applications, especially related to energy storage, the high cost of the precursor does not really pay off.

More valuable methodologies are thus those in which the ionic liquids can be used as a recyclable solvents. The lack of vapor pressure that characterizes ionic liquids is what provides their "green" character and makes them interesting alternatives for replacing highly volatile organic solvents in synthetic processes. In addition, their excellent solubilization properties especially for biomass-derived components should make them very suitable for the production of various materials from biomass. All these, in combination with their very high thermal stability, should make them ideal "solvents" for high-temperature carbonization reactions in ionic liquids.

Cooper *et al.* [95] were the first to report a new type of solvothermal synthesis in which ionic liquids were used as both the solvent and the structure-directing agent in the synthesis of zeolites. This methodology has been termed *ionothermal synthesis* and, since then, this methodology has become one of the most widely used synthetic strategies among the zeolite community. It was also extended to the synthesis of metal organic frameworks, covalent organic frameworks, polymer organic frameworks, porous silicas, nanoparticles, polymers and others [96–99].

In the context of this chapter, ionic liquids have been also used for "ionothermal carbonizations." Titirici and Taubert reported that metal-containing ionic liquids can play

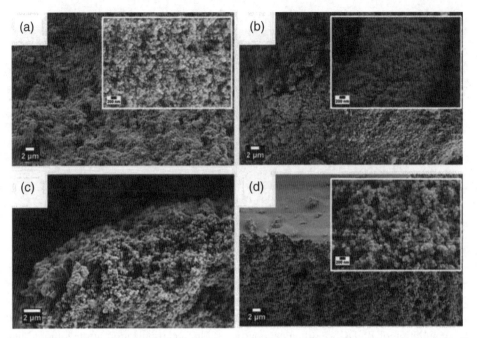

Figure 1.14   *SEM micrographs of as-synthesized ionothermal carbons: (a) C-glucose, (b) C-fructose, (c) C-xylose, and (d) C-starch (scale bar in inset: 200 nm).*

the simultaneous roles of structure-directing agent, catalyst for carbonization, and solvent [100]. Interestingly, the ionic liquids can be fully recovered at the end of each carbonization process without any effect on their chemical structure. A variety of carbohydrates were used as a carbon source (i.e., D-glucose, D-fructose, D-xylose, and starch) with 1-butyl-3-methylimidazolium tetrachloroferrate(III) ([Bmim][FeCl$_4$]) as a reusable solvent and catalyst. The carbon materials derived from these different carbohydrates were similar in terms of particle size (Figure 1.14) and chemical composition, possessing relatively high surface areas from 44 to 155 m^2 g^{-1} after ionothermal processing, which could be significantly increased to more than 350 m^2 g^{-1} by further thermal treatment (e.g., post-carbonization at 750 °C). CO$_2$ and nitrogen sorption analysis, combined with mercury intrusion porosimetry, revealed a promising hierarchical pore structuring to these carbon materials. The ionic liquid [Bmim][FeCl$_4$] has a triple role: it acted as a soft template to generate the characterized pore structuring, as a solvent, and as a catalyst, resulting in enhanced ionothermal carbon yields. Importantly from a process point of view, the ionic liquid could be successfully recovered and reused.

The group of Dai *et al.* used a protic ionic liquid (*N,N*-dimethyl-*N*-formylammonium bis(trifluoromethylsulfonyl)imide) ([DMFH][Tf$_2$N]) for the synthesis of ionothermal carbons from glucose and fructose at low temperature and ambient pressure. The observed results were similar with those of Titirici where the ionic liquid induces porosity in the resulting carbons while the carbonization yield is significantly increased [101].

Although only these two publications exist currently in the literature, this methodology could represent an interesting direction for the synthesis of sustainable carbon materials. The carbonization can take place at atmospheric pressure in a standard flask while real biomass precursors can be employed. In particular, cellulose is known to be solubilized by protic ionic liquids. In addition, it was already reported that the ionic liquids can efficiently catalyze the production of hydroxymethylfurfural (HMF) from biomass [102], which can then significantly increase the HTC yield. By designing tailor-made ionic liquids it should be possible to control the porosity as well as introduce various functions into the resulting ionocarbons. Metal-containing ionic liquids should offer the possibility to produce various interesting nanocomposites under the appropriate synthetic conditions. Furthermore, the use of ionic liquids will be far more justified in such a procedure compared with their irreversible conversion in various materials (i.e., carbons, polymers).

A related class of ionic liquids named deep eutectic solvents (DESs) is obtained by complexion of quaternary ammonium salts with hydrogen-bond donors [103]. DESs share many characteristics of conventional ionic liquids (e.g., nonreactive with water, nonvolatile, and biodegradable) while offering certain advantages. For instance, the preparation of eutectic mixtures in a pure state is accomplished more easily than for ionic liquids. There is no need for postsynthesis purification as the purity of the resulting DES will only depend on the purity of its individual components. The low cost of those eutectic mixtures based on readily available components [104, 105] makes them particularly desirable (more so than conventional ionic liquids) for large-scale synthetic applications. However, a close inspection of the recent literature revealed that their use in materials synthesis is sporadic compared to ionic liquids. This situation is currently changing and different authors consider DESs as the next generation of ionic liquids because, besides the above-mentioned interesting features, they can act as true solvent–template–reactant systems, where the DES is at the same time the precursor, template, and reactant medium for the fabrication of a desired material with a defined morphology or chemical composition. A very comprehensive review on the topic was recently published by the group of Francisco del Monte [106].

Carriazo *et al.* reported the preparation of a DES based on the mixture of resorcinol and choline chloride. Polycondensation with formaldehyde resulted in the formation of monolithic carbons with a bimodal porosity comprising of both micropores and large mesopores of around 10 nm (Figure 1.15) [107]. The morphology of the resulting carbons consisted of a bicontinuous porous network built out of highly cross-linked clusters that aggregated and assembled into a stiff, interconnected structure. This morphology is typical for carbons obtained via spinodal decomposition [108]. Carriazo *et al.* hypothesized that one of the components forming the DES (e.g., resorcinol) acts as a precursor of the polymer phase whereas the second component (e.g., choline chloride) acts as a structure-directing agent following a synthetic mechanism based on DES rupture and controlled delivery of the segregated structure-directing agent into the reaction mixture.

The wide range of DESs that can be prepared provides a remarkable versatility to this synthetic approach. For instance, Carriazo *et al.* also reported that the use of ternary DESs composed of resorcinol, urea, and choline chloride results in carbons with surface areas of nearly 100 m^2 g^{-1} higher then those of carbons obtained from binary DESs composed of resorcinol and choline chloride. Urea was here partially incorporated into the resorcinol/formaldehyde network (upon its participation in polycondensation reactions),

(a)   (b)

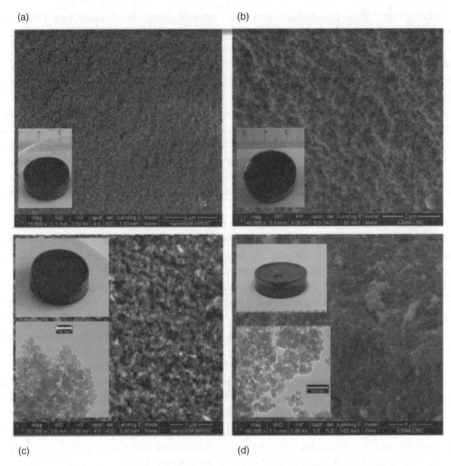

(c)   (d)

Figure 1.15 *SEM micrographs of RFRC-DES (a, bar: 5 mm) and RFRUC-DES (b, bar: 2 mm) gels. Insets show a picture of the monolithic RF gels. SEM micrographs of CRC-DES (c, bar: 1 mm) and CRUC-DES (d, bar: 1 mm) monoliths. Insets show a picture of the respective monolithic carbons and TEM micrographs of CRC-DES (left, bar: 50 nm) and CRUC-DES (right, bar: 150 nm). R = resorcinol; C = chlorine chloride; U = urea; F = formaldehyde. (Reproduced with permission from [107]. © 2010 American Chemical Society.)*

the release of which (during carbonization) resulted in the above-mentioned increase in the surface area.

The same group used DESs composed of resorcinol, 3-hydroxypyridine, and choline chloride [109]. In this case, DESs played multiple roles in the synthetic process: liquid medium ensuring reagent homogenization, structure-directing agent responsible for the achievement of the hierarchically porous structure, and source of carbon and nitrogen (Figure 1.16). The formation of a polymer-rich phase upon resorcinol and 3-hydroxypyridine polycondensation promotes DES rupture and choline chloride segregation into a spinodal-like decomposition process. Interestingly, the resulting carbons exhibited a combination of surface areas and nitrogen contents (from around 550 to 650 m^2 g^{-1} and from around 13 to 5 at% for carbonization temperatures ranging from 600 to 800 °C, respectively) that, unless

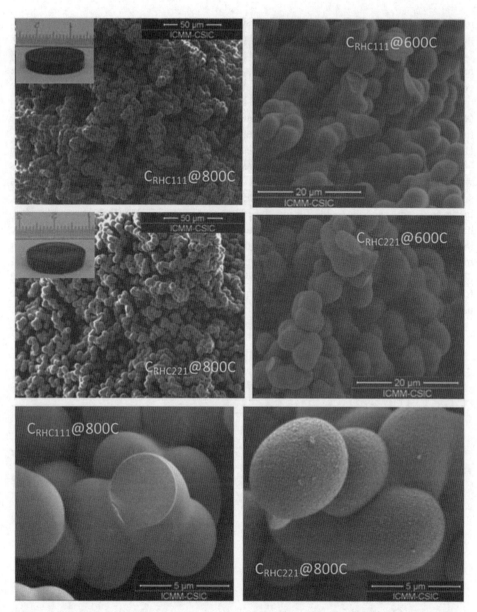

Figure 1.16 *SEM micrographs of CRHC-DES and CRHC-DES obtained after thermal treatments at 600 and 800 °C. Insets show pictures of the CRHC-DES and CRHC-DES monoliths obtained after thermal treatment at 800 °C. C = chlorine chloride; R = resorcinol; H = 3-hydroxypyridine. (Reproduced with permission from [109]. © 2011 Royal Society of Chemistry.)*

traditional structure-directing agents are also used, had never been attained by synthetic processes carried out in ionic liquids.

It is finally worth noting the "green" character of this process because of the absence of residues and/or byproducts eventually released after the synthetic process; that is, one of the components forming the DES (e.g., resorcinol, mixtures of resorcinol and hydroxypyridine or urea, and furfuryl alcohol) becomes the material itself with high yields of conversion (within the 60–80% range), whereas the second component (e.g., single choline chloride in resorcinol-based synthesis) is fully recovered and can be reused in subsequent reactions. However, using DES based on carbohydrates [110] should improve even further the green character of this methodology for the future production of carbon materials. Such DESs were found to serve some basic functions in living cells and organisms. They include sugars, some amino acids, choline, and some organic acids such as malic acid, citric acid, lactic acid, and succinic acid. Taking the plant metabolomics data that Verpoorte *et al.* have collected over recent years, they could clearly see similarities with the synthetic ionic liquids. The above-mentioned major cellular constituents are perfect candidates for making ionic liquids and DES. The authors made various combinations of these candidates (Table 1.1 [110]). Such "natural DESs" (NADES) could be a potential interesting source for new and existing sustainable carbon materials that will be surely exploited in the near future.

Table 1.1   *List of natural ionic liquids and DES. (Reprinted with permission from [110]. © 2011 American Society of Plant Biologists.)*

Combination	Molar ratio
Citric acid/choline chloride	1 : 2, 1 : 3
Malic acid/choline chloride	1 : 1, 1 : 2, 1 : 3
Maleic acid/choline chloride	1 : 1, 1 : 2, 1 : 3
Aconitic acid/choline chloride	1 : 1
Glc/choline chloride/water	1 : 1 : 1
Fru/choline chloride/water	1 : 1 : 1
Suc/choline chloride/water	1 : 1 : 1
Citric acid/Pro	1 : 1, 12, 1 : 3
Malic acid/Glc	1 : 1
Malic acid/Fru	1 : 1
Malic acid/Suc	1 : 1
Citric acid/Glc	2 : 1
Citric acid/trehalose	2 : 1
Citric acid/Suc	1 : 1
Maleic acid/Glc	4 : 1
Maleic acid/Suc	1 : 1
Glc/Fru	1 : 1
Fru/Suc	1 : 1
Glc/Suc	1 : 1
Suc/Glc/Fru	1 : 1 : 1

1.2.6 Hydrothermal Carbon Materials (HTC)

The last procedure to produce green carbon materials is HTC and this forms the subject of this book. Its versatility, end-products, applications, and limitations will be described within the following chapters. In my opinion, it is by far the most suitable and up-scalable technology at the moment to process waste biomass and transform it into end-products with remarkable properties.

Before going deeper into the secrets of HTC-derived materials, some important advantages of this unconventional carbonization technique should be emphasized here:

- Carbonization temperatures are low – typically in the range 130–250 °C.
- Carbonization takes place in water under self-generated pressures, thus avoiding precursors drying costs.
- Typically, spherical microsized particles are generally obtained.
- Controlled porosity can be easily introduced using nanocasting procedures, natural templates, or activation procedures/thermal treatments (see Chapters 2 and 3).
- Carbonaceous materials can be combined with other components (e.g., inorganic nanoparticles) to form composites with special physicochemical properties (see Chapter 4).
- The resulting carbon particles have (polar) oxygenated groups residing at the surface that can be in turn used in postfunctionalization strategies (see Chapter 5).
- The surface chemistry and electronic properties can be easily controlled via additional thermal treatment, while the morphology and porosity are maintained (see Chapter 6).
- The materials have a wide range of timely technological applications (see Chapter 7).
- The synthesis can be described as "carbon-negative," meaning that it has the potential to bind the CO_2 fixed by from original plant precursor (see Chapters 8 and 9).

Here, we will provide a short history of the HTC technology.

1.3 Brief History of Hydrothermal Carbons

As early as 1911, Friedrich Bergius was researching topics that are currently of extreme importance for finding alternative fuels to the fossil-based fuels. Back then, Bergius was convinced that it should be possible to carry out the "water gas reaction" and produce hydrogen gas according to the formula $C + 2H_2O = CO_2 + H_2$, if the right temperature and pressure conditions are satisfied [111]. The aim was to inhibit the troublesome formation of CO. In order to achieve such goals, Bergius was working at temperatures below 600 °C, at which steam practically ceases to act on coal. His intention was to discover whether, the reaction velocity might be increased sufficiently to adjust the equilibrium. Indeed, Bergius managed to oxidize coal when liquid water was reacting with it at 200 bar, producing CO_2 and hydrogen, in particular when transitional metal catalysts were present in the system. However, oxidation velocities, which would have made such reaction of commercial interest, failed to be achieved.

On the other hand, another very important observation was made: when peat was used as coal material, it was observed that exceptionally large amounts of CO_2 formed and that the carbonaceous residue remaining in the vessel had the same elemental composition of natural coal. This observation prompted Bergius to study the decomposition process of plant substances more closely. He thought the process would be similar to the process of metamorphism that such plant-based compounds undergo in nature over the course of millions of years during their gradual transition into coal.

In those times, many researchers had already attempted to convert biomass (i.e., wood, which in addition to cellulose also contains lignin) into coal by heating; however, the cellulose was decomposed. What Bergius *et al.* did differently was that they managed to prevent superheating and thus prevented the decomposition of cellulose by introducing steam. The resulting carbonization products had a very similar composition to natural coal [112].

The secret that Bergius discovered was that the biomass precursor had to be in intimate contact with the liquid water that, at those mild temperatures (200 °C) in the high-pressure vessel, could not decompose into gases.

The apparatus that Bergius used to produce the very first hydrothermal carbon ever is presented in Figure 1.17. The temperatures were between 200 and 330 °C in the presence of liquid water at pressures up to 200 bar.

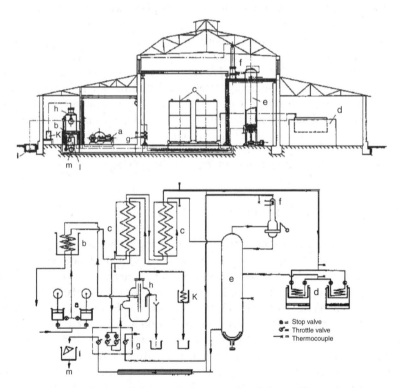

Figure 1.17 *Cross-section and schematic diagram of cellulose effluent decarbonization: (a) pump, (b) preheater, (c) heat exchanger, (d) lead bath furnace, (e) reaction vessel, (f) column, (g) control post, (h) tar separator/tar receiver, (k) cooler for alcohol vapor, (l) scales for coal slurry, and (m) screw conveyor. (Reprinted with permission from Chemical reactions under high pressure, Nobel Lecture, May 21, 1932 by Friedrich Bergius. © The Nobel Foundation 1931.)*

The first report on the formation of what today I call hydrothermal carbon, while others call "humins" [113], citing from Bergius was "per 2 parts cellulose, there were formed 2 parts CO_2 and 5 parts water, and a powdery substance which corresponded approximately to $C_{10}H_8O$" [112].

This important discovery was followed by numerous studies of Bergius and his assistant Hugo Specht on the hydrogenation of this artificial coal [114]. Their experiments become the basis for the production of liquid and soluble compounds from coal – a process they named "coal liquefaction" [115].

For all his studies on the production of synthetic coal as well as hydrogenation studies and his contributions in general to high-pressure reactions [116], Bergius was awarded the Nobel Prize in 1931.

The enormous discovery of Bergius was later followed by other studies. For example, Berl and Schmidt in 1932 varied the biomass source and treated the different samples, in the presence of water, at temperatures between 150 and 350 °C [117]. The latter authors summarized, via a series of papers in 1932, the state-of-the-art knowledge about the emergence of coal [118]. Later, Schuhmacher *et al.* analyzed the influence of pH on the outcome of the HTC reaction and found serious differences in the decomposition schemes, as identified by the C/H/O composition [119]. A review of the current knowledge on coal structure and its origin is found in Haenel [120].

A renaissance of such experiments was started with reports on the low-temperature hydrothermal synthesis of carbon spheres (around 200 °C) using sugar or glucose as precursors in 2001 [121, 122]. At the same time, Antonietti and Shu Hong Yu found that the presence of metal ions can effectively accelerate the HTC of starch, which shortened the reaction time to several hours and directed the synthesis towards various metal/carbon nanoarchitectures, such as Ag@C nanocables [123], CNFs [124], and spheres [120].

This was the state-of-the-art in 2005 when I joined the Max Planck Institute, first as a post-doctoral fellow and later on as a group leader. Ever since, myself together with a team of very talented PhD students and postdocs (most of them are coauthors of various chapters in this book) worked intensively on this technique, its fundamentals, the production of porous HTC carbons, hybrid materials, and their applications.

It has to be mentioned that HTC is today a very-well-established technique. Many research groups have embarked on discovering and revealing the new secrets this method still has to offer. In addition to the academic interest, several private companies have started large-scale HTC production. This is because, on the one hand, the resulting char has coal-like properties and is expected to exhibit favorable behavior with respect to combustion, gasification, and, on the other hand, because of thermal conversion processes for decentralized applications. The energy content of HTC-derived chars was measured to be between 25 and 35 MJ kg^{-1}, which is about 40% higher then that of the staring biomass precursor [125]. Some of these more industrial aspects of HTC will be described in Chapter 9.

Maybe one of the most appealing large-scale applications of the HTC process is, as suggested by Seifritz in a pioneering dissertation in 1993 [126], the fact that anthropogenic CO_2 emissions could be mitigated by converting biomass into charcoal. Thus, converting fast-growing biomass into hydrothermal carbon, the CO_2 bound in the parent biomass will be no longer liberated via the photosynthesis process or atmospheric decomposition of the biomass residue but bound to the final solid carbonaceous structure. This represents an efficient way of taking the CO_2 out of the carbon cycle and thus a solution for reducing the already alarming amounts of CO_2 present in the atmosphere, known to be responsible for climate change. The concept of biochar as an approach to carbon sequestration, as well as increasing soil fertility, raising agricultural productivity, and reducing pressure on forests, has received increasing attention over the past few years [127]. Several papers have already been published where the tremendous potential of HTC as a CO_2-negative solution for soil improvement is discussed [128–130]. This is also described in Chapter 8.

In addition to the applications of HTC associated with large cost production, HTC can be used on a much lower scale in nanotechnology for energy- and environmental-related applications (Chapter 7). This is the author's field of expertise as well as the main focus of this book. For such applications, laboratory-scale production of hydrothermal carbon (500 g to 5 kg) in commercially available reactors is sufficient.

Whether this technology will really represent one solution to the many problems our society is confronted these days regarding depletion of resources, global warming, and energy is still to be determined in the coming years.

References

(1) van Wyk, J.P.H. (2001) *Trends in Biotechnology*, **19**, 172–177.

(2) Lichtenthaler, F.W. and Peters, S. (2004) *Comptes Rendus Chimie*, **7**, 65–90.

(3) Rahmstort, S., Morgan, J., Levermann, A., and Sach, K. (2010) in *Global Sustainability – A Nobel Cause* (eds. M. Molina, H.J. Schellnhuber, N. Stern, V. Huber, and S. Kadner), Cambridge University Press, Cambridge, p. 68.

(4) Green, M.L., Espinal, L., Traversa, E., and Amis, E.J. (2012) *MRS Bulletin*, **37**, 303–309.

(5) Koopmans, R.J. (2006) *Soft Matter*, **2**, 537–543.

(6) Bobleter, O. (1994) *Progress in Polymer Science*, **19**, 797–841.

(7) (a) Gray, K.A., Zhao, L.S., and Emptage, M. (2006) *Current Opinion in Chemical Biology*, **10**, 141–146; (b) Hendriks, A.T.W.M. and Zeeman, G. (2009) *Bioresource Technology*, **100**, 10–18; (c) Lee, J. (1997) *Journal of Biotechnology*, **56**, 1–24; (d) Lin, Y. and Tanaka, S. (2006) *Applied Microbiology and Biotechnology*, **69**, 627–642; (e) Lynd, L.R. (1996) *Annual Review of Energy and the Environment*, **21**, 403–465; (f) Lynd, L.R., van Zyl, W.H., McBride, J.E., and Laser, M. (2005) *Current Opinion in Biotechnology*, **16**, 577–583; (g) Lynd, L.R., Weimer, P.J., van Zyl, W.H., and Pretorius, I.S. (2002) *Microbiology and Molecular Biology Reviews*, **66**, 506; (h) Ni, M., Leung, D.Y.C., and Leung, M.K.H. (2007) *International Journal of Hydrogen Energy*, **32**, 3238–3247; (i) Sanchez, O.J. and Cardona, C.A. (2008) *Bioresource Technology*, **99**, 5270–5295; (j) Sun, Y. and Cheng, J.Y. (2002) *Bioresource Technology*, **83**, 1–11; (k) Wyman, C.E. (1999) *Annual Review of Energy and the Environment*, **24**, 189–226; (l) Zaldivar, J., Nielsen, J., and Olsson, L. (2001) *Applied Microbiology and Biotechnology*, **56**, 17–34.

(8) (a) Bridgwater, A.V. (1995) *Fuel*, **74**, 631–653; (b) Demirbas, A. (2007) *Progress in Energy and Combustion Science*, **33**, 1–18; (c) Devi, L., Ptasinski, K.J., and Janssen, F. (2003) *Biomass & Bioenergy*, **24**, 125–140; (d) Matsumura, Y., Minowa, T., Potic, B. *et al.* (2005) *Biomass & Bioenergy*, **29**, 269–292; (e) McKendry, P. (2002) *Bioresource Technology*, **83**, 37–46; (f) McKendry, P. (2002) *Bioresource*

Technology, **83**, 55–63; (g) Peterson, A.A., Vogel, F., Lachance, R.P. *et al.* (2008) *Energy & Environmental Science*, **1**, 32–65; (h) Sutton, D., Kelleher, B., and Ross, J.R.H. (2001) *Fuel Processing Technology*, **73**, 155–173.

(9) (a) Alonso, D.M., Bond, J.Q., and Dumesic, J.A. (2010) *Green Chemistry*, **12**, 1493–1513; (b) Lin, Y.-C. and Huber, G.W. (2009) *Energy & Environmental Science*, **2**, 68–80; (c) Nigam, P.S. and Singh, A. (2011) *Progress in Energy and Combustion Science*, **37**, 52–68; (d) Saxena, R.C., Seal, D., Kumar, S., and Goyal, H.B. (2008) *Renewable & Sustainable Energy Reviews*, **12**, 1909–1927; (e) Sivasamy, A., Cheah, K.Y., Fornasiero, P. *et al.* (2009) *ChemSusChem*, **2**, 278–300; (f) Tong, X., Ma, Y., and Li, Y. (2010) *Applied Catalysis A – General*, **385**, 1–13; (g) Van de Vyver, S., Geboers, J., Jacobs, P.A., and Sels, B.F. (2011) *ChemCatChem*, **3**, 82–94; (h) Yung, M.M., Jablonski, W.S., and Magrini-Bair, K.A. (2009) *Energy & Fuels*, **23**, 1874–1887.

(10) Anastas, P.T. and Kirchhoff, M.M. (2002) *Accounts of Chemical Research*, **35**, 686–694.

(11) (a) Adamczak, M., Bornscheuer, U.T., and Bednarski, W. (2009) *European Journal of Lipid Science and Technology*, **111**, 800–813; (b) Blanch, H.W., Simmons, B.A., and Klein-Marcuschamer, D. (2011) *Biotechnology Journal*, **6**, 1086–1102; (c) Chundawat, S.P.S., Beckham, G.T., Himmel, M.E., and Dale, B.E. (2011) *Annual Review of Chemical and Biomolecular Engineering*, **2**, 121–145; (d) Hoell, I.A., Vaaje-Kolstad, G., and Eijsink, V.G.H. (2010) *Biotechnology and Genetic Engineering Reviews*, **27**, 331–366.

(12) Hong, C.K. and Wool, R.P. (2005) *Journal of Applied Polymer Science*, **95**, 1524–1538.

(13) Nampoothiri, K.M., Nair, N.R., and John, R.P. (2010) *Bioresource Technology*, **101**, 8493–8501.

(14) Gharbi, S., Andreolety, J.P., and Gandini, A. (2000) *European Polymer Journal*, **36**, 463–472.

(15) Mitiakoudis, A. and Gandini, A. (1991) *Macromolecules*, **24**, 830–835.

(16) Boufi, S., Belgacem, M.N., Quillerou, J., and Gandini, A. (1993) *Macromolecules*, **26**, 6706–6717.

(17) Pavier, C. and Gandini, A. (2000) *European Polymer Journal*, **36**, 1653–1658.

(18) (a) Brandelli, A., Daroit, D.J., and Riffel, A. (2010) *Applied Microbiology and Biotechnology*, **85**, 1735–1750; (b) Briens, C., Piskorz, J., and Berruti, F. (2008) *International Journal of Chemical Reactor Engineering*, **6**, 1–49; (c) Fischer, R. and Emans, N. (2000) *Transgenic Research*, **9**, 279–299; (d) Harun, R., Singh, M., Forde, G.M., and Danquah, M.K. (2010) *Renewable & Sustainable Energy Reviews*, **14**, 1037–1047; (e) Mamman, A.S., Lee, J.-M., Kim, Y.-C. *et al.* (2008) *Biofuels Bioproducts & Biorefining-Biofpr*, **2**, 438–454; (f) Pandey, A. and Soccol, C.R. (1998) *Brazilian Archives of Biology and Technology*, **41**, 379–389; (g) Pulz, O. and Gross, W. (2004) *Applied Microbiology and Biotechnology*, **65**, 635–648; (h) Singhania, R.R., Patel, A.K., Soccol, C.R., and Pandey, A. (2009) *Biochemical Engineering Journal*, **44**, 13–18; (i) Sipkema, D., Osinga, R., Schatton, W. *et al.* (2005) *Biotechnology and Bioengineering*, **90**, 201–222; (j) Tokiwa, Y. and Caiabia, B.P. (2008) *Canadian Journal of Chemistry – Revue Canadienne De Chimie*, **86**, 548–555.

(19) Baughman, R.H., Zakhidov, A.A., and de Heer, W.A. (2002) *Science*, **297**, 787–792.

(20) Su, D.S. (2009) *ChemSusChem*, **2**, 1009–1020.

(21) Dupuis, A.-C. (2005) *Progress in Materials Science*, **50**, 929–961.

(22) Su, D.S. and Chen, X.-W. (2007) *Angewandte Chemie International Edition*, **119**, 1855–1856.

(23) Endo, M., Takeuchi, K., Kim, Y.A. *et al.* (2008) *ChemSusChem*, **1**, 820–822.

(24) Rinaldi, A., Zhang, J., Mizera, J. *et al.* (2008) *Chemical Communications*, 6528–6530.

(25) Kawasaki, S., Shinoda, M., Shimada, T. *et al.* (2006) *Carbon*, **44**, 2139–2141.

(26) Su, D.S., Chen, X., Weinberg, G. *et al.* (2005) *Angewandte Chemie International Edition*, **44**, 5488–5492.

(27) Chen, X.-W., Su, D.S., Hamid, S.B.A., and Schlögl, R. (2007) *Carbon*, **45**, 895–898.

(28) Chen, X.-W., Timpe, O., Hamid, S.B.A. *et al.* (2009) *Carbon*, **47**, 340–343.

(29) (a) Dosodia, A., Lal, C., Singh, B.P. *et al.* (2009) *Fullerenes Nanotubes and Carbon Nanostructures*, **17**, 567–582; (b) Jieshan, Q., Yongfeng, L., Yunpeng, W. *et al.* (2003) *Carbon*, **41**, 2170–2173; (c) Mathur, R.B., Lal, C., and Sharma, D.K. (2007) *Energy Sources Part A – Recovery Utilization and Environmental Effects*, **29**, 21–27; (d) Qiu, J.S., Li, Y.F., Wang, Y.P., and Li, W. (2004) *Fuel Processing Technology*, **85**, 1663–1670; (e) Qiu, J.S., Zhang, F., Zhou, Y. *et al.* (2002) *Fuel*, **81**, 1509–1514; (f) Qiu, J.S., Zhou, Y., Wang, L.N., and Tsang, S.C. (1998) *Carbon*, **36**, 465–467; (g) Qiu, J.S., Zhou, Y., Yang, Z.G. *et al.* (2000) *Fuel*, **79**, 1303–1308; (h) Qiu, J.S., Zhou, Y., Yang, Z.G. *et al.* (2000) *Molecular Materials*, **13**, 377–384; (i) Wang, M.Z. and Li, F. (2005) *New Carbon Materials*, **20**, 71–82; (j) Wilson, M.A., Patney, H.K., and Kalman, J. (2002) *Fuel*, **81**, 5–14; (k) Zhao, X.-f., Qiu, J.-s., Sun, Y.-x. *et al.* (2009) *New Carbon Materials*, **24**, 109–113; (l) Zhi, W., Bin, W., Qianming, G. *et al.* (2008) *Materials Letters*, **62**, 3585–3587.

(30) Kumar, M., Okazaki, T., Hiramatsu, M., and Ando, Y. (2007) *Carbon*, **45**, 1899–1904.

(31) Kang, Z.H., Wang, E.B., Mao, B.D. *et al.* (2005) *Nanotechnology*, **16**, 1192.

(32) Gogotsi, Y., Libera, J.A., and Yoshimura, M. (2000) *Journal of Materials Research*, **15**, 2591–2594.

(33) Calderon Moreno, J.M., Fujino, T., and Yoshimura, M. (2001) *Carbon*, **39**, 618–621.

(34) Moreno, J.M.C. and Yoshimura, M. (2001) *Journal of the American Chemical Society*, **123**, 741–742.

(35) Pol, V.G. and Thiyagarajan, P. (2010) *Journal of Environmental Monitoring*, **12**, 455–459.

(36) Pol, S.V., Pol, V.G., Sherman, D., and Gedanken, A. (2009) *Green Chemistry*, **11**, 448–451.

(37) Sevilla, M. and Fuertes, A.B. (2010) *Chemical Physics Letters*, **490**, 63–68.

(38) Ashokkumar, M., Narayanan, N.T., Mohana Reddy, A.L. *et al.* (2012) *Green Chemistry*, **14**, 1689–1695.

(39) Mermin, N.D. (1968) *Physical Review*, **176**, 250–254.

(40) Orlita, M., Faugeras, C., Plochocka, P. *et al.* (2008) *Physical Review Letters*, **101**, 267601.

(41) Balandin, A.A., Ghosh, S., Bao, W. *et al.* (2008) *Nano Letters*, **8**, 902–907.

(42) Lee, C., Wei, X., Kysar, J.W., and Hone, J. (2008) *Science*, **321**, 385–388.

(43) Park, S., An, J., Jung, I. *et al.* (2009) *Nano Letters*, **9**, 1593–1597.

(44) Park, S., An, J., Piner, R.D. *et al.* (2008) *Chemistry of Materials*, **20**, 6592–6594.

(45) Novoselov, K.S., Jiang, D., Schedin, F. *et al.* (2005) *Proceedings of the National Academy of Sciences of the United States of America*, **102**, 10451–10453.

(46) Novoselov, K.S., Geim, A.K., Morozov, S.V. *et al.* (2004) *Science*, **306**, 666–669.

(47) Berger, C., Song, Z., Li, X. *et al.* (2006) *Science*, **312**, 1191–1196.

(48) Kim, K.S., Zhao, Y., Jang, H. *et al.* (2009) *Nature*, **457**, 706–710.

(49) Jiao, L., Zhang, L., Wang, X. *et al.* (2009) *Nature*, **458**, 877–880.

(50) Stankovich, S., Dikin, D.A., Piner, R.D. *et al.* (2007) *Carbon*, **45**, 1558–1565.

(51) Cote, L.J., Kim, F., and Huang, J. (2008) *Journal of the American Chemical Society*, **131**, 1043–1049.

(52) Stankovich, S., Piner, R.D., Chen, X. *et al.* (2006) *Journal of Materials Chemistry*, **16**, 155–158.

(53) Wang, G., Yang, J., Park, J. *et al.* (2008) *Journal of Physical Chemistry C*, **112**, 8192–8195.

(54) Guo, H.L., Wang, X.F., Qian, Q.Y. *et al.* (2009) *ACS Nano*, **3**, 2653–2659.

(55) Ding, Y.H., Zhang, P., Zhuo, Q. *et al.* (2011) *Nanotechnology*, **22**.

(56) Li, X.-H., Chen, J.-S., Wang, X. *et al.* (2012) *ChemSusChem*, **5**, 642–646.

(57) Zhu, C.Z., Guo, S.J., Fang, Y.X., and Dong, S.J. (2010) *ACS Nano*, **4**, 2429–2437.

(58) Fan, Z.-J., Kai, W., Yan, J. *et al.* (2011) *ACS Nano*, **5**, 191–198.

(59) Liu, Y., Li, Y., Zhong, M. *et al.* (2011) *Journal of Materials Chemistry*, **21**, 15449–15455.

(60) Sui, Z., Zhang, X., Lei, Y., and Luo, Y. (2011) *Carbon*, **49**, 4314–4321.

(61) Kai, W., Tao, F., Min, Q. *et al.* (2011) *Applied Surface Science*, **257**, 5808–5812.

(62) Khanra, P., Kuila, T., Kim, N.H. *et al.* (2012) *Chemical Engineering Journal*, **183**, 526–533.

(63) Wang, Y., Shi, Z., and Yin, J. (2011) *ACS Applied Materials & Interfaces*, **3**, 1127–1133.

(64) Wang, G., Qian, F., Saltikov, C.W. *et al.* (2011) *Nano Research*, **4**, 563–570.

(65) Salas, E.C., Sun, Z., Lüttge, A., and Tour, J.M. (2010) *ACS Nano*, **4**, 4852–4856.

(66) Liu, K., Zhang, J.-J., Cheng, F.-F. *et al.* (2011) *Journal of Materials Chemistry*, **21**, 12034–12040.

(67) Nursanto, E.B., Nugroho, A., Hong, S.-A. *et al.* (2011) *Green Chemistry*, **13**, 2714–2718.

(68) Li, X., Cai, W., An, J. *et al.* (2009) *Science*, **324**, 1312–1314.

(69) Reina, A., Jia, X., Ho, J. *et al.* (2008) *Nano Letters*, **9**, 30–35.

(70) Kondo, D., Sato, S., Yagi, K. *et al.* (2010) *Applied Physics Express*, **3**, 025102.

(71) Ago, H., Ito, Y., Mizuta, N. *et al.* (2010) *ACS Nano*, **4**, 7407–7414.

(72) Kang, B.J., Mun, J.H., Hwang, C.Y., and Cho, B.J. (2009) *Journal of Applied Physics*, **106**, 104309.

(73) Sun, Z., Yan, Z., Yao, J. *et al.* (2010) *Nature*, **468**, 549–552.

(74) Ruan, G., Sun, Z., Peng, Z., and Tour, J.M. (2011) *ACS Nano*, **5**, 7601–7607.

(75) Primo, A., Atienzar, P., Sanchez, E. *et al.* (2012) *Chemical Communications*, **48**, 9254–9256.

(76) Ruiz-Hitzky, E., Darder, M., Fernandes, F.M. *et al.* (2011) *Advanced Materials*, **23**, 5250–5255.

(77) (a) Ahmad, T., Rafatullah, M., Ghazali, A. *et al.* (2011) *Journal of Environmental Science and Health Part C – Environmental Carcinogenesis & Ecotoxicology Reviews*, **29**, 177–222; (b) Chen, Y., Zhu, Y., Wang, Z. *et al.* (2011) *Advances in Colloid and Interface Science*, **163**, 39–52; (c) Demirbas, A. (2009) *Journal of Hazardous Materials*, **167**, 1–9; (d) Ioannidou, O. and Zabaniotou, A. (2007) *Renewable & Sustainable Energy Reviews*, **11**, 1966–2005; (e) Lin, S.-H. and Juang, R.-S. (2009) *Journal of Environmental Management*, **90**, 1336–1349; (f) Mayhew, M. and Stephenson, T. (1997) *Environmental Technology*, **18**, 883–892; (g) Mohamed, A.R., Mohammadi, M., and Darzi, G.N. (2010) *Renewable & Sustainable Energy Reviews*, **14**, 1591–1599; (h) Rodriguez, G., Lama, A., Rodriguez, R. *et al.* (2008) *Bioresource Technology*, **99**, 5261–5269; (i) Skodras, G., Diamantopouiou, I., Zabaniotou, A. *et al.* (**2007**) *Fuel Processing Technology*, **88**, 749–758; (j) Uraki, Y. and Kubo, S. (2006) *Mokuzai Gakkaishi*, **52**, 337–343; (k) Yu, H., Covey, G.H., and O'Connor, A.J. (2008) *International Journal of Environment and Pollution*, **34**, 427–450.

(78) (a) Frackowiak, E. and Beguin, F. (2001) *Carbon*, **39**, 937–950; (b) Pandolfo, A.G. and Hollenkamp, A.F. (2006) *Journal of Power Sources*, **157**, 11–27; (c) Simon, P. and Gogotsi, Y. (2008) *Nature Materials*, **7**, 845–854; (d) Zhang, L.L. and Zhao, X.S. (2009) *Chemical Society Reviews*, **38**, 2520–2531.

(79) (a) Choi, S., Drese, J.H., and Jones, C.W. (2009) *ChemSusChem*, **2**, 796–854; (b) Lozano-Castello, D., Alcaniz-Monge, J., de la Casa-Lillo, M.A. *et al.* (2002) *Fuel*, **81**, 1777–1803.

(80) White, R.J., Budarin, V., Luque, R. *et al.* (2009) *Chemical Society Reviews*, **38**, 3401–3418.

(81) White, R.J., Luque, R., Budarin, V.L. *et al.* (2009) *Chemical Society Reviews*, **38**, 481–494.

(82) White, R.J., Budarin, V.L., and Clark, J.H. (2008) *ChemSusChem*, **1**, 408–411.

(83) Budarin, V., Clark, J.H., Hardy, J.J.E. *et al.* (2006) *Angewandte Chemie International Edition*, **45**, 3782–3786.

(84) White, R.J., Antonio, C., Budarin, V.L. *et al.* (2010) *Advanced Functional Materials*, **20**, 1834–1841.

(85) White, R.J., Budarin, V.L., and Clark, J.H. (2010) *Chemistry – A European Journal*, **16**, 1326–1335.

(86) Budarin, V., Luque, R., Macquarrie, D.J., and Clark, J.H. (2007) *Chemistry – A European Journal*, **13**, 6914–6919.

(87) Luque, R., Budarin, V., Clark, J.H., and Macquarrie, D.J. (2009) *Green Chemistry*, **11**, 459–461.

(88) Paraknowitsch, J.P., Zhang, J., Su, D. *et al.* (2010) *Advanced Materials*, **22**, 87.

(89) Lee, J.S., Wang, X., Luo, H., and Dai, S. (2010) *Advanced Materials*, **22**, 1004–1007.

(90) Kuhn, P., Forget, A., Hartmann, J. *et al.* (2009) *Advanced Materials*, **21**, 897–901.

(91) Lee, J.S., Wang, X., Luo, H. *et al.* (2009) *Journal of the American Chemical Society*, **131**, 4596–4597.

(92) Wang, X. and Dai, S. (2010) *Angewandte Chemie International Edition*, **49**, 6664–6668.

(93) Paraknowitsch, J.P., Thomas, A., and Antonietti, M. (2010) *Journal of Materials Chemistry*, **20**, 6746–6758.

(94) Paraknowitsch, J.P., Zhang, Y., and Thomas, A. (2011) *Journal of Materials Chemistry*, **21**, 15537–15543.

(95) Cooper, E.R., Andrews, C.D., Wheatley, P.S. *et al.* (2004) *Nature*, **430**, 1012–1016.

(96) Parnham, E.R. and Morris, R.E. (2007) *Accounts of Chemical Research*, **40**, 1005–1013.

(97) Taubert, A. and Li, Z. (2007) *Dalton Transactions*, 723–727.

(98) Morris, R.E. (2009) *Chemical Communications*, 2990–2998.

(99) Ma, Z., Yu, J., and Dai, S. (2010) *Advanced Materials*, **22**, 261–285.

(100) Xie, Z.-L., White, R.J., Weber, J. *et al.* (2011) *Journal of Materials Chemistry*, **21**, 7434–7442.

(101) Lee, J.S., Mayes, R.T., Luo, H., and Dai, S. (2010) *Carbon*, **48**, 3364–3368.

(102) (a) Binder, J.B. and Raines, R.T. (2009) *Journal of the American Chemical Society*, **131**, 1979–1985; (b) Hu, S., Zhang, Z., Zhou, Y. *et al.* (2008) *Green Chemistry*, **10**, 1280–1283; (c) Lansalot-Matras, C. and Moreau, C. (2003) *Catalysis Communications*, **4**, 517–520; (d) Moreau, C., Finiels, A., and Vanoye, L. (2006) *Journal of Molecular Catalysis A – Chemical*, **253**, 165–169; (e) Zhao, H., Holladay, J.E., Brown, H., and Zhang, Z.C. (2007), *Science*, **316**, 1597–1600.

(103) Abbott, A.P., Capper, G., Davies, D.L. *et al.* (2003) *Chemical Communications*, 70–71.

(104) Abbott, A.P., Boothby, D., Capper, G. *et al.* (2004) *Journal of the American Chemical Society*, **126**, 9142–9147.

(105) Nkuku, C.A. and LeSuer, R.J. (2007) *Journal of Physical Chemistry B*, **111**, 13271–13277.

(106) Carriazo, D., Serrano, M.C., Gutierrez, M.C. *et al.* (2012) *Chemical Society Reviews*, **41**, 4996–5014.

(107) Carriazo, D., Gutiérrez, M.C., Ferrer, M.L., and del Monte, F. (2010) *Chemistry of Materials*, **22**, 6146–6152.

(108) Takenaka, M., Izumitani, T., and Hashimoto, T. (1993) *Journal of Chemical Physics*, **98**, 3528–3539.

(109) Katuri, K., Ferrer, M.L., Gutierrez, M.C. *et al.* (2011) *Energy & Environmental Science*, **4**, 4201–4210.

(110) Choi, Y.H., van Spronsen, J., Dai, Y. *et al.* (2011) *Plant Physiol*, **156**, 1701–1705.

(111) Bergius, F. (1915) *Zeitschrift fur Komprimierte und Flussige Gase*, **17**.

(112) Bergius, F. (1928) *Naturwissenschaften*, **16**, 1–10.

(113) Rice, J.A. (2001) *Soil Science*, **166**, 848–857.

(114) Bergius, F. (1913) *Journal of the Society of Chemical Industry*, **32**.

(115) Bergius, F. (1925) *Zeitschrift Des Vereines Deutscher Ingenieure*, **69**, 1359–1362.

(116) Bergius, F. (1912) *Zeitschrift fur Elektrochemie*, **18**.

(117) Berl, E. and Schmidt, A. (1928) *Justus Liebigs Annalen Der Chemie*, **461**, 192–220.

(118) Berl, E., Schmidt, A., and Koch, H. (1932) *Angewandte Chemie*, **45**, 517–519.

(119) (a) Schuhmacher, J.P., Vanvucht, H.A., Groenewege, M.P. *et al.* (1956) *Fuel*, **35**, 281–290; (b) Schuhmacher, J.P., Huntjens, F.J., and Vankrevelen, D.W. (1960) *Fuel*, **39**, 223–234.

(120) Haenel, M.W. (1992) *Fuel*, **71**, 1211–1223.

(121) Qing Wang, H.L., Chen, L., and Huang, X. (2001) *Carbon*, **39**, 2211–2214.

(122) Sun, X. and Li, Y. (2004) *Angewandte Chemie International Edition*, **43**, 597–601.

(123) Yu, S.H., Cui, X.J., Li, L.L. *et al.* (2004) *Advanced Materials*, **16**, 1636–1640.

(124) Qian, H.S., Yu, S.H., Luo, L.B. *et al.* (2006) *Chemistry of Materials*, **18**, 2102.

(125) Hoekman, S.K., Broch, A., and Robbins, C. (2011) *Energy Fuels*, **25**, 1802–1810.

(126) Seifritz, W. (1993) *International Journal of Hydrogen Energy*, **18**, 405.

(127) (a) Chan, K.Y., Van Zwieten, L., Meszaros, I. *et al.* (2007) *Australian Journal of Soil Research*, **45**, 629–634; (b) Kuzyakov, Y., Subbotina, I., Chen, H. *et al.* (2009) *Soil Biology & Biochemistry*, **41**, 210–219; (c) Warnock, D.D., Lehmann, J., Kuyper, T.W., and Rillig, M.C. (2007) *Plant and Soil*, **300**, 9–20.

(128) Sevilla, M., Macia-Agullo, J.A., and Fuertes, A.B., *Biomass and Bioenergy*, **35**, 3152–3159.

(129) Libra, J.A., Ro, K.S., Kammann, C. *et al.* (2011) *Biofuels*, **2**, 89–124.

(130) Titirici, M.M., Thomas, A., and Antonietti, M. (2007) *New Journal of Chemistry*, **31**, 787–789.

2

Porous Hydrothermal Carbons

Robin J. White[1], Tim-Patrick Fellinger[2], Shiori Kubo[3], Nicolas Brun[2], and Maria-Magdalena Titirici[4]

[1] *Institute for Advanced Sustainability Studies, Earth, Energy and Environment Cluster, Germany*
[2] *Max Planck Institute of Colloids and Interfaces, Germany*
[3] *Absorption and Decomposition Technology Research Group, National Institute of Advanced Industrial Science and Technology, Japan*
[4] *School of Engineering and Materials Science, Queen Mary, University of London, UK*

2.1 Introduction

Beginning with their use as adsorbents for the decolorization of alcohol, water, and sugar or for the removal of moisture from architecture in ancient times, porous carbon materials have been widely used in our society [1]. More recent applications are dedicated to their use in industry as adsorbents (e.g., activated carbon) for drinking water, wastewater, and gas purification, or as catalysts or catalyst supports [2–5]. The well-developed surface area and porosity of such materials are critical to these applications, typically enhancing the adsorption capacity/selectivity or catalytic activity. Porous carbon materials are becoming of increasing interest in the developing application fields of energy storage [6] (e.g., electrodes for Li-ion batteries or supercapacitors), fuel cells (e.g., novel catalysts or catalyst supports for the oxygen reduction reaction) [7–9], or chromatography technologies [10–12], which is expected to amount to a market worth US$90 billion in 2012 (Figure 2.1) [13]. When compared to conventional electrolytic capacitors or to batteries, carbon-based supercapacitors offer good specific energy (1–10 Wh kg^{-1}), as recently demonstrated,

---

Sustainable Carbon Materials from Hydrothermal Processes, First Edition. Edited by Maria-Magdalena Titirici.
© 2013 John Wiley & Sons, Ltd. Published 2013 by John Wiley & Sons, Ltd.

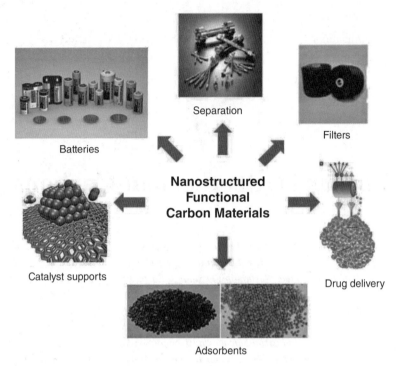

Figure 2.1   Schematic representation of the main application fields of porous carbon materials. (Reprinted with permission from [15]. © 2011 S. Kubo, University of Potsdam (Potsdam, Germany).)

for example, by carbon aerogels with high porosity and surface area (greater than 50%, 400–1000 m^2 g^{-1}) [14].

Due to the extreme versatility of carbon materials available (e.g., crystalline, amorphous, bulk, nanostructured, activated, functionalized), a great deal of literature research has discussed the use of various electrode configurations for supercapacitors or secondary batteries (e.g., see [6]). This is a good example of how material porosity directly affects electrochemical performance. In order to improve even further the performance of carbon materials, novel synthetic approaches as well as a developed fundamental understanding of their properties are strongly desirable from a materials chemistry point of view.

In this context, as we have seen in the previous introductory chapter, the hydrothermal carbonization (HTC) process can be considered as a very interesting, green, and versatile technique to produce carbon materials [16, 17]. However, as they are synthesized directly from biomass or carbohydrates at low temperatures (130–250 °C), they essentially present very limited surface area and pore volumes (see Section 6.10), thereby inhibiting their use in the aforementioned application fields.

This chapter focuses on the development of synthetic routes to introduce well-defined porosity into hydrothermal carbons. In this chapter, various synthetic routes towards porous HTC-derived materials or composites will be introduced and discussed, focusing on the synthesis of different pore systems and morphologies directed by the presence of various structure-directing agents (e.g., organic and inorganic templates). Attention will also be given to the utilization of naturally occurring biocomposites (e.g., crustacean shells) and their use in the preparation of porous HTC-derived materials. The presented methods allow tailoring of the final structure via the tools of colloid and polymer science, leading to selectable material morphology for a wide range of applications.

2.2 Templating – An Opportunity for Pore Morphology Control

In relation to the introduction of porosity into carbon materials, Knox *et al.* [10] was one of the first to synthesize nanoporous carbon materials with a variety of pore structures using the appropriate nanostructured silica materials as sacrificial templates [10]. This procedure reflects the "hard (exo) templating" approach to structure replication [18, 19]. Generally, in hard templating, a carbon precursor (e.g., phenol/formaldehyde resin) infiltrates a sacrificial hard template structure and is carbonized within the pores (normally at high temperatures; e.g., greater than 700 °C). The template is then removed to give the inverse nanostructures of the original inorganic templates (Figure 2.2a)

More recently, Zhao *et al.* and Dai *et al.* have independently showed that under the correct synthesis conditions, the use of a sacrificial inorganic template is not required, demonstrating a direct route to ordered porous carbon materials via the self-assembly of block copolymers and suitable aromatic carbon precursors (e.g., phloroglucinol or resorcinol/formaldehyde) [20–25]. This method, widely recognized as one of the classical methods to produce inorganic porous materials by sol–gel chemistry, is called "soft (endo) templating" (Figure 2.2b).

The hydrothermal process has clear advantages when compared to the classical carbonization methods (e.g., pyrolysis) since it takes place in the aqueous phase and thus can,

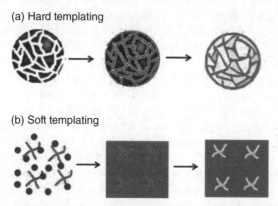

Figure 2.2 *Schematic drawing of the (a) hard templating and (b) soft templating approach to synthesize porous and high-surface area materials. For soft (endo) templating, the templating species are occluded in the forming solid; for hard (exo) templating, a rigid porous solid is infiltrated with the precursor for the final solid. Endo templates as well as exo templates are removed from the composite to yield the porous or high-surface area materials. (Adapted with permission from [18]. © 2003 WILEY-VCH Verlag GmbH & Co. KGaA, Weinheim.)*

in principle, be easily combined with classical templating procedures. Pure carbohydrates are in this case the most convenient precursors since they are easily dissolved in water. Additionally, polar functionalities are present on the materials surface, allowing relatively facile further modification to be performed, which is somewhat difficult in the case of carbon materials with a highly developed aromatic/graphitic character usually obtained from the conventional carbonization procedure (e.g., synthesized using a phenol/formaldehyde resin precursor) [17]. The surface chemistry of the resulting HTC-derived materials can also easily be controlled and in turn carbon condensation/aromaticity (e.g., by post-thermal treatment) offers the opportunity to direct surface or pore wall chemistry for specific application needs. More details on the surface chemistry of hydrothermal carbons are provided in Chapter 6, dealing with the characterization of these materials.

2.2.1   Hard Templating in HTC

Following the same principle as the structural replication presented by Knox *et al.* [10], mesoporous hydrothermal carbons have been produced by conducting HTC at 180 °C in the presence of nanostructured sacrificial silica bead templates (LiChrospher® Si100; average particle size 10 μm, pore size 10 nm) [26]. Significantly, it was found that it is important to match the template surface polarity with that of the carbon precursor. Hydrophobization of the silica template can be performed via methylation or thermal treatment. A variety of morphologies are obtained using silica templates with different polarities and 2-furaldehyde as a carbon source (Figure 2.3). The use of silica templates with a very hydrophobic surface character leads to the generation of macroporous carbon casts (Figure 2.3a), whereas hollow carbon spheres are produced when using a moderately hydrophobic surface (Figure 2.3b). Dehydroxylated silica templates filled with 60 wt% 2-furaldehyde resulted in mesoporous carbonaceous microspheres (Figure 2.3d), thus demonstrating successful templating, whereas application of a 30 wt% carbon precursor gave only small carbon spherules (6–10 nm diameter) owing to the lack of interconnectivity between growing particles in the coating. Furthermore, when rehydroxylated silica was used, demixing occurred due to the enhanced hydrophilic character of the template surface essentially inhibiting impregnation of the 2-furaldehyde.

Complete impregnation was achieved by controlling the degree of hydrophobicity of the template and by controlling the concentration of the carbon precursor in an aqueous solution. In addition to the matching of the surface polarities, it was also illustrated that the carbon nanocoating procedure operates in a "patchwise" manner, via the generation of stable colloidal intermediates. For nonporous templates, hollow carbonaceous spheres with a robust carbon coating are observed (Figure 2.3c). Further investigation was conducted using mesoporous silica beads (LiChrospher Si300) with an average particle size and pore diameter of 10 μm and 30 nm, respectively, to yield spherically shaped carbonaceous particles [15, 26]. Scanning electron microscopy (SEM) images of the obtained carbon replica of Si300 reveals that the both particle size (around 10 μm) and morphology uniformity were maintained (Figure 2.4a and b).

One of the main advantages of the combined HTC/hard templating method is the accessibility of nanostructured materials possessing useful surface functionalities. By further thermal treatment under an inert atmosphere of the obtained composites (i.e., 350, 550, and

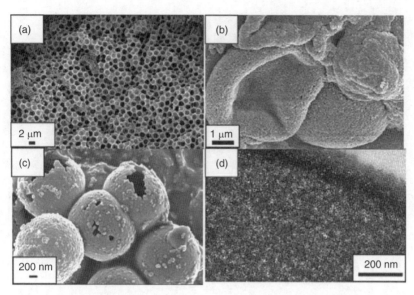

Figure 2.3 *Morphologies obtained via HTC of 2-furaldehyde in the presence of silica templates with different polarities. (a) Macroporous carbon cast obtained in the presence of nonporous silica with a very hydrophobic surface, (b) mesoporous-shell hollow spheres obtained from mesoporous silica templates with moderate hydrophobicity, (c) hollow spheres obtained using nonporous silica spheres with moderate hydrophobicity, and (d) mesoporous hydrothermal carbon replica obtained using a calcined mesoporous silica template. (Reproduced with permission from [26]. © 2007 WILEY-VCH Verlag GmbH & Co. KGaA, Weinheim.)*

Figure 2.4 *SEM micrographs of (a and b) an as-synthesized carbon replica (HTC at 180 °C) and (c and d) a carbon replica after postcarbonization at 750 °C. (Reprinted with permission from [15]. © 2011 S. Kubo, University of Potsdam (Potsdam, Germany).)*

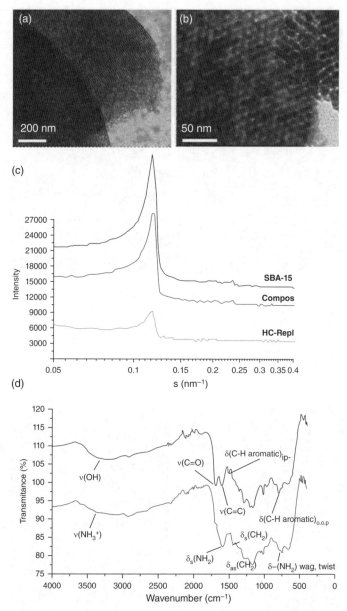

Figure 2.5   *(a and b) TEM micrographs of the ordered mesoporous carbon obtained by replication of the SBA-15 silica template. (c) SAXS patterns of the silica template, corresponding composites and ordered HTC-derived materials). (d) FTIR spectra of the ordered carbon replica before and after amino functionalization. (Reproduced with permission from [27]. © 2007 Royal Society of Chemistry.)*

750 °C) followed by template removal, carbon spheres with different surface functionalities and pore properties were obtained. This postcarbonization treatment had no significant impact on the spherical material morphology (Figure 2.4c and d). Elemental analysis, revealed an increase in carbon content of the synthesized materials from 64 to 81 wt% (350 to 750 °C), thus showing an increasing carbon-rich material character.

Similarly to the production of CMK-type materials pioneered by Ryoo *et al.* [19], hexagonally ordered mesoporous SBA-15 silica has also been used as a sacrificial inorganic template for the preparation of ordered carbonaceous replicas under hydrothermal conditions [27]. Transmission electron microscopy (TEM) images of the carbonaceous duplication also demonstrates that the ordered hexagonal pattern is maintained in the carbon product (Figure 2.5a and b). Small-angle X-ray scattering (SAXS) analysis of the carbon replica presented Bragg reflections of (100), (110), and (200), characteristic of the two-dimensional hexagonal arrangement (Figure 2.5c), further confirming the maintenance of the structural regularity. Using the functionality generated by the HTC process, relatively simple chemical modification strategies can employed to further modify the SBA-15 carbon replica pore wall and surface groups. The introduction of surface amino groups was confirmed by elemental analysis (around 4 mmol N g^{-1}) and Fourier transform infrared spectroscopy (FTIR) spectroscopy as indicated by bands at 1615 and 772 cm^{-1} characteristic of amino ($-NH_2$) groups (Figure 2.5d).

Analogous to the structural replication of mesoporous silica, the nanostructure of a macroporous anodic aluminum oxide membrane was also achieved via a combined HTC/hard templating strategy to yield tubular carbons [28]. Again, the surface character of the tubular carbonaceous nanostructure was controlled by employing postsynthesis thermal treatments. TEM images reveal that the micrometer-long tubular carbons possess an open-ended structure with smooth surfaces constituted of a hollow internal diameter of around 125 ($\pm$25) nm and around 40 nm thick carbonaceous tube walls (Figure 2.6), with pore wall thickness observed to marginally decrease as the treatment temperature passed 550 °C (i.e., as the carbon network condenses).

High-resolution TEM images of before (Figure 2.6e) and after carbonization (Figure 2.6f), revealed the level of local order improved with temperature indicative of increasing aromatic/pseudographitic character. In addition, the surface area and the pore volume were observed to increase upon employing higher postcarbonization temperature as was observed in the case of carbon spheres again due to the development of microporosity (see Chapter 6 for details). Furthermore, the useful material chemistry of these tubular structures can be modified with stimuli-responsive polymers for controlled dispersion applications [28].

2.2.2 Soft Templating in HTC

The use of soft, polymeric templates of defined size and shape (e.g., polymer nanoparticles) can also be employed in the synthesis of nanostructured HTC-derived materials. White *et al.* have used aqueous dispersions of polystyrene latexs (PSL) nanoparticle (e.g., $D = 100$ nm) as a templates to synthesize functional hollow carbon nanospheres (Figure 2.7a) [29]. Using the hydrothermal conversion of glucose at 180 °C in the presence of the PSL template results in a controllable coating of hydrothermal carbon of the polymer body. Thermal treatment of the resulting composite, above the decomposition point of PSL (e.g., above 500 °C), results

Figure 2.6 *SEM micrographs of (a) as-synthesized tubular carbons and (b) tubular carbons postcarbonized at 750 °C. TEM micrographs of (c) as-synthesized tubular carbons and (d) tubular carbons postcarbonized at 750 °C. High-resolution TEM micrographs of (e) as-synthesized tubular carbons and (f) tubular carbons postcarbonized at 750 °C. (Reproduced with permission from [28]. © 2010 American Chemical Society.)*

in formation of hollow carbonaceous nanospheres. Latex nanoparticles of any desired diameter and narrow size distribution can be readily prepared allowing, in principle, subtle control of the hollow carbon nanosphere size over a wide range of diameters. Details on the macroporosity of these materials determined by mercury intrusion and the perfection of replication with regard to the size of the initial latex templates is provided in Section 6.10.

Thermal treatment at 1000 °C generates a turbostratic-type of carbon layering in the nanospheres wall with a thickness of around 20 nm (Figure 2.7b), suitable for the application as anode materials in Li^+ [30], Na^+ [31] as well as cathode materials in Li–S batteries [32], as discussed in Chapter 7 dedicated to the applications of HTC.

It is proposed in materials science that enhanced application performance (e.g., catalysis) can be generated by the synthesis of an organized, uniformly sized and shaped material pore texture at the nanoscale, in tandem with controlled pore wall chemistry. As mentioned

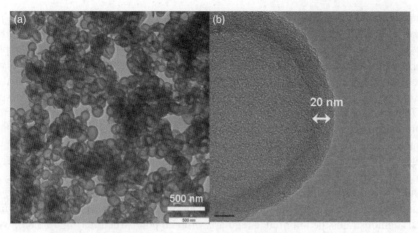

Figure 2.7   *(a) TEM images of MPG-C*100 nm *HTC-derived composite material (i.e., before template removal). (b) MPG-C*100 nm-*550 °C demonstrating polystyrene latex template removal. (c and d) High-resolution TEM images of MPG-C*100 nm-*1000 °C. (Reproduced with permission from [29].* © *2010 American Chemical Society.)*

earlier, Zhao *et al.* and Dai *et al.* have recently demonstrated a soft templating route, a direct route to ordered porous carbons via organic–organic self-assembly of a block copolymer (e.g., Pluronic® block copolymers or polystyrene-*b*-poly(4-vinylpyridine)) and a suitable aromatic carbon precursor (e.g., phloroglucinol [21] or phenol/formaldehyde [22, 23]). Hydrogen-bond interactions play a significant role in orientating the resulting polyaromatic network, whilst pore structuring was conventionally controlled mainly by polymeric template (e.g., via concentration and structure), copolymer/precursor ratio, and pH. The polymeric template can be then removed either by solvent extraction or heat treatment around its decomposition temperature. Carbons prepared using such strategies typically present rather chemically condensed (e.g., graphitic-like) pore walls/surfaces due to the high-temperature syntheses, which inhibit facile postchemical modification and limit surface hydrophobicity/polarity modification. Therefore, performing the same type of soft templating process under HTC conditions becomes of great interest. One of the main advantages is the fact it offers direct access to ordered porous carbons with useful surface functionality, whilst it will be a key point and highly challenging to achieve the stable formation of a polymeric template/precursor assembly under HTC conditions. In this context, it was found that the selection of a suitable carbon precursor and HTC temperature leads to successful "soft" templating of hydrothermal carbons.

In the first attempt, HTC of D-fructose was performed at 130 °C in the presence of block copolymer Pluronic F127 (ethylene oxide$_{106}$–propylene oxide$_{70}$–ethylene oxide$_{106}$) to yield a carbonaceous precipitate with an ordered nanostructure [33].

This composite was further heat treated under nitrogen gas to 550 °C, resulting in template removal and yielding a carbonaceous product (denoted as C-MPG1-*micro*), composed of 82.6 wt% C, 14.2 wt% O, and 3.2 wt% H, respectively. Here, the use of D-fructose is considered to play a crucial role in successful templating of the supramolecular assembly of the block copolymer template since it undergoes HTC at 130 °C – a significantly lower temperature than that previously reported for other sugars (e.g., D-glucose: 180 °C) [33]. Typically, block copolymer micelles are not stable at the high temperature used for the HTC process [34]. Therefore, the use of D-fructose allows access to a more stable micellar phase and opens the opportunity for soft templating. The complete removal of block copolymer species from the template/carbon composite was confirmed by thermogravimetric analysis, representing a significant mass loss regime at around 400 °C (i.e., the F127 decomposition) followed by a smaller secondary charring step at around 600 °C (i.e., further condensation/aromatization of the carbon structure). SEM image analysis of C-MPG1-*micro* indicated a layer-by-layer growth mode yielding relatively uniform cuboctahedron-like particles in the size range 1–10 μm. Whilst of a similar size to nontemplated HTC-derived material, particles synthesized via this templating approach present a faceted edge/layered morphology, indicative of the growing direction of the hydrothermal carbon network by the templating phase as well as of the formation of near-single-crystalline particles. TEM image analysis demonstrates the long-range regularly ordered pore structure (Figure 2.8a), whilst synchrotron SAXS analysis of C-MPG1-*micro* indicates the formation of a near perfect cubic *Im3m* symmetry (Figure 2.8b). Moreover, the obtained two-dimensional scattering

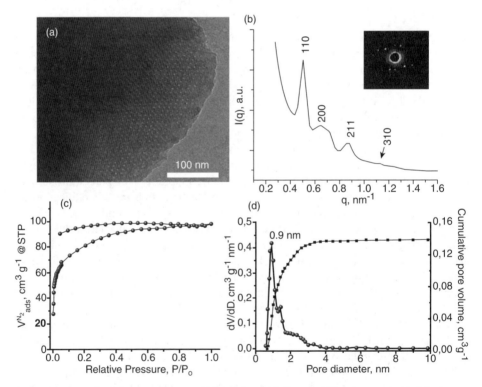

Figure 2.8 *(a) TEM, (b) synchrotron SAXS pattern (right-up: diffraction pattern), (c) nitrogen sorption isotherm, and (d) pore size distribution (quenched solid density functional theory (QSDFT)) of as-synthesized C-MPG1-micro material. (Reproduced with permission from [33]. © 2011 American Chemical Society.)*

pattern (Figure 2.8b, insert) is indicative of a single-crystalline nature of this material and presents well-resolved peaks at q-spacing values of 0.51, 0.72, and 0.87 nm^{-1}, respectively, corresponding to a d-spacing of 12.3 nm and equating to a unit cell parameter of 17.4 nm. The calculated synchrotron SAXS unit cell parameter agrees well with the presented TEM image, while unusually both synchrotron SAXS and TEM indicate the generation of a very thick pore wall feature for C-MPG1-*micro* with dimensions of between 7 and 10 nm.

In order to investigate the pore property of the obtained material, nitrogen sorption analysis was carried out generating a microporous type I isotherm with a nonreversible nature due to structural changes in pore wall dimensions during sorption processes (Figure 2.8c) [35]. We provide a detailed explanation regarding which models we have to apply in order to get a correct interpretation of such an adsorption isotherm when we will describe the gas sorption applied to HTC-derived materials in Chapter 6. Specific surface area and total pore volume were calculated as 257 m^2 g^{-1} and 0.14 cm^3 g^{-1}, respectively. The pore size distribution presents a sharp peak at a diameter of 0.9 nm and a less discrete broader shoulder centered around 2 nm (Figure 2.8d). By comparison, the F127/HTC carbon composite material presented very limited pore volume and surface area, indicating that indeed pores are opened upon removal of the block copolymer template via calcination. TEM image analysis of this composite indicated an F127 micelle diameter of around 11 nm and carbon wall thickness of around 7 nm (Figure 2.9a). SAXS analysis of the F127/HTC carbon composite demonstrated a well-resolved pattern with the d-spacing value of the first peak equivalent to 16.8 nm, corresponding to a unit cell parameter of 23.6 nm, which is considerably larger than pore dimensions observed after template removal (Figure 2.9b). A possible formation mechanism was proposed. It is believed that as HTC proceeds via the dehydration/polycondensation reactions to generate a polyfuran-like network [36] (see details in Section 6.9). The organized block copolymer micellar phase is essentially "templated," whereby in the initial steps D-fructose

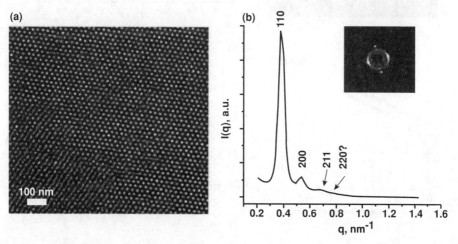

Figure 2.9 (a) TEM micrograph and (b) synchrotron SAXS pattern of C-MPG1-micro before template removal (insert: two-dimensional scattering pattern). (Reproduced with permission from [34]. © 1989 VCH Verlag GmbH, Weinheim.)

is absorbed via hydrogen-bonding interactions into the hydrophilic poly(ethylene oxide) (PEO) moiety of the F127. Here, the resulting polyfuran-like network is comparatively hydrophobic and therefore the possibility of "templating" of a reversed phase through a hydrophobic interaction still cannot be discounted. Upon removal of the template at 550 °C, the resulting pores are smaller than that before template removal due to structural shrinkage (e.g., carbon network condensation) or possible partial carbonization of the block copolymer.

Importantly, further work revealed that the ordered pore phase dimensions of this C-MPG1-*micro* material could be shifted into the mesoporous domain by the addition of a pore-swelling agent, trimethylbenzene (TMB) to the F127/D-fructose reaction mixture (denoted as C-MPG1-*meso*). Thermal template removal at 550 °C generates carbonaceous particles again with faceted ages similar in morphology to C-MPG1-*micro*. Close examination of the pore structuring via TEM image analysis indicated the maintenance of the well-ordered pore structuring upon addition of the pore-swelling agent with a pore diameter and wall thickness of around 5 nm and around 10 nm, respectively (Figure 2.10a). Synchrotron SAXS analysis of C-MPG1-*meso* presented well-resolved peaks at inverse

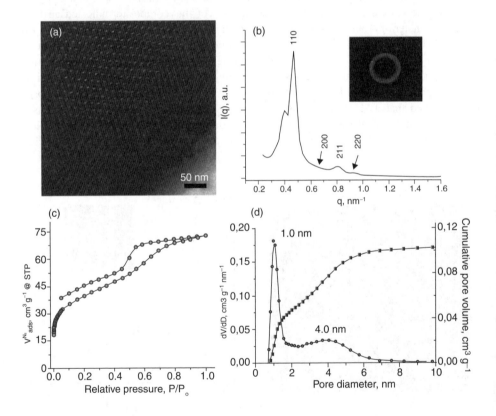

Figure 2.10   (a) TEM, (b) synchrotron SAXS pattern (right-up; diffraction pattern), (c) nitrogen sorption isotherm, and (d) pore size distribution (QSDFT) of as-synthesized C-MPG1-meso material. (Reproduced with permission from [33]. © 2011 American Chemical Society.)

nanometer values of $q = 0.47, 0.65, 0.80$, and 0.94 nm^{-1}, and a further peak at higher q values (Figure 2.10b), indicative of *Im3m* symmetry and a unit cell of 18.9 nm showing the increased regular unit. The peak at 0.47 nm^{-1} is accompanied with a small shoulder feature indicating some structural heterogeneity (e.g., the presence of a mixed phase) as also demonstrated by the increased polycrystalline nature of this material (two-dimensional scattering pattern; Figure 2.10b, insert).

The transition from a type I (i.e., for C-MPG1-*micro*) to a type IV nitrogen sorption profile for C-MPG1-*meso* was observed with associated capillary condensation feature at $p/p_0 \sim 0.45$, indicative of a shift in pore structuring into the mesopore domain (Figure 2.10c). Consequently, this resulted in a reduction in surface area ($S_{BET} = 116$ m^2 g^{-1}) and total pore volume ($V_{total} = 0.10$ cm^3 g^{-1}). The addition of the swelling agent resulted in a bimodal pore size distribution with a new maximum at 4.0 nm in the mesopore domain and a discrete micropore peak at 1.0 nm (e.g., from template removal in mesopore walls) (Figure 2.10d). Correspondingly, mesopore volume was enhanced as a proportion of the total pore volume, increasing from around 20% in C-MPG1-*micro* to around 60% upon addition of TMB, for C-MPG1-*meso*. Mechanistically, the added TMB is believed to interact with hydrophobic poly(propylene oxide) moieties, thus swelling the spatial volume of the hydrophobic region, which in turn results in the observed pore expansion into the mesopore range as observed in more classical inorganic templating examples [37, 38].

It is presumed that carbonaceous crystal growth is determined by the rate of hexose dehydration (i.e., the generation of hydroxymethylfurfural (HMF)) and subsequent polymerization reactions. Relatively slow dehydration and strong hydrogen-bonding interactions between HMF and the PEO segments of the F127 template favor the growth of the templated, almost perfect single-crystalline carbonaceous structures.

2.2.3 Naturally Inspired Systems: Use of Natural Templates

The organization of organic and inorganic matter at the nanometer scale is exemplified best by biological materials systems and the development of novel materials mimicking or deriving inspiration from these natural systems is a dominant theme in materials chemistry, with one aim to produce material systems with the same efficiency and selectivity as those provided by nature [39]. In this regard, biominerals are typically natural inorganic/organic (e.g., polysaccharide) composites that perform as structural supports, mineral storage, and protective surfaces [40]. Self-assembly of polysaccharides into organized structures with a cooperative inorganic partner such as CaCO$_3$ produces lightweight and dimensionally strong natural composites, as demonstrated by the exoskeletons of crustaceans and insects [41], where the polysaccharide content is sufficiently high to enable follow-up conversion chemistry. The inorganic and organic components are often organized into nanoporous-sized domains and replication/mimicking of these systems can potentially lead to the design of novel advanced materials. White *et al.* have reported on the HTC of crustacean exoskeletons, namely prawn and lobster shells, into nanostructured (nitrogen-doped) porous carbonaceous materials (Figure 2.11) [42].

Marine-derived crustacean shells are composed of the nitrogen-containing polysaccharide chitin (poly-β-(1 $\rightarrow$ 4)-*N*-acetyl-D-glucosamine) and an inorganic component (e.g., CaCO$_3$). Such high-volume industry shell waste represents an inexpensive, accessible

Figure 2.11 *Prawn shell-derived carbons via HTC processing and post-thermal treatment at 750 °C; acetic acid [H⁺] is used to remove the CaCO₃ component. (Reproduced with permission from [42]. © 2009 Royal Society of Chemistry.)*

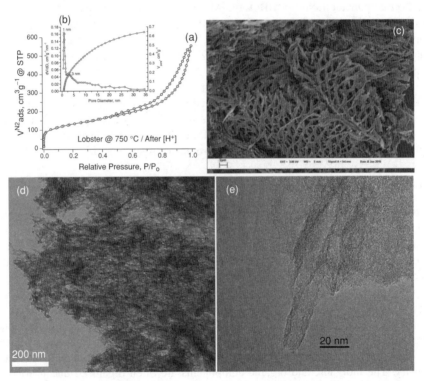

Figure 2.12 *Lobster shell-derived hydrothermal carbon materials (and thermal treatment at 750 °C): (a) nitrogen sorption isotherm, (b) corresponding nonlocal density functional theory pore size distribution, (c) SEM image of the macromorphology, (d) TEM, and (e) high-resolution TEM images of the carbon nanostructure. (Adapted with permission from [46]. © 2012 Royal Society of Chemistry.)*

source of nitrogen-containing biomass that does not directly impact the food chain. These exoskeleton biocomposites represented useful precursor biomasses for the preparation of porous carbons, whereby the organic component is carbonized in the presence of the (structure-donating) inorganic component $CaCO_3$, which finally could be removed with the use importantly from a process point, a weak (waste) Brønsted acid, acetic acid (e.g., vinegar). These acids are to be compared to the more aggressive reagents usually employed in inorganic template removal in conventional mesoporous carbon synthesis, for instance HF/NH_4F (aq.) or caustic NaOH conditions [43].

The use of nitrogen-containing polysaccharide-derived biomass is an important aspect, as it directly allows the preparation of especially useful nitrogen-doped carbon materials with improved performance range (e.g., for electrodes for supercapacitors [44] or fuel cells [45]). We provide detailed aspects of such improved application performance in Chapter 5 and especially Chapter 7.

Nitrogen-doped carbon synthesis from crustacean shells such as lobster shells [46] would not only be beneficial from an economic standpoint, but also allow a simple route for the introduction of material dopant within carbon structures that would be almost impossible to reproduce synthetically (Figure 2.12) – particularly advantageous in high-value end applications. Nitrogen doping of carbons is, for instance, known to enhance material electrical conductivity, as the introduction of the nitrogen heteroatom into the graphitic structure contributes further electron carriers in the conduction band [47]. Further discussion regarding the heteroatom doping of hydrothermal carbon structures is developed in Chapter 5, while the use of these materials as electrodes in supercapacitors or heterogeneous catalyst supports is discussed in Chapter 7.

2.3　Carbon Aerogels

Aerogels are known in a great variety of compositions and are used in numerous high-end applications, including chromatography, adsorption, separation, gas storage, detectors, heat insulation, and as supports and ion exchange materials [48, 49]. Without doubt, silica aerogels are the most prominent and developed materials in this field, as they have revolutionized high-pressure liquid chromatography (HPLC), allowing the highest resolution and fast separation at the same time [50]. Silica, however, also has some chemical disadvantages and, in 1987, Pekala *et al.* described for the first time organic aerogels as an interesting alternative [51]. Aerogels are defined by the International Union of Pure and Applied Chemistry as nonfluid networks composed of interconnected colloidal particles as the dispersed phase in a gas that is usually air. Typically, an aerogel is formed via the supercritical drying of a wet colloidal gel. Their characteristics include interconnected large-diameter mesopores, low material density ($\rho \sim 0.004$–0.5 g cm^{-3}). and high specific surface area. The formation of colloidal gels is a "bottom-up" synthetic process following the well-established sol–gel chemistry where colloidal particles form at first, and then are aligned and condensed by thermodynamic forces [52]. Therefore, aerogel synthesis, achieved via the HTC processing of sugars, is of significant interest, not only from the viewpoint of green chemistry, but also with application potential in mind [53].

The exciting new HTC system was further investigated, especially in terms of particle size and porosity control. The results should be related to the well-known trends in the resorcinol/formaldehyde system. Hierarchically structured but also inexpensive porous organic aerogels are capable of competing with their inorganic (e.g., silica) counterparts in sorption and insulation applications [53]. The most common organic aerogels are the aforementioned resorcinol/formaldehyde systems [52]. Here, the formation of the gel phase occurs through the catalyzed polycondensation of the reactive precursors in water. The whole gel formation can take several days. A well-developed microstructure is also crucial for the following drying procedure to achieve the organic aerogels. The solvent has to be removed from the pores, while the structure has to withstand capillary forces induced by evaporation. An introduction to this topic has been provided by Job *et al.* [54]. So far, supercritical drying is still the most prominent procedure used to minimize the effects of capillary collapse.

In addition to being highly porous and lightweight, another advantage of organic aerogels is the possibility to introduce electric conductivity by converting them into carbon aerogels, thus accessing electrical/electrochemical applications like batteries [55], supercapacitors [56], and as conductive supports as cathodes for fuel cell applications [57]. For those applications, it is important to have additional control over micro- and mesoporosity development. It is possible to subtly influence and design material nanoscale structuring by chemical means, such that carbon aerogels can also be classified as nanostructured carbon. To convert polymers into carbon aerogels, pyrolysis is usually employed (e.g., $T > 500\,°C$). The carbonization process leads to a loss of oxygen and hydrogen functionalities, and therefore more condensed carbon structures [57]. Finally, depending on the pyrolysis temperature and the respective precursor chemistry ("graphitizability") of the former organic structure, an electrically conductive carbon network is achieved and can be used for a wide range of applications. A review article on resorcinol/formaldehyde-based organic and carbon aerogels was published by Al-Muhtaseb *et al.* [58]. In the following sections, four promising routes to the formation of HTC aerogel materials are introduced and discussed.

2.3.1   Ovalbumin/Glucose-Derived HTC-Derived Carbogels

Moving away from the use of templates, the introduction of secondary biomass precursors (e.g., proteins) has been found to have interesting consequences on the porosity, morphology, and texture of the resulting HTC-derived materials. Baccile *et al.* first reported that throughout addition of the glycoprotein (ovalbumin), at variation concentrations, carbonaceous nanoparticles or continuous nanosponges with high specific surface areas can be efficiently produced [59]. Taking this approach one step further, it was found that under the right HTC processing conditions of temperature, time, drying technique (e.g., supercritical CO_2 extraction), and ovalbumin concentrations, it is possible to synthesis highly functional, low-density, nitrogen-doped HTC-derived monolith materials, termed "Carbogels" (Figure 2.13) [60]. Commonly available inexpensive proteins such as ovalbumin are interesting additives for the HTC process, being H_2O soluble, nitrogen rich, and having the potential to act as surface-stabilizing/structure-directing agents during the formation of the evolving hydrothermal carbon network. After supercritical drying of the recovered gel product, a low-density monolith product is produced (Figure 2.13). The material was then characterized by solid-state nuclear magnetic resonance (NMR), FTIR and X-ray photoelectron spectroscopy (XPS).

Qualitative bulk ^{13}C NMR analysis indicates distinct resonances in the $\delta = 0–60$ ppm region for sp^3 aliphatic "C," and at $\delta = 29.8$ and 52.9 ppm for methyl and/or methoxyester

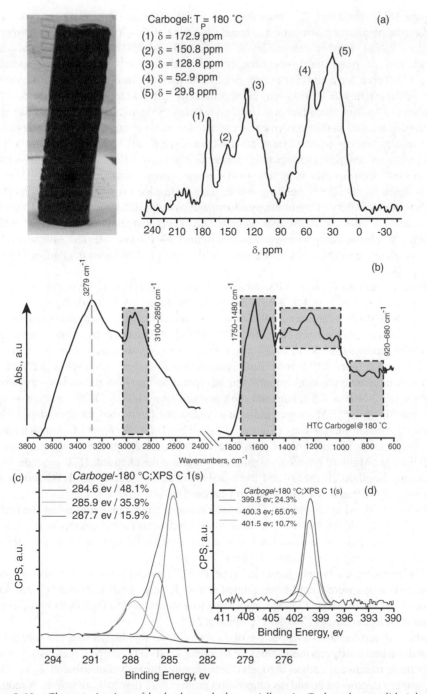

Figure 2.13 Characterization of hydrothermal glucose/albumin Carbogel monolith (photograph) product by (a) ^{13}C cross-polarization/magic angle spinning NMR, (b) attenuated total reflectance FTIR, and high-resolution XPS analysis of the (c) C 1(s) and (d) N 1(s) photoelectron envelopes. (Adapted with permission from [60]. © 2009 Royal Society of Chemistry.)

groups [61], whilst sp^2 "C" resonances (δ = 90–160 ppm) are relatively pronounced, indicative of a strong aromatic Carbogel character at this low preparation temperature (Figure 2.13a). Further resonances at δ = 128.8 and 150.8 ppm relate to all-carbon aromatics and poly(furan)-type networks, respectively [36]. The intense resonance at δ = 172.9 ppm for carboxylate-type groups and partially resolved higher resonances for ketones/aldehydes (δ > 200 ppm) are significantly different to previously reported NMR analyses of hydrothermal carbons. Notably, resonances are relatively sharp and decoupled, indicating a rather mobile (polymeric) nature to material structure. FTIR analysis indicates carbonyl/carboxylic features (at 1750–1480 cm^{-1} (e.g., v(C=O) $\sim$ 1626 cm^{-1})) mixed with nitrogen-containing groups (e.g., amines at 1580 cm^{-1}) (Figure 2.13b) [62]. A band at 1666 cm^{-1} corresponds to amide C=O groups whilst a band at around 1076 cm^{-1} is assigned to the C—N bending mode. There are also possible contributions in this region from a variety of pyridinic/pyridonic and pyrrolic-like structures [63]. The band at 1516 cm^{-1} is reflective of conjugated/condensed oxygen-containing groups (e.g., C=C—O; "furans"). A broad composite spectral band in the 900–720 cm^{-1} region, assigned to C—H (out-of-plane) bending modes, reflects a wide range of (nitrogen-containing) aromatic groups [63].

High-resolution C 1(s) XPS analysis reveals three distinct binding energies of 284.6 (C—C/—CH$_x$; 48.1% area), 285.9 (C—N bond: 35.9 % area), and 287.7 eV (—C=O/—C(O)N— (amide); 15.9% area) (Figure 2.13c). N 1(s) envelope analysis indicated a high pyrrolic surface character (400.3 eV; 65.0% area), contributions from primary/secondary amines (399.5 eV; 24.3% area), and pyridinic/quaternary-type motifs (401.5 eV; 10.7% area) (Figure 2.13d) [64]. Both bulk elemental analysis and surface XPS quantification demonstrate the uniform nitrogen incorporation into both the carbon structure and surface (e.g., N% = 7.5 (elemental (combustion) analysis)/6.8 (XPS survey scan quantification)), whilst SEM images indicate a continuous interlinked nitrogen-doped fibrous monolithic macromorphology (Figure 2.14a and b). The early onset of aromatization and relative absence of keto groups was believed to be a consequence of the reaction with protein (e.g., Maillard reaction), significantly altering the chemical HTC cascade. Details regarding Maillard chemistry and the reactions occurring during the HTC formation of nitrogen-doped carbons are provided in Chapter 5.

Using the dried hydrothermal Carbogel, derivatives were prepared using thermal carbonization (T_p) in the 350–900 °C range to manipulate surface chemistry (e.g., carbon or nitrogen condensation state) and demonstrate material flexibility (see Chapter 5 for further discussion relating to heteroatom doping).

The resulting Carbogels presented type IV/H3 reversible sorption isotherms, where hysteresis loops were found to be limited at low T_p (e.g., 350 °C or below), developing through the intermediate T_p range, stabilizing to a more classical type IV/H3 profile with a more open hysteresis, at T_p = 900 °C (Figure 2.15).

Material surface areas irrespective of T_p are considerably higher (S_{BET} > 240 m^2 g^{-1}) than those previously reported values for (nontemplated) HTC-derived materials (but lower than more traditional carbon aerogels), here passing through a maxima at T_p = 550 °C as strongly micropore-bound decomposition products are thermally removed. A reduction in V_{micro} at higher T_p then occurs as the carbon network undergoes further chemical condensation and micropore closure. Mesoporosity as a proportion of total porosity stabilizes after T_p = 550 °C, the result of network contraction, reaching promising values of

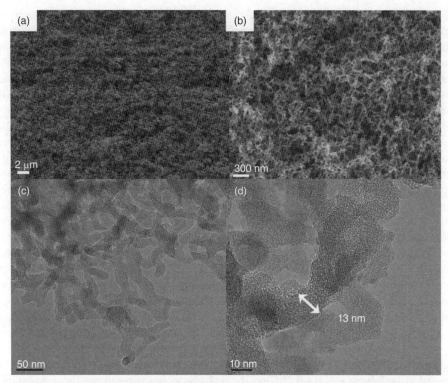

Figure 2.14 *(a and b) SEM images of the aerogels, and (c and d) high-resolution TEM images of aerogels thermally treated at 750 and 900 °C, respectively. (Reproduced with permission from [60]. © 2009 Royal Society of Chemistry.)*

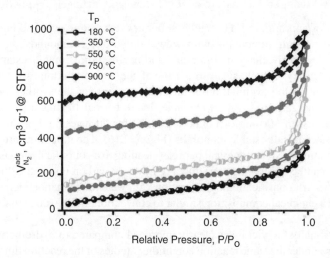

Figure 2.15 *Nitrogen sorption isotherms for the Carbogel/180 °C and carbonaceous derivative therefrom prepared at increasing carbonization temperature (T_p).*

$V_{meso} > 0.45$ cm^3 g^{-1}. For all materials, broad pore size distributions were observed with partially defined diameter maxima in the micropore range (pore diameter around 1 nm) and lower mesopore region (pore diameter around 3 nm) although significantly volume contributions are made from all mesopore size domains. High-resolution TEM images of the resulting carbonized materials maintained the continuous highly porous nature of the Carbogel parent (Figure 2.14c and d). Material in all cases presented highly (meso)porous nanomaterial structuring, with image analysis demonstrating elegantly the unusual coral-like continuous carbonaceous hierarchical nano-architecture. The hyperbranched network has walls of disordered graphitic-like sheets of around 10–15 nm thickness composed of around two or three short carbon layers stacking locally in no preferential long-range (greater than 100 nm) orientation.

It is proposed that the protein (and associated Maillard products) acts as a surface-stabilizing/structure-directing agent(s) to create a stable heteroatom (nitrogen)-doped carbon scaffolds possessing cocontinuous flexible porosity, tunable chemistry, and structural size (Figure 2.14). The reaction of the reducing sugar and free amino groups of the protein is provided to dramatically alter the HTC cascade reaction, leading to branched Carbogel networks with different morphologies from previously reported HTC-derived materials. Given the structural similarities with corresponding monolithic silicas [50], it was speculated that a similar underlying formation mechanism resulted in the formation of the continuous porous structure. This involves sugar dehydration where carbon precursors demix from the aqueous phase in a spinodal fashion, which in this case is stopped from further ripening towards droplets by an early and efficient reaction with the water-based proteins. As discussed in Chapter 5, material chemistry (e.g., nitrogen-bonding motifs) can simply be directed via selection of a postsynthesis thermal treatment temperature. The applications of these monoliths in catalysis is discussed in Chapter 7.

2.3.2   Borax-Mediated Formation of HTC-Derived Carbogels from Glucose

In the context of appropriate HTC catalysis, borates (e.g., boric acid and borax) are well-known for their strong interaction with vicinal-dihydroxy compounds, such as glucose and fructose. Negatively charged borate-diol and borate-didiol complexes are formed that change the reactivity and physical properties of the sugar solutions [65, 66]. We have recently reported on the exploitation of these interactions for the synthesis of hydrothermal carbonaceous gels [67]. As an example, hydrothermal treatment of 20 ml of 30% glucose solution with 500 mg of borax at 200 °C for 20 h resulted in the formation of dark brown, mechanically stable monoliths (Figure 2.16). The monoliths are composed of aggregated nanometer-sized spherical particles, leading to a colloidal hydrogel. Macro- and mesoporosity can be observed generating a lightweight carbon material exhibiting hierarchical porosity. After purification by extraction with water and ethanol, elemental analysis demonstrated a chemical composition similar to conventional HTC (i.e., around 64% C and around 5% H).

Since additionally the yield was clearly improved compared to hydrothermal treatment of pure glucose under the same reaction conditions, studies of the reaction time were carried out (solid yield improved from 26 to 44%, carbon yield from 42 to 73% for the HTC of 20 ml 30% glucose solution at 200 °C for 19 h). Aliquots of 20 ml of 30% aqueous glucose

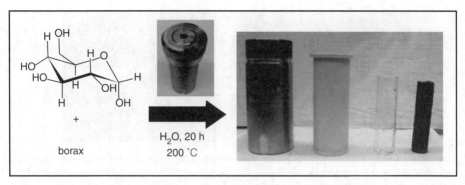

Figure 2.16 *Scheme of the hydrothermal reaction of glucose in the presence of borax resulting in a black monolith. (Reproduced with permission from [67]. © 2012 WILEY-VCH Verlag GmbH & Co. KGaA, Weinheim.)*

solutions with and without 500 mg borax were heat treated at 200 °C for 1, 2, 3, and 4 h for comparison (Figure 2.17).

The accelerated carbonization by borax (i.e., higher yields in shorter times with an onset at lower temperatures) could be explained by catalysis of glucose dehydration. Referring to the reactivity of borax with sugars [68, 69], one can expect that the interaction of borax with the vicinal diols leads to negatively charged boron-diol or boron-didiol complexes, enhancing the reactivity in the dehydration of glucose to HMF. Concurrent to our research, Riisager *et al.* have presented a study describing the impact of boric acid on the hydrothermal dehydration of glucose, confirming the proposition that borax does in fact catalyze the dehydration of glucose [70]. Further studies indicate that catalytic activity may arise from the stabilization of intermediates appearing within the isomerization of glucose to fructose mediated by negatively charged boron-diol complexes with their results correlating with the lowered activation energy to the corresponding coordinated intermediate structures [68]. However, under the conditions used during HTC, the Lobry de Bruyn–Alberda van Ekstein isomerization already converts glucose into fructose [69]. Compared to the dehydration

Figure 2.17 *Comparison of HTC products from glucose with and without the addition of borax after different reaction times. Shown are results from 20 ml glucose solutions (30%) with (cat.) and without (ref.) the addition of 500 mg of borax for 1, 2, 3, and 4 h at 200 °C. (Reproduced with permission from [67]. © 2012 WILEY-VCH Verlag GmbH & Co. KGaA, Weinheim.)*

Figure 2.18 *Comparison of HTC products from fructose with and without the addition of borax after different reaction times. Shown are results from 20 ml glucose solutions (30%) with (cat.) and without (ref.) the addition of 500 mg of borax for 2, 3, and 4 h at 140 °C.*

of fructose, HTC is a rather slow process and therefore should not be affected by the accelerated isomerization.

To further investigate the role of borax, a HTC system employing borax and fructose as precursor was investigated, whereby 20 ml of 30% aqueous fructose solutions with and without 500 mg borax were heat treated at 140 °C for 2, 3, and 4 h for comparison (Figure 2.18). It is clear that in this case, borax also accelerates the HTC of fructose. In addition to the improved dehydration of glucose, the subsequent carbonization/resinification reactions occurring in a later stage are therefore catalyzed, as well.

Particle aggregation indicates that borax also promotes cross-linking, which is essential for the gelation process and monolith evolution. It has to be mentioned that HTC of pure HMF with borax as additive leads to perfect gels as well. Regarding the mechanism of catalysis, borates are known to catalyze esterification [71] and decarboxylation [72]. However, these properties at least do not obviously affect the HTC reaction pathways. One characteristic, also in the context of HTC, is the formation of boron-dioxo complexes (Figure 2.19). It is proposed that borax coordinates sugars and thereby competes with the

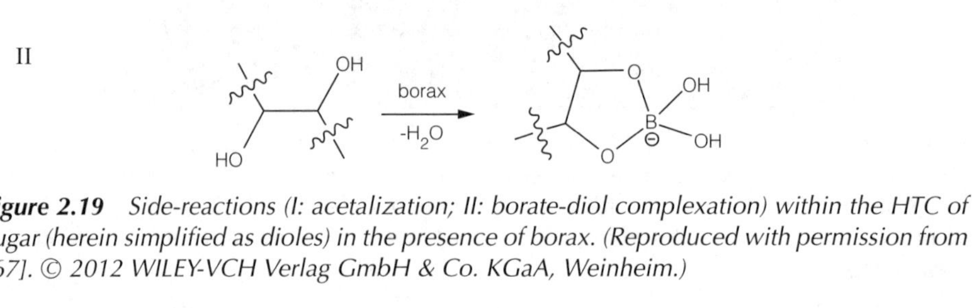

I

Hydroxymethylfurfural (HMF)

II

Figure 2.19 *Side-reactions (I: acetalization; II: borate-diol complexation) within the HTC of sugar (herein simplified as dioles) in the presence of borax. (Reproduced with permission from [67]. © 2012 WILEY-VCH Verlag GmbH & Co. KGaA, Weinheim.)*

aldehyde-protecting acetalization reaction of sugars with HMF, eventually promoting aldol condensations of the "free" HMF.

Representative of our reported system, four Carbogel examples were synthesized at different sugar/borax ratios denoted as Carbogels-X, where X corresponds to the used amount of borax in milligrams while the amount of water and glucose was kept the same (20 ml of 30% glucose solutions). The Carbogels were produced by hydrothermal treatment of the solutions at 180 °C for 8 h. Shorter reaction times resulted in the formation of unstable gels or sol-like suspensions. Under the reported reaction conditions, low-density (e.g., $\rho \sim 0.12$ g cm^{-3}) Carbogels were produced of around 3.1 g mass with a carbon content of around 64% (thus resulting in 80% carbon yield with respect to glucose; the volume fraction of pores is $\Phi \sim 0.94$).

SEM and TEM image analysis of the resulting aerogels revealed the presence of very small, spherical HTC nanoparticles, aggregating to create a gel structure with hierarchical porosity. (Figure 2.20). With increasing system borax concentration, the microstructure becomes increasingly finer due to decrease in primary particle size (from a to d in Figure 2.20). The decreasing particle size with decreasing sugar/borax ratio is in accordance with the structure control trends in the classical resorcinol/formaldehyde system. Both SEM and TEM images reveal the Carbogel network to be composed of uniform-sized, aggregated, spherical nanoparticles generating the desired hierarchical porous network. It is worth mentioning that the particle size control is not limited to the sub-50-nm range. Experiments with less borax and/or longer reaction times lead to gels with particle sizes greater than 100 nm and therefore very likely cover the whole possible size range underneath classical HTC particle dimensions.

Thermal treatment of the dried hydrophilic gels under nitrogen atmosphere at 550 and 1000 °C (heating rate $= 10$ K min^{-1} followed by an isothermal period of 5 h) leads to uniform dimensional shrinkage of the monolith macrostructure with preservation of the overall cylindrical shape.

As observed via TEM, at higher sugar/borax ratios the local nano- and microstructure in terms of particle shape and connectivity are also retained upon heating to these carbonization temperatures (Figure 2.21). However, aerogels composed of very fine, sub-10-nm particles do not completely withstand the carbonization process and shrink to more condensed systems. Determination of the particle size via TEM (Figure 2.21, Carbogel-750@1000 °C) indicates that also the primary particle size decreases upon carbonization, in accordance with the mass loss throughout carbonization.

In addition, nitrogen sorption measurements were carried out to investigate the porosity of the as-formed hierarchical carbon structures (Figure 2.22). With increasing sugar/borax ratio, the aerogels show an increasing gas uptake and a higher specific surface area due to the presence of smaller primary nanoparticles. Carbogel-750 breaks this trend, as the S_{BET} of 209 m^2 g^{-1} is decreased compared to that of 233 m^2 g^{-1} for Carbogel-500. In accordance with the TEM images (Figure 2.21), this reflects the collapse of the very fine structure within the freeze-drying process (i.e., 7-nm disordered carbon cannot withstand the capillary forces of a pore of similar size). This is well known from corresponding, even rigid polymer structures [73].

Regarding gas sorption behavior for the postcarbonized systems, a strong increase of the specific surface area especially after thermal treatment at 550 °C/N$_2$ is observed. High surface area materials exhibiting around 600 m^2 g^{-1} (Carbogel-250@550 °C) are accessible

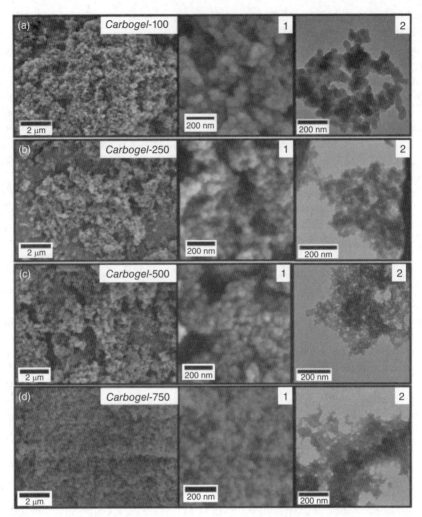

Figure 2.20 *SEM and TEM images of Carbogels-X prepared from different amounts X of borax (in mg), with the same amount of water and glucose. SEM overview images are presented on the left side, while higher-resolved SEM and TEM images are for each Carbogel indicated as 1 and 2. (Reproduced with permission from [67]. © 2012 WILEY-VCH Verlag GmbH & Co. KGaA, Weinheim.)*

even without chemical activation using the relatively simple synthetic route and conditions. The carbonized Carbogel-100 and Carbogel-250 samples show a trend of increasing surface area with increasing sugar/borax ratio – the result of a reduction in the primary particle size. The high borax structures are again already too fine to withstand carbonization completely, resulting in partial structure collapse and a marked reduction of the accessible surface and overall porosity. The interplay of enhancement of surface area and fragility of the architecture is also nicely reflected in the high temperature data: carbonization at higher

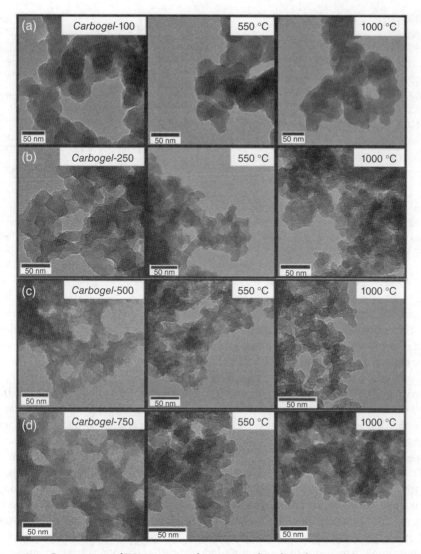

Figure 2.21 *Comparison of TEM images of as-prepared Carbogels-X and the respective post-carbonized samples at 550 and 1000 °C, where X stands for the amount of borax used within the synthesis protocol. (Reproduced with permission from [67]. © 2012 WILEY-VCH Verlag GmbH & Co. KGaA, Weinheim.)*

temperatures (i.e., 1000 °C) produces lower surface area materials, although the primary particle size has been dramatically decreased.

A more quantitative porosimetry analysis of the so-far optimized products, Carbogel-250 and the corresponding postcarbonized Carbogels, was performed (Figure 2.22). Only marginal hysteresis and lack of defined plateau region as relative pressure reaches unity (i.e., $p/p_0 \sim 1$) of the sorption profile reflect the combination of a slit-type micropore

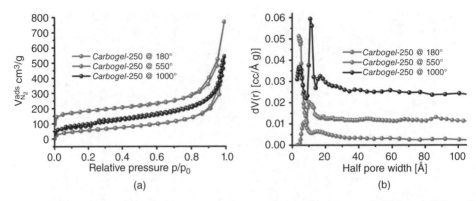

Figure 2.22 *Comparison of nitrogen sorption analysis of as-prepared Carbogel-250 and the respective products from postcarbonization at 550 and 1000 °C. (a) Nitrogen sorption isotherms. (b) Pore size distributions. (Reproduced with permission from [70]. © 2010 Royal Society of Chemistry.)*

structure combined with a high external surface area due to small, interconnected primary carbon nanoparticles. The "exploding" sorption capacity at relative pressures near unity reflects the presence of excessive large diameter mesoporosity and macroporosity, which can be expected from a material with 94 vol% overall porosity.

As-prepared Carbogels show in general limited microporosity ($V_{micro} \leq 0.10$ cm^3 g^{-1}), which is very typical for ductile, polymer-like behavior of the walls [73]. Heat treatment at 550 °C significantly increases this micropore content to about 0.25 cm^3 g^{-1}, which means that the material now behaves as a rigid carbon scaffold, postcarbonized carbon aerogels showing isotherms (left) and pore size distributions (right). Also the mesopore volume is observed to increase nicely upon carbonization at 550 °C ($V_{meso} > 0.5$ cm^3 g^{-1} at 550 °C), most likely as the result of material contraction/condensation processes opening up interstitial porosity between the primary particles, which was closed ahead by ductile deformation. Interestingly, nitrogen sorption at Carbogel-250@1000 °C shows a loss both of micro- and mesoporosity. This means that the structure rearrangements observed by TEM on a larger scale are also found on the molecular and low-nanometer scale: the whole structure simply sinters due to temperature-induced carbon–carbon rearrangements.

To investigate the applicability in electrochemical applications (i.e., as electrode materials in various systems), the electrical conductivities of pulverized and pressed Carbogel-250 before and after thermal treatment under inert atmosphere was measured. Electrochemical measurements were carried out with a Gamry Reference 600 potentiostat (Gamry Instruments) and Gamry EIS 300/Physical Electrochemistry software. The electrical conductivity was evaluated applying a simple resistor model on potentiostatic impedance spectroscopy at 1–1000 Hz using a two-electrode setup. The specific conductivities increase with increasing carbonization temperature reaching 290 S m^{-1} at 1000 °C. These values are, considering the high porosity of the materials, comparably high and absolutely sufficient to support an electrode application. The hierarchical porosity of the monolithic Carbogels suggests the use of the material as electrodes for fast processes, were mass transport takes a crucial role.

2.3.3 Carbogels from the Hydrothermal Treatment of Sugar and Phenolic Compounds

Ryu *et al.* [74] have shown that the HTC of sugars (i.e., xylose or fructose) in the presence of phenolic compounds can largely enhance the yield of the typical hydrothermal carbon microspheres up to 20-fold through the addition of phloroglucinol. In this study, the authors suggest a possible mechanism where the phenolic compounds act as a cross-linker via intermolecular condensation with dehydrated sugars (i.e., HMF or furfural), and/or hydrothermal carbon nuclei [74]. The higher reactivity of phloroglucinol, related to its electron density in the 2, 4, 6 ring positions, seems to involve a more efficient cross-linking without any additive [74], while the use of resorcinol to design HTC-derived materials implies the presence of a suitable catalyst [75]. According to the investigation by Katsoulidis *et al.* [76] on the solvothermal polymerization of phloroglucinol and aldehyde derivatives leading to highly microporous polymers, electrophilic aromatic substitutions can also be considered as a possible mechanism pathway. This second route involves the covalent linking of a carbonyl group on dehydrated sugar with two phloroglucinol moieties by eliminating a water molecule [76].

Taking this "phenolic/sugar approach" one step further, it has been found that highly porous carbon aerogels can be obtained from phloroglucinol/monosaccharide mixtures without any catalyst or additive in hydroalcoholic media [77]. Since monosaccharides can be isolated from the cellulosic fraction of lignocellulosic biomass and phloroglucinol can be extracted from the bark of fruit trees, this approach constitutes an interesting renewable synthetic pathway.

As shown in Figure 2.23a, after hydrothermal treatment of fructose and phloroglucinol monomers at 180 °C for 20 h, smooth carbon-based wet gels were produced. After further CO_2 supercritical drying, the as-synthesized monolithic brown aerogels (Figure 2.23b) depict an airy continuous interlinked macromorphology made of small aggregated particles (Figure 2.23c). TEM micrographs display a consistent hyperbranched structure with a pore diameter up to 100 nm (Figure 2.23d) and a wall thickness of about 15 nm (Figure 2.23e). Nitrogen sorption experiments performed on these aerogels depict consistent mixed isotherms (Figure 2.24). While the first steep rise at low relative pressures (i.e., from 0 to about 0.05) is characteristic of microporous materials, the gradual increase in sorption volume from 0.1 suggests a multilayer adsorption within the bigger pores. Since the nitrogen sorption curves show unclear hysteresis, we can assume type IV (H3)/type II mixed isotherms, displaying both adsorption on open surfaces of macropores and volume filling of smaller mesopores, associated with the capillary condensation phenomenon. This feature is in good agreement with TEM observations, highlighting a broad pore size distribution from few nanometers up to 100 nm (Figure 2.23d and e).

Depending on the synthetic pathway, the carbon-based aerogels present different porous characteristics (Figure 2.24). Indeed, two categories of aerogels were identified. The first category is related to gels made from a lower phenolic cross-linker loading or depicting a poor carbon yield (typically, gels made from glucose or at a low hydrothermal temperature, 130 °C). As a direct consequence, these aerogels exhibit a poor cross-linking level and therefore unsteady framework stability, leading to a significant drop in specific surface area after drying process. These aerogels present BET (Brunauer, Emmett, and Teller) surface areas lower than 610 m^2 g^{-1} with a microporous contribution around 300 m^2 g^{-1}.

Figure 2.23   *Pictures of a gel obtained from a mixture of phloroglucinol and fructose at a molar ratio of 1 : 2 and HTC temperature of 180 °C: (a) before and (b) after CO₂ supercritical drying. (c) SEM micrograph and (d and e) TEM micrographs of the corresponding supercritical-dried aerogel. (Reproduced with permission from [32]. © 2012 Royal Society of Chemistry.)*

Nevertheless, after further thermal treatment under inert atmosphere (typically at 950 °C under nitrogen), an increase in the total surface area, mainly related to a large rise in the microporous contribution, was systematically noticed.

The second category corresponds to materials obtained from highly reactive monosaccharide/phloroglucinol systems (i.e., typically from fructose and xylose made at high hydrothermal temperature, 180 °C). These aerogels exhibit a higher cross-linking level, involving a better preservation of the micromesoporous character of the scaffold after supercritical drying, leading to higher specific surface areas (up to $1100 \, m^2 \, g^{-1}$), with a surprisingly high microporous contribution, around $650 \, m^2 \, g^{-1}$. After further thermal treatment at higher temperatures, a significant drop of the specific surface area can be noticed, suggesting a large modification of the porous framework. Indeed, the noticeable macroscopic shrinkage of the carbon-based scaffold is also effective at the micro- and mesoscopic levels. After further pyrolysis, most of the shrunk micropores are probably too small to be detected by nitrogen sorption. However, the BET and mesoporous surface areas, about 650 and

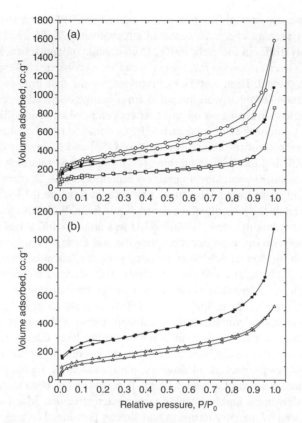

Figure 2.24 *(a) Nitrogen sorption isotherms of aerogels obtained from hydrothermal treatment of a phloroglucinol/monosaccharide mixture at a molar ratio of 1 : 2 and temperature of 180 °C from (○) xylose, (■) fructose, and (□) glucose. (b) Nitrogen sorption isotherms of aerogels obtained from hydrothermal treatment of phloroglucinol/fructose mixture at 180 °C with a molar ratio of (■) 1 : 2 and (△) 1 : 7.*

$150 \ m^2 \ g^{-1}$, respectively, are still noteworthy. These Carbogels are attractive as potential materials for separation science, adsorption, or heterogeneous catalysis.

2.3.4 Emulsion-Templated "Carbo-HIPEs" from the Hydrothermal Treatment of Sugar Derivatives and Phenolic Compounds

When the desired application of porous carbons involves the immobilization of bulky catalysts, such as proteins or bacteria, larger macropores are essential to increase the accessibility of the supported biocatalysts, while reducing the low diffusion kinetics. This need is greatly increased when the application involves the use of monolithic supports [78]. A rational pathway to increase the pore diameter of monoliths to the micro- or even millimeter scale is to combine sol–gel polymerization with a soft templating approach, such as air–liquid foaming [79] or liquid–liquid emulsification [80].

Particularly, the use of high internal phase emulsion (HIPE) as a template to design macroporous polymers has been the center of attention since Barby *et al.* [81] developed the design of polyHIPEs in the early 1980s, from organic monomers such as styrene and divinylbenzene. This formulation involving a water-in-oil concentrated emulsion remains, by far, the most studied. Brun and Titirici reported for the first time on "carbo-HIPEs" ("carbo-" for carbonized) that were prepared from biomass-derived precursors (i.e., furfural and phloroglucinol) using an oil-in-water concentrated emulsion [82]. These foams represent the first emulsion-templated carbo-HIPEs obtained from saccharide derivatives, leaving the beaten path of resorcinol/formaldehyde [83] and divinylbenzene-based polymers [84]. Considering the HTC conditions ($T > 130$ °C and relatively high pressures, around 10 bar), an oil-in-water direct emulsion cannot be stable. A prepolymerization of the monomers within the continuous phase was necessary before the hydrothermal treatment, in order to retain the integrity of the liquid–liquid foam. With this aim, phloroglucinol was used as a cross-linking agent, while Fe(III) was employed as a Lewis acid catalyst, promoting the polycondensation between phenolic and furanic compounds, following a mechanism similar to the one described previously for the "phenolic/sugar approach." As shown in Figure 2.25a, after a one-step synthesis, HTC-derived macroporous monoliths were obtained. These materials depict a typical macromorphology made up of aggregated hollow spheres (Figure 2.25b), called *cells* [80a]. This macrostructure is directly induced by the polymerization of the hydroalcoholic continuous phase of the direct emulsion. These cells, corresponding to the removed oil droplets, exhibit large diameters from 10 up to 50 μm. Furthermore, due to the interstices left in the continuous phase or by direct contact between two nearby packed oil droplets, these monolithic foams present narrower macropores, called *connecting windows* [80a]. This feature has been underlined by SEM (Figure 2.25c), showing a highly interconnected macrostructure. Macroscopic pore size distribution obtained by mercury intrusion porosimetry performed on a carbo-HIPE after a further thermal treatment at 950 °C displays a main contribution centered at 3 μm, while two weak contributions, centered at 300 nm and 4 μm, can also be noticed. As a direct consequence of a high connectivity, the carbo-HIPEs depict a large macroporosity of about 98%, together with total macroscopic cumulative volume and surface area respectively up to 18 cm^3 g^{-1} and 300 m^2 g^{-1} [82].

Beyond a *cellular* macrostructure, these carbo-HIPEs depict, after further thermal treatment at higher temperature, highly micro- and mesoporous frameworks (Figure 2.26) [82]. While the first contribution can be attributed to the physical activation induced by carbonization, the mesopore contribution is related to the catalytic effect of the iron-based particles toward the graphitization of amorphous carbons [85]. As shown in Figure 2.25h, at high temperature these metallic nanoparticles diffuse within the carbonaceous framework, while catalyzing graphitization preferentially on their surface, leading to hollow graphitic rings. As a direct consequence, mesopore volumes up to 0.1 cm^3 g^{-1} were determined by the QSDFT approach and BET surface areas of about 500 m^2 g^{-1} were reached.

Another important feature related to the macrocellular porous structure of the carbo-HIPE foams concerns the mechanical properties. To assess these properties, mechanical tests under compression have been performed (Figure 2.27). The first linear portion of the stress–strain curve (Figure 2.27b), from 0 to 5% strain, corresponds to the elastic portion, from which can be extracted the Young's modulus, of about 3 MPa. This Young's modulus is rather low compared with those usually obtained for nonreinforced organic polyHIPEs

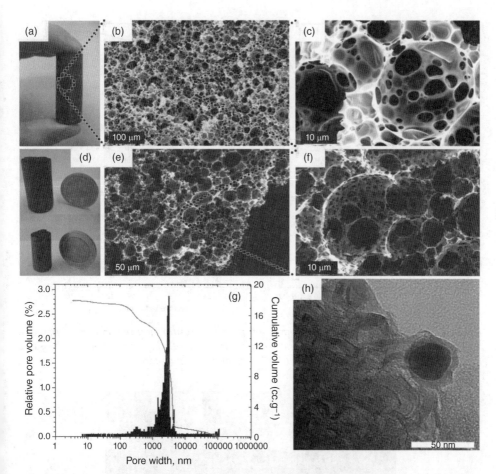

Figure 2.25 *(a) Picture of an as-synthesized carbo-HIPE made at 130 °C during 24 h, after Soxhlet extraction and drying at 80 °C. (b and c) SEM micrographs of a carbo-HIPE made at 130 °C during 24 h and synthesized with an oil volume fraction of 0.8. Micrograph (c) is a higher magnification of micrograph (b). (d) Picture of an as-synthesized carbo-HIPE, after Soxhlet extraction and drying at 80 °C (top) and of the same monolith after further thermal treatment at 950 °C under nitrogen (bottom). (e and f) SEM micrographs of a carbo-HIPE after further thermal treatment at 950 °C, synthesized with an oil volume fraction of 0.8. Micrograph (f) is a higher magnification of micrograph (e). Pore size distribution obtained by mercury intrusion porosimetry (g) and TEM micrograph (h) of a carbo-HIPE after further thermal treatment at 950 °C under nitrogen.*

[86], from 20 up to 65 MPa, revealing a low cross-linking level. Then, after reaching the yield strength (0.16 MPa), a nonlinear plastic behavior was observed, related to permanent deformations of the material. Finally, a densification of the monolith, typical behavior of foam, was noticed. Nevertheless, up to 60% strain, the ductile foam did not collapse and kept its monolithic aspect (Figure 2.27a) together with its macrocellular morphology. However, the first stress–strain cycle induced a noticeable permanent volume drop of 11%.

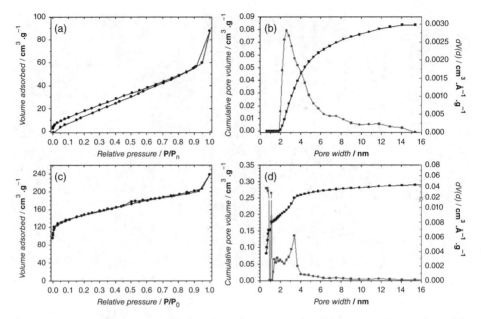

Figure 2.26 *Nitrogen sorption isotherms and pore size distribution of carbo-HIPE made at 130 °C during 24 h, after Soxhlet extraction and drying at 80 °C (a and b) and of the same foam after further thermal treatment at 950 °C under nitrogen (c and d).*

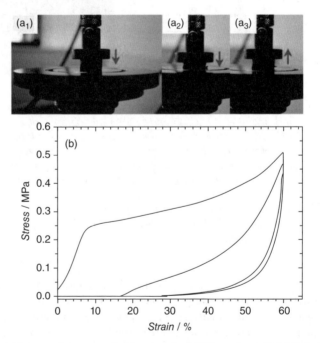

Figure 2.27 *(a) Picture of an as-synthesized carbo-HIPE made at 130 °C during 24 h, after Soxhlet extraction and drying at 80 °C. (b and c) SEM micrographs of a carbo-HIPE made at 130 °C during 24 h and synthesized with an oil volume fraction of 0.8.*

After the immobilization of a catalytic mixture made of glucose oxidase and redox polymer, these carbo-HIPE foams were successfully used as bioelectrodes for enzymatic fuel cells through the electrooxidation of glucose [82]. These macrocellular foams can be considered as promising candidates in many applications involving the use of continuous flow process, particularly in the field of separation science and heterogeneous catalysis [78a, 87].

2.4 Summary and Outlook

In this chapter, we have evaluated the recent main research directions related to the introduction of porosity to HTC-based carbons.

The combined HTC/templating technique has been revealed to be a useful technique for us to gain access to functional nanostructured carbon materials. From the presented examples, we have learned that this combined HTC/templating strategy allows simultaneous control of material morphology, surface character, and pore properties, thus widening the possible applications fields of promising porous HTC-derived materials. Furthermore, the flexibility of the presented approach allows access to a wide range of functional materials with surface functionality/chemistry tunable via selection of postcarbonization treatment temperature, thus allowing access to materials with application-specific surface/bulk properties (e.g., surface functionality-rich character towards increased aromatic/pseudographitic character).

Carbon beads and ordered mesoporous carbons were produced through the combined HTC/hard templating technique using the corresponding silica sacrificial templates. In a very similar manner, tubular carbons could be synthesized using the macroporous anodic alumina membrane template. The presented materials are suitable for various applications such as heterogeneous catalysis, separation/adsorption, or electrochemistry. Here, the possibility of controlling surface chemistry from an oxygenated functionality-rich character towards an increased aromatic/pseudographitic character via postcarbonization step provided an added advantage. As such, these highly flexible materials can be very interesting candidates as separation media, allowing the separation not only of nonpolar analytes, but also of polar analytes (e.g., neutral sugars, sugar alcohols). Furthermore, introduction of other important chemical moieties using the existing oxygenated surface functional groups will be of great interest.

When HTC is combined with the soft templating technique, direct access to functional ordered micro/meso porous materials was possible via a direct "templating" of block copolymer micellar templates. The cooperative self-assembly of block copolymer species and D-fructose (i.e., a biomass-derived carbon source) occurring in an aqueous phase at 130 °C is considered responsible for the stable formation of the ordered block copolymer/carbon composite. Importantly, pore size could be shifted into the mesopore region simply through the addition of an organic pore-swelling agent. Further studies on the formation mechanism (e.g., via *in situ* SAXS/small-angle neutron scattering studies or ^{13}C NMR study of the reaction solution) and the corresponding synthesis optimization are desirable for this highly interesting soft templating system. Such information would then allow us to ultimately manipulate the formed ordered phases, opening access to materials with enhanced porous features (e.g., increased pore size and pore volume) necessary for successful real-world applications. Additionally, incorporation/doping of other important chemical species will also offer an added advantage. As such, these materials are highly versatile, potentially finding application in fields such as electrochemistry (e.g., batteries and supercapacitors) or drug delivery systems (e.g., encapsulation of drug molecules). Regularly ordered pore structuring confers particular properties such as controlled linear electronic conduction pathways and fast charge/discharge properties or could facilitate fast encapsulation and release of drug molecules.

Bioresources (e.g., crustacean shells) were also introduced as potential HTC-derived material sources, whereby the naturally occurring bioinorganic/organic composites were converted to extremely promising, highly textured, hierarchically porous nitrogen-doped carbon networks. Furthermore, we have alluded to the first step towards the true transfer of traditional sol–gel principles typically used in inorganic material preparation to this new rapidly developing area of HTC to allow, in principle, access to a whole new range of nanostructured functional carbonaceous material platforms. In this sense the previous limitations in many applications (i.e., a lack of developed porosity in HTC-derived material) has been overcome, and it is expected that associated improvements in catalytic efficient and adsorption capacity will be markedly improved as a result.

Furthermore, in the final section, we also discussed four different pathways for the successful preparation of HTC-derived monoliths with tunable morphologies, pore properties, and chemistries, which seem to lend themselves directly to applications in flash and analytical chromatography and the emerging catalytic and separation challenges associated with the twenty-first century concept of the "biorefinery."

If porous sustainable HTC-derived materials can be utilized successfully as separation or catalytic devices in this new chemical production concept then this would represent a step towards a complete, holistic, synergistic biorefinery and, in principle, a near-closed CO_2 loop in chemical manufacture.

References

(1) Kondo, S., Ishikawa, T., and Abe, I. (2001) *Science of Adsorption*, Maruzen, Tokyo.
(2) Eltekova, N.A., Berek, D., Novák, I., and Belliardo, F. (2000) *Carbon*, **38**, 373–377.
(3) Steel, K.M. and Koros, W.J. (2003) *Carbon*, **41**, 253–266.
(4) Makowski, P., Demir Cakan, R., Antonietti, M. *et al.* (2008) *Chemical Communications*, 999–1001.
(5) Serp, P. and Figueiredo, J.L. (2009) *Carbon Materials for Catalysis*, John Wiley & Sons, Inc., Hoboken, NJ.
(6) Su, D.S. and Schlögl, R. (2010) *ChemSusChem*, **3**, 136–168.
(7) Antolini, E. (2009) *Energy & Environmental Science*, **2**, 915–931.
(8) Joo, S.H., Kwon, K., You, D.J. *et al.* (2009) *Electrochimica Acta*, **54**, 5746–5753.
(9) Othman, R., Dicks, A.L., and Zhu, Z. (2012) *International Journal of Hydrogen Energy*, **37**, 357–372.
(10) Knox, J.H., Kaur, B., and Millward, G.R. (1986) *Journal of Chromatography A*, **352**, 3–25.
(11) Hanai, T. (2003) *Journal of Chromatography A*, **989**, 183–196.

(12) White, R.J., Antonio, C., Budarin, V.L. *et al.* (2010) *Advanced Functional Materials*, **20**, 1834–1841.
(13) Lu, G.Q. and Zhao, X.S. (2005) *Nanoporous Materials: Science and Engineering*, Imperial College Press, London.
(14) Pandolfo, A.G. and Hollenkamp, A.F. (2006) *Journal of Power Sources*, **157**, 11–27.
(15) Kubo, S. (2011) PhD Thesis, University of Potsdam, Potsdam, Germany.
(16) Titirici, M.-M., White, R.J., Falco, C., and Sevilla, M. (2012) *Energy & Environmental Science*, **5**, 6796–6822.
(17) Titirici, M.-M. and Antonietti, M. (2010) *Chemical Society Reviews*, **39**, 103–116.
(18) Schüth, F. (2003) *Angewandte Chemie International Edition*, **42**, 3604–3622.
(19) Ryoo, R., Joo, S.H., Kruk, M., and Jaroniec, M. (2001) *Advanced Materials*, **13**, 677–681.
(20) Liang, C., Hong, K., Guiochon, G.A. *et al.* (2004) *Angewandte Chemie International Edition*, **43**, 5785–5789.
(21) Liang, C. and Dai, S. (2006) *Journal of the American Chemical Society*, **128**, 5316–5317.
(22) Zhang, F., Meng, Y., Gu, D. *et al.* (2005) *Journal of the American Chemical Society*, **127**, 13508–13509.
(23) Huang, Y., Cai, H., Yu, T. *et al.* (2007) *Angewandte Chemie International Edition*, **46**, 1089–1093.
(24) Meng, Y., Gu, D., Zhang, F. *et al.* (2005) *Angewandte Chemie International Edition*, **44**, 7053–7059.
(25) Meng, Y., Gu, D., Zhang, F. *et al.* (2006) *Chemistry of Materials*, **18**, 4447–4464.
(26) Titirici, M.M., Thomas, A., and Antonietti, M. (2007) *Advanced Functional Materials*, **17**, 1010–1018.
(27) Titirici, M.-M., Thomas, A., and Antonietti, M. (2007) *Journal of Materials Chemistry*, **17**, 3412–3418.
(28) Kubo, S., Tan, I., White, R.J. *et al.* (2010) *Chemistry of Materials*, **22**, 6590–6597.
(29) White, R.J., Tauer, K., Antonietti, M., and Titirici, M.-M. (2010) *Journal of the American Chemical Society*, **132**, 17360–17363.
(30) Tang, K., White, R.J., Mu, X. *et al.* (2012) *ChemSusChem*, **5**, 400–403.
(31) Tang, K., Fu, L., White, R.J. *et al.* (2012) *Advanced Energy Materials*, **2**, 873–877.
(32) Brun, N., Sakaushi, K., Yu, L. *et al.* (2013) *Physical Chemistry Chemical Physics*, **15**, 6080–6087.
(33) Kubo, S., White, R.J., Yoshizawa, N. *et al.* (2011) *Chemistry of Materials*, **23**, 4882–4885.
(34) Bloss, P., Hergeth, W.D., Döring, E. *et al.* (1989) *Acta Polymerica*, **40**, 260–265.
(35) Sing, K.S.W. (1985) *Pure and Applied Chemistry*, **57**, 603.
(36) Baccile, N., Laurent, G., Babonneau, F. *et al.* (2009) *Journal of Physical Chemistry C*, **113**, 9644–9654.
(37) Blin, J.L. and Su, B.L. (2002) *Langmuir*, **18**, 5303–5308.
(38) Huo, Q., Margolese, D.I., and Stucky, G.D. (1996) *Chemistry of Materials*, **8**, 1147–1160.
(39) White, R.J., Budarin, V., Luque, R. *et al.* (2009) *Chemical Society Reviews*, **38**, 3401–3418.

(40) Addadi, L. and Weiner, S. (1992) *Angewandte Chemie International Edition in English*, **31**, 153–169.

(41) Sugawara, A., Nishimura, T., Yamamoto, Y. *et al.* (2006) *Angewandte Chemie International Edition*, **45**, 2876–2879.

(42) White, R.J., Antonietti, M., and Titirici, M.-M. (2009) *Journal of Materials Chemistry*, **19**, 8645–8650.

(43) Ryoo, R., Joo, S.H., and Jun, S. (1999) *Journal of Physical Chemistry B*, **103**, 7743–7746.

(44) Su, F., Poh, C.K., Chen, J.S. *et al.* (2011) *Energy & Environmental Science*, **4**, 717–724.

(45) Gong, K., Du, F., Xia, Z. *et al.* (2009) *Science*, **323**, 760–764.

(46) Soorholtz, M, White, R.J., Zimmermann, T. *et al.* (2013) *Chemical Communications*, **49**, 240–242.

(47) Czerw, R., Terrones, M., Charlier, J.C. *et al.* (2001) *Nano Letters*, **1**, 457–460.

(48) Kabbour, H., Baumann, T.F., Satcher, J.H. *et al.* (2006) *Chemistry of Materials*, **18**, 6085–6087.

(49) Lu, X., Arduini-Schuster, M.C., Kuhn, J. *et al.* (1992) *Science*, **255**, 971–972.

(50) Nakanishi, K., Minakuchi, H., Soga, N., and Tanaka, N. (1997) *Journal of Sol–Gel Science and Technology*, **8**, 547–552.

(51) Pekala, R.W. (1989) *Journal of Materials Science*, **24**, 3221–3227.

(52) Tamon, H. and Ishizaka, H. (2000) *Journal of Colloid and Interface Science*, **223**, 305–307.

(53) Gutierrez, M.C., Pico, F., Rubio, F. *et al.* (2009) *Journal of Materials Chemistry*, **19**, 1236–1240.

(54) Job, N., Théry, A., Pirard, R. *et al.* (2005) *Carbon*, **43**, 2481–2494.

(55) Dayong, G., Jun, S., Nianping, L. *et al.* (2011) *Journal of Reinforced Plastics and Composites*, **30**, 827–832.

(56) Saliger, R., Fischer, U., Herta, C., and Fricke, J. (1998) *Journal of Non-Crystalline Solids*, **225**, 81–85.

(57) Marie, J., Chenitz, R., Chatenet, M. *et al.* (2009) *Journal of Power Sources*, **190**, 423–434.

(58) Al-Muhtaseb, S.A. and Ritter, J.A. (2003) *Advanced Materials*, **15**, 101–114.

(59) Baccile, N., Antonietti, M., and Titirici, M.-M. (2010) *ChemSusChem*, **3**, 246–253.

(60) White, R.J., Yoshizawa, N., Antonietti, M., and Titirici, M.-M. (2011) *Green Chemistry*, **13**, 2428–2434.

(61) Barron, P.F. and Wilson, M.A. (1981) *Nature*, **289**, 275–276.

(62) Jia, Y.F., Xiao, B., and Thomas, K.M. (2001) *Langmuir*, **18**, 470–478.

(63) Lua, A.C. and Yang, T. (2004) *Journal of Colloid and Interface Science*, **274**, 594–601.

(64) Pérez-Cadenas, M., Moreno-Castilla, C., Carrasco-Marín, F., and Pérez-Cadenas, A.F. (2008) *Langmuir*, **25**, 466–470.

(65) Conner, J.M. and Bulgrin, V.C. (1967) *Journal of Inorganic and Nuclear Chemistry*, **29**, 1953–1961.

(66) Levy, M. and Doisy, E.A. (1929) *Journal of Biological Chemistry*, **84**, 749–762.

(67) Fellinger, T.-P., White, R.J., Titirici, M.-M., and Antonietti, M. (2012) *Advanced Functional Materials*, **22**, 3254–3260.

(68) Ståhlberg, T., Rodriguez-Rodriguez, S., Fristrup, P., and Riisager, A. (2011) *Chemistry – A European Journal*, **17**, 1456–1464.

(69) Bobleter, O. (1994) *Progress in Polymer Science*, **19**, 797–841.

(70) Hansen, T.S., Mielby, J., and Riisager, A. (2011) *Green Chemistry*, **13**, 109–114.

(71) Houston, T.A., Wilkinson, B.L., and Blanchfield, J.T. (2004) *Organic Letters*, **6**, 679–681.

(72) Wehrli, P.A. and Chu, V. (1973) *The Journal of Organic Chemistry*, **38**, 3436–3436.

(73) Thomas, A., Goettmann, F., and Antonietti, M. (2008) *Chemistry of Materials*, **20**, 738–755.

(74) Ryu, J., Suh, Y.W., Suh, D.J., and Ahn, D.J. (2010) *Carbon*, **48**, 1990–1998.

(75) Zhang, W.L., Tao, H.X., Zhang, B.H. *et al.* (2011) *Carbon*, **49**, 1811–1820.

(76) Katsoulidis, A.P. and Kanatzidis, M.G. (2011) *Chemistry of Materials*, **23**, 1818–1824.

(77) Brun, N., Garcia-Gonzalez, C.A., Smirnova, I., and Titirici, M.M. (2013) *RSC Advances*, in press.

(78) (a) Brun, N., Babeau-Garcia, A., Achard, M.F. *et al.* (2011) *Energy & Environmental Science*, **4**, 2840–2844; (b) Brun, N., Garcia, A.B., Deleuze, H. *et al.* (2010) *Chemistry of Materials*, **22**, 4555–4562; (c) Flexer, V., Brun, N., Backov, R., and Mano, N. (2010) *Energy & Environmental Science*, **3**, 1302–1306; (d) Flexer, V., Brun, N., Courjean, O. *et al.* (2011) *Energy & Environmental Science*, **4**, 2097–2106.

(79) (a) Fujiu, T., Messing, G.L., and Huebner, W. (1990) *Journal of the American Ceramic Society*, **73**, 85–90; (b) Wu, M.X., Fujiu, T., and Messing, G.L. (1990) *Journal of Non-Crystalline Solids*, **121**, 407–412; (c) Chandrappa, G.T., Steunou, N., and Livage, J. (2002) *Nature*, **416**, 702–702.

(80) (a) Brun, N., Ungureanu, S., Deleuze, H., and Backov, R. (2011) *Chemical Society Reviews*, **40**, 771–788; (b) Cameron, N.R. and Sherrington, D.C. (1996) *Biopolymers Advances in Polymer Science*, **126**, 163–214; (c) Zhang, H.F. and Cooper, A.I. (2005) *Soft Matter*, **1**, 107–113; (d) Carn, F., Colin, A., and Backov, R. (2004) *MRS Proceeedings*, **847**, EE9.4.

(81) Barby, D. and Haq, Z. (1982) European Patent 0060138.

(82) Brun, N., Edembe, L., Gounel, S. *et al.* (2012) *ChemSusChem*, doi: 10.1002/cssc.201200692.

(83) (a) Gross, A.F. and Nowak, A.P. (2010) *Langmuir*, **26**, 11378–11383; (b) Thongprachan, N., Yamamoto, T., Chaichanawong, J. *et al.* (2011) *Adsorption – Journal of the International Adsorption Society*, **17**, 205–210.

(84) (a) Wang, D., Smith, N.L., and Budd, P.M. (2005) *Polymer International*, **54**, 297–303; (b) Cohen, N. and Silverstein, M.S. (2011) *Polymer*, **52**, 282–287.

(85) (a) Kicinski, W., Szala, M., and Nita, M. (2011) *Journal of Sol–Gel Science and Technology*, **58**, 102–113; (b) Maldonado-Hodar, F.J., Moreno-Castilla, C., Rivera-Utrilla, J. *et al.* (2000) *Langmuir*, **16**, 4367–4373.

(86) (a) Lepine, O., Birot, M., and Deleuze, H. (2008) *Colloid and Polymer Science*, **286**, 1273–1280; (b) Youssef, C., Backov, R., Treguer, M. *et al.* (2010) *Journal of Polymer Science Part A – Polymer Chemistry*, **48**, 2942–2947.

(87) El Kadib, A., Chimenton, R., Sachse, A. *et al.* (2009) *Angewandte Chemie International Edition*, **48**, 4969–4972.

3

Porous Biomass-Derived Carbons: Activated Carbons

Dolores Lozano-Castelló[1], Juan Pablo Marco-Lozar[1], Camillo Falco[2], Maria-Magdalena Titirici[3], and Diego Cazorla-Amorós[1]

[1] *Materials Institute and Inorganic Chemistry Department, Universidad de Alicante, Spain*
[2] *Institute for Advanced Sustainability Studies, Earth, Energy and Environment Cluster, Germany*
[3] *School of Engineering and Materials Science, Queen Mary, University of London, UK*

3.1 Introduction to Activated Carbons

Activated carbons are highly porous carbon materials, exhibiting appreciable apparent surface area and micropore volume (MPV), which can present a wide variety of pore size distributions (PSDs) and micropore size distributions (MPSDs) [1–4]. They are solids that can be prepared in different forms, such as powders, granules, pellets, fibers, cloths, and others. Due to these features and their special chemical characteristics, they can be used for very different applications, such as liquid- and gas-phase treatments and energy storage [1–7].

Activated carbons are not present in nature. In order to produce them, the right selection of precursor and preparation process is crucial. Several precursors, such as wood, coals, pitches, polymers, residues with a high amount in carbon, and also different preparation methods have been used. These two factors have a great importance as they determine the final porous structure of the activated carbons.

Sustainable Carbon Materials from Hydrothermal Processes, First Edition. Edited by Maria-Magdalena Titirici.
© 2013 John Wiley & Sons, Ltd. Published 2013 by John Wiley & Sons, Ltd.

Considering the preparation process, the main stage determining the porous structure is the method of activation. The objective during the activation is both to increase the number of pores and to increase the size of the existing pores, so that the resulting activated carbon has a high adsorption capacity.

The different activation processes are divided into two different groups: chemical and physical activation [1]. The differences between them are the procedure and the type of activating agents used.

The preparation of activated carbons by physical activation [1, 8] includes a controlled gasification of the carbonaceous material that has been previously carbonized, although occasionally the activation of the precursor can be done directly. Thus, the samples are heat treated up to 800–1000 °C under an oxidant gas atmosphere, so that carbon atoms are selectively removed. The removal of the outer and less-ordered carbon atoms of the graphitic microcrystals leads to the creation of new micropores and/or the widening of their size in the char, which results in an increase of its pore volume.

Thus, the PSD in the activated carbon depends on the precursor, preparation conditions (mainly temperature, time, and gas flow), activating agent used, and presence of catalysts.

CO_2 and steam are the mostly commonly used activated agents. Their reaction with carbon is endothermal. These gases react with the carbon atoms in the precursor according to the following reactions:

$$C + CO_2 \leftrightarrow 2\,CO \qquad \Delta H = 159.0 \text{ kJ mol}^{-1}$$
$$C + H_2O \leftrightarrow CO + H_2 \qquad \Delta H = 118.5 \text{ kJ mol}^{-1}$$

Nevertheless, the activation process is not so easy. In order to have an efficient activation process, the reaction must predominantly take place inside the particles and the higher the amount of carbon removed, the higher the porosity development. The extension of the activation, which must be as selective and controlled as possible, is usually expressed as weight loss percentage, which is called the "burn-off percentage." Therefore, to obtain an appropriate porosity development an accurate selection of the experimental conditions is needed, mainly the temperature, gas flow, and precursor weight. This selection is very different depending on the precursor.

The chemical activation process consists of contacting a carbonaceous precursor with a chemical activating agent, followed by a heat treatment stage, and finally by a washing step to remove the chemical agent and the inorganic reaction products [1, 3, 6]. In the literature the use of several activating agents, such as phosphoric acid (H_3PO_4) [9–12], $ZnCl_2$ [13–17], alkaline carbonates [18, 19], KOH [20–22], and NaOH [17, 20, 21, 23, 24], has been reported. In the case of chemical activation, it is better not to use the term burn-off degree, as for physical activation, but it is better to talk about the degree of carbon reacted (or degree of activation).

Chemical activation offers well-known advantages [3, 13, 21, 22] over the physical approach, which can be summarized as follows: (i) it uses lower temperatures and heat treatment times, (ii) it usually consists of one stage, and (iii) the yields obtained are typically higher. However, chemical activation presents some disadvantages [3, 6, 22], such as the need for an additional washing stage after the heat treatment and a more corrosive behavior of the chemical agents used in comparison to CO_2 or steam. Traditionally, chemical activation has been carried out using two activating agents: H_3PO_4 or $ZnCl_2$.

In the case of chemical activation with H_3PO_4, lignocellulosic materials are preferred as precursors [6, 10]. At low degrees of activation, the activated carbons do not have a highly developed area and they are essentially microporous, whereas at higher activation degree, the surface area and the MPV increase, but there is also a remarkable increase in the mesopore volume and a widening of the MPSD [25, 26]. Therefore, in the case of activation with H_3PO_4 both high adsorption capacity and narrow MPSD cannot be achieved. However, for activated carbons that require a well-developed mesoporosity (e.g., for gasoline removal [5]), H_3PO_4 activation is a very suitable activation method [27].

The activated carbons prepared by chemical activation with $ZnCl_2$ are essentially microporous [13, 15]. The loading of zinc has an important effect on the porosity: samples activated at high zinc loadings present high porosity development and MPV, but also a more heterogeneous MPSD [15]. Although higher MPVs can be obtained by $ZnCl_2$ activation than by physical activation or by H_3PO_4 activation, the increase in porosity development is also concurrent with a widening of the microporosity. The main disadvantage of this activating agent is that the emission of zinc may cause serious environmental problems, which strongly limits its present use.

The purpose of developing activated carbons with tailored porosity over the whole range of microporosity has motivated continued research towards the use of other activating agents, such as alkaline hydroxides. Since the pioneering patent of Wennerberg *et al.* [28], the production of very-high-surface-area ("superactive") activated carbons using alkaline hydroxide activation started its commercialization first by Amoco and then by Kansai Coke and Chemical Company (Japan) [28, 29]. In addition, a considerable number of studies have been carried out over the years focusing on chemical activation with hydroxides [14, 17, 20, 21–23, 28]. A systematic study, developed by our research work, has underlined that the reactivity of the carbon precursor, carbon/hydroxide ratio, heat treatment temperature, and nitrogen gas flow rate are the crucial variables determining the porous character of the activated carbons [30]. In some cases their careful control has allowed the production of highly porous activated carbons ($S_{BET} > 3500$ m^2 g^{-1}) with narrow MPSDs.

Both activation processes previously described (physical activation and chemical activation) can be used for preparing activated carbons with a suitable porous texture for gas-phase pollutant abatement, high-pressure gas storage (e.g., CO_2, hydrogen, natural gas), and as advanced electrodes for supercapacitors with improved volumetric and gravimetric energy storage. Nowadays, there are two major guidelines in material research: (i) increased environmental concerns and (ii) need for cost-effective competitive products. As a consequence, many efforts in this field are directed towards the production of materials that can be synthesized from cheap natural precursors and through efficient processes.

Thus, this chapter deals with the production of activated carbons with a highly developed microporosity from lignocellulosic biomass. Two different synthetic routes are covered: (i) carbonization followed by (or together with) chemical activation with H_3PO_4, $ZnCl_2$, and KOH/NaOH, and (ii) hydrothermal carbonization (HTC) of biomass followed by chemical activation with KOH/NaOH.

3.2 Chemical Activation of Lignocellulosic Materials

In the literature it has been shown that when physical activation is used to prepare activated carbons from lignocellulosic materials, meso- and macroporosity usually exist in activated carbons, reflecting the botanic texture of the precursor [31]. As explained previously, in the case of physical activation, microporosity is developed by activation with gases (mainly steam or CO_2) and in the case of lignocellulosic precursor the microporosity is created by widening the rudimentary porosity of the parent char [32]. It has been observed that variables such as temperature, pressure, heating rate, and so on, do not have a strong influence on the MPSD [33]. Thus, to obtain activated carbons with a highly developed microporosity from lignocellulosic precursors, well-controlled chemical activations are usually required because the development of porosity is substantially modified.

A typical chemical composition of lignocellulosic materials is 48 wt% C, 6 wt% H, and 45 wt% O, the inorganic matter being a minor component. The corresponding atomic ratios are: $O/C = 0.73$, $H/C = 1.5$, and $H/O = 2.07$. As the transformation to the char requires the removal of oxygen and hydrogen, the degree of conversion to carbon (carbonization yield) varies widely as a function of the amount of carbon being removed with oxygen and hydrogen (as CO_x, H_2, or hydrocarbons). A typical carbonization process yields around 20–30 wt%, lower than the expected value (up to 48 wt%) if water only was the result of the removal of hydrogen and oxygen. This suggests that a chemical activation, where the activating agent is a dehydrating compound, will increase the yield, but will also change the thermal degradation of the precursor, leading to a subsequent change in the evolution of porosity. Several authors have studied the carbonization of lignocellulosic materials [15, 34] and found that the main degradation takes place at 200–350 °C, with evolution of H_2O, CO, CO_2, CH_4, aldehydes, and so on. Distillation of heavier hydrocarbons (tar) takes place in the range 350–500 °C; above 500 °C, there is little weight loss, thus indicating that the basic structure of the char has already been formed.

Impregnation of the lignocellulosic precursor with the activating agents ($ZnCl_2$, H_3PO_4, and KOH) may lead to fragmentation of cellulose and other components of the botanic precursor such as hemicellulose and lignin. Although the three activating agents react with the precursor, some clear differences exist between them, as has been described in the literature [35]: $ZnCl_2$ and H_3PO_4, in the interior of the particles, produce a dehydrating effect on the cellulose, hemicellulose, and lignin components during heat treatment. Dehydration is possible because the chemical is a liquid at the temperature of the process, thus facilitating the bonding to the precursor being thermally degraded. The precursor is thus able to transfer water to the reactant in the reacting mass to form a hydrated compound that then loses water with increasing temperature. The dehydration produced by both $ZnCl_2$ and H_3PO_4 is strong, whereas that of KOH does not seem to affect carbonization.

In the following sections, some examples of chemical activation of different biomass materials taken from the literature are included, showing the most relevant parameters affecting the final porosity of the materials.

3.2.1   H₃PO₄ Activation of Lignocellulosic Precursors

H_3PO_4 acts both as an acid catalyst to promote bond cleavage and the formation of cross-links via cyclization and condensation reactions, and also can combine with organic species to form phosphate and polyphosphate bridges that connect and cross-link biopolymer fragments. The addition (or insertion) of phosphate groups drives a process of dilation that, after removal of the acid, leaves the matrix in an expanded state with an accessible pore structure. Essentially, this is the activation process. At temperatures above 450 °C, a secondary contraction of the structure occurs when the phosphate linkages become thermally unstable. The reduction in cross-link density allows the growth and alignment of polyaromatic clusters, producing a more densely packed structure with some reduction in porosity.

A suitable selection of the experimental conditions of the activation process allows preparing activated carbon materials with textural properties tailored to specific applications (i.e., pore volume, surface area, PSD). In order to obtain these activated carbons, it is first necessary to select the carbon precursor and also the preparation method. Regarding the former, any carbonaceous material with high carbon content could be used for the preparation of activated carbons. In the particular case of the use of H_3PO_4 as activating agent, lignocellulosic materials are widely used as carbon precursors [10, 35–37]. The reason for this selection is based on the advantages that this activating agent offers such as nonpolluting character (e.g., compared to $ZnCl_2$) and ease of elimination by leaching with water, thereby recovering H_3PO_4 that can be recycled for further uses. It has also been published that in the case of lignocellulosic precursors, H_3PO_4 also produces an increase in the char yield [38].

In general, mixing H_3PO_4 and the carbonaceous precursor is performed by impregnation using (i) wet impregnation, utilizing a high excess of solution, or (ii) incipient wetness impregnation, adding a volume of solution slightly higher than the pore volume of the solid. For this, H_3PO_4 solutions of various concentrations can be added to the raw material to vary the content of impregnating agent, which is expressed as impregnation ratio (X_P, wt%), defined as (gram of H_3PO_4 per gram of precursor) $\times$ 100. After the impregnation process, the H_3PO_4/precursor mixture is usually submitted to a soft thermal treatment to dry the sample and to improve the contact between the activating agent and the precursor material. The next step after the impregnation process is the carbonization. As in any chemical activation process, after cooling, a washing step of the obtained material is necessary in order to remove the byproducts produced during the activation procedure (e.g., distilled water). Once the clean material is dried, the activated carbon has been obtained.

Many parameters of the H_3PO_4 activation process affect the final properties of the obtained activated carbons. However, the most relevant variables for the chemical activation by H_3PO_4 are: (i) H_3PO_4/precursor ratio, (ii) activation temperature, and (iii) activation time.

The influence that the H_3PO_4/precursor ratio (X_P) has on the final textural properties is analyzed next, taking as an example a lignocellulosic material such as apple pulp impregnated with aqueous solutions of H_3PO_4 [39]. Impregnation ratios used were: 21, 30, 43, 50, 64, 75, 85, 100, 125, and 150 wt%. For this study, an activation temperature of 450 °C (temperature value most widely used in the H_3PO_4 activation) and 1 h of activation time were selected. Table 3.1 summarizes the characterization results obtained from the activated samples.

In general, looking to the textural results obtained, it can be observed that the higher the chemical ratio, the higher the porosity development, allowing us to achieve specific surface area values around 1000 $m^2\,g^{-1}$. However, a more detailed analysis allows us to see different trends. For example, at low impregnation ratios ($X_P < 43$ wt%), the obtained activated carbons are essentially microporous as can be concluded from the small differences between the total MPV ($V_{DR}(N_2)$) and the narrow MPV ($V_{DR}(CO_2)$). For intermediate impregnation ratios (50–85 wt%), the obtained activated carbons are still microporous materials, but this microporosity becomes wider (increase in the difference between the $V_{DR}(N_2)$ and the $V_{DR}(CO_2)$). Finally, for the samples with the highest impregnation ratios (100–150 wt%),

Table 3.1   *Textural properties of the activated carbons obtained from apple pulp and using different H_3PO_4/precursor ratios.*

X_P (wt%)	S_{BET} (m^2 g^{-1})	$V_{DR}(N_2)$ (cm^3 g^{-1})	$V_{DR}(CO_2)$ (cm^3 g^{-1})	S_{ext} (m^2 g^{-1})^a
0	3	—	0.10	—
21	454	0.19	0.15	16
30	591	0.23	0.16	16
43	837	0.31	0.21	19
50	854	0.32	0.20	19
64	1010	0.37	0.19	28
75	1004	0.36	0.19	31
85	1022	0.37	0.18	37
100	1011	0.36	0.16	37
125	954	0.34	0.16	98
150	914	0.32	0.16	128

^{a}Obtained by using the α_σ method.

the microporosity is slightly reduced with the increase in the chemical ratio, indicating that the microporosity development is almost negligible. Under these conditions, H_3PO_4 promotes the development of wider porosity. Similar tendencies with increasing amounts of H_3PO_4 used in the activation process have been reported by other authors [35,40].

According to the activation mechanism proposed by Jagtoyen and Derbyshire [37, 38], porosity is generated by the H_3PO_4 that is inserted in the internal structure of lignocellulosic materials, preventing their shrinkage. By increasing the amount of H_3PO_4 used in the impregnation, the volume filled by this activating agent obviously increases, resulting in larger pore volume and pore size.

Another important variable in this activation process with H_3PO_4 is the activation temperature (also called carbonization temperature). Activation temperature, together with the activating agent/precursor ratio, is probably the most widely studied parameter. Typically, the temperature range applied for H_3PO_4 activation is between 350 and 650 °C. Figure 3.1 shows the effect of the activation temperature on the textural properties. These activated carbons were prepared from white oak using a weight ratio of H_3PO_4 to wood on an as-received basis of 145 wt%. As can be seen in Figure 3.1, microporosity increases to temperature values of 350 °C. From this value, MPV begins to decrease. On the other hand, mesopore volume development evolves slowly up to about 350 °C, then increases and reaches a maximum around 500 °C. For higher temperatures, a contraction of the porous structure seems to occur provoking a reduction in porosity. Taking into account these results it seems that the H_3PO_4 treatment promotes the expansion of the lignocellulosic structure over the temperature range 250–450 °C, provoking the porosity development and the creation of an extensive surface area.

As an example of the influence of the activation time on the textural properties of H_3PO_4-activated carbons, Table 3.2 summarizes the characterization results of activated carbons obtained from the activation of sugar cane residues by H_3PO_4 [42]. Experimental conditions used in this work were the following: activation temperature of 400 °C and a weight H_3PO_4/precursor ratio of 200 wt%.

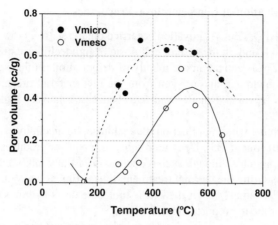

Figure 3.1 *Porosity development in H_3PO_4-activated carbons. Effect of the activation temperature. (Redrawn with permission from [41]. © 1995 Elsevier.)*

For short periods (lower than 1 h), the obtained results suggest that textural properties do not achieve a full development of porosity. Maximum pore volumes are reached for carbonization times of 1 h where microporous activated carbons are obtained. Increasing the carbonization time beyond 1 h produces a significant reduction in the porosity development, as reflected by a decreases in the BET (Brunauer, Emmett, and Teller) surface area values and the MPV. However, mesoporosity increases with the activation time, indicating an important pore widening. These results have been corroborated by many other authors [40].

This porosity development for H_3PO_4-activated carbons has been related to the different reactions between the H_3PO_4 and the chemical composition of the lignocellulosic materials (cellulose, hemicellulose, lignin) [41, 43], in which H_3PO_4 acts as an acid catalyst, promoting bond cleavage and the formation of cross-links via cyclization and condensation, and also reacting with organic species to form phosphate and polyphosphate bridges that connect and cross-link biopolymer fragments. These phosphate groups drive a process of dilation that after removal of the acid, leaves the matrix in an expanded state with an accessible pore structure. This mechanism would explain that, for example, at high temperatures these linkages become unstable, producing a more densely packed structure with some reduction in porosity or that for high activation times, a breakdown of cross-links within the carbon structures would be produced, resulting in a reduction of the microporosity [35, 39, 41].

Table 3.2 *Influence of the carbonization time on surface properties.*

Carbonization time (min)	S_{BET} (m^2 g^{-1})	V_{micro} (cm^3 g^{-1})	V_{meso} (cm^3 g^{-1})
0	479	0.25	0.05
30	673	0.35	0.14
60	779	0.41	0.18
180	440	0.23	0.21

3.2.2   ZnCl₂ Activation of Lignocellulosic Precursors

As in the case of H_3PO_4, $ZnCl_2$ activation is carried out at relatively low temperatures (i.e., 450–600 °C) [15, 44] and, in the case of cellulosic or lignocellulosic precursors, reactions that take place in $ZnCl_2$ activation are mainly of dehydration. In this sense, $ZnCl_2$ favors the elimination of hydrogen and oxygen from the raw material precursor, resulting in a relatively low loss of volatile matter and tars; hence, a relatively high reaction yield is obtained.

$ZnCl_2$ has a high dehydrating effect on the cellulose, hemicellulose, and lignin components during heat treatment [35]. Dehydration of the precursor produces a reduction in the dimensions of the particle, although such reduction is partially inhibited because the reactant remains inside during the thermal treatment, thus acting as a template for the creation of microporosity. The small size of the $ZnCl_2$ molecule or its hydrates explains the small and uniform size of the micropores created.

As described previously, any chemical activation procedure begins with the mixing of the activating agent and the precursor. In the particular case of $ZnCl_2$, the mixing method is performed by impregnation. For this purpose, $ZnCl_2$ solutions of different concentrations (depending on the desired activating agent/precursor ratio) are added to the pristine material. After that, the mixture is submitted to heating (about 85 °C) in order to improve the contact between activating agent and precursor, which facilitates the access of the $ZnCl_2$ into the internal structure of the material. This stage takes several hours. Once the impregnation step is finished, the mixture is carbonized using an inert gas flow (usually nitrogen). After the heat treatment, the obtained treated sample needs to be washed to eliminate the byproducts produced during the $ZnCl_2$ activation (e.g., diluted acid chloride and distilled water). Finally, cleaned samples are subsequently dried.

Attending to the chemical activation procedure, parameters such as $ZnCl_2$/precursor ratio (X_{Zn} = gram of Zn per gram of precursor), activation temperature, activation time or precursor can strongly affect the final textural properties of the activated carbon obtained.

Probably the influence of the chemical ratio ($ZnCl_2$/precursor) has been the parameter most widely studied for the $ZnCl_2$ activation. As an example, Figure 3.2 shows the effect of

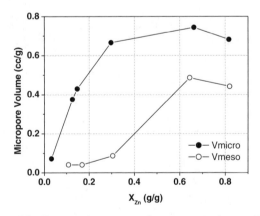

Figure 3.2   *Influence of the X_{Zn} on micropore and mesopore volumes. (Redrawn with permission from [15]. © 1991 Elsevier.)*

the chemical ratio ($X_{Zn} = 0$–1.2) on the evolution of the textural properties of activated carbons obtained from a peach stone precursor [15]. The unactivated material ($X_{Zn} = 0$) reveals a slight microporosity development produced by the carbonization process. However, even when low X_{Zn} values are introduced, important changes in the porosity are observed. For $X_{Zn} < 0.4$, the porosity development is mainly focused on the microporosity obtaining a homogenous MPSD (mesoporous development is almost negligible). For higher amounts of zinc incorporated to the lignocellulosic material ($X_{Zn} > 1$), the porosity continues growing, but the PSD becomes more heterogeneous (important increase of the mesopore volumes), indicating the widening of the microporosity previously created.

It has been proposed that the mechanism of porosity creation by $ZnCl_2$ is due to the spaces left by the activating agent after the washing step (template behavior). This hypothesis is based on the comparison between the volume of $ZnCl_2$ (using the density of 2.9 g cm^{-3} for solid $ZnCl_2$) and the obtained micropore and mesopore volumes from the prepared activated carbons. Thus, for $X_{Zn} < 0.4$, where the micropores are the only porosity that is being developed, MPVs correspond to the reactant volume, which indicates the template behavior of the $ZnCl_2$. The small size of the $ZnCl_2$ molecule or its hydrates may explain the small and uniform size of the micropores created. For higher values ($X_{Zn} > 0.4$), where the mesoporosity development is important, there is a deviation between the volumes (pore volume is lower than the activating agent volume). This tendency has been explained arguing the severe attack of the pristine material by the $ZnCl_2$.

Another important variable on the $ZnCl_2$ process that has a significant effect on the textural properties of the final material is the selection of the activation temperature (carbonization temperature). This aspect has been also widely investigated in the literature [14, 15]. In general, for temperatures in the range between 450 and 800 °C, the obtained results highlight the reduction on the surface area and MPV with the increase in the activation temperature for this activating agent. Figure 3.3 shows the influence of this parameter on the surface area and the MPV.

The trends observed in Figure 3.3 show clearly the important decrease in the textural parameters with the increase in the activation temperature. This behavior is attributed to the shrinkage in the carbon structure with the increase of the temperature, resulting in

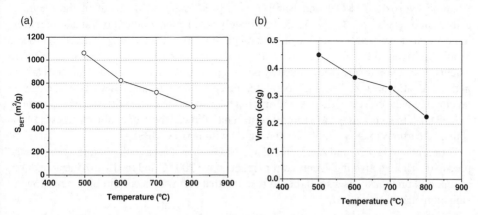

Figure 3.3   *Effect of the activation temperature on the surface area and MPV of ZnCl$_2$-activated carbons. (Reproduced with permission from [14]. © 1996 Elsevier.)*

Table 3.3   *Effect of the activation time on the textural properties of different ZnCl$_2$ chemically activated carbons. (Extracted with permission from [48, 50]. © 2005 and 1998 Elsevier.)*

Time (h)	BET surface area (m^2 g^{-1})	Total pore volume (cm^3 g^{-1})	Reference
0.5	750	0.37	[50]
1	820	0.41	[50]
2	791	0.39	[50]
4	744	0.38	[50]
0.5	1538	0.76	[48]
1	1635	0.83	[48]
2	1576	0.80	[48]
3	1475	0.75	[48]

a reduction in porosity since slight contraction of the particles has been observed after carbonization. These results indicate that the role of the ZnCl$_2$ on the porosity creation takes place at low temperatures (below 500 °C). It has been demonstrated that, at higher temperatures, ZnCl$_2$ does not have any additional effect on the porous structure. For this reason, an activation temperature of around 500 °C is recommended for ZnCl$_2$ activation [14, 45, 46].

The influence of the carbonization time on ZnCl$_2$ activation, in general, has two different tendencies [47–50] (see Table 3.3). For low activation times, with increasing the time, the porosity developed becomes higher probably due to the release of volatile matter. However, for long activation times, porosity seems to decrease continuously. This effect has been explained due to the softening of the volatile compounds present in the pristine material, which may produce the closing of porosity.

3.2.3   KOH and NaOH Activation of Lignocellulosic Precursors

Alkaline hydroxides (KOH and NaOH) have been analyzed in detail in the chemical activation of coals [20–23, 29, 51, 52] and lately also in some other precursors, including lignocellulosic materials [19, 53–56]. Several parameters of the chemical activation process by alkaline hydroxides affect the main features of the produced activated carbons to different extents. A systematic study, developed by our research group, has underlined that the carbon/hydroxide ratio, heat treatment temperature, nitrogen flow rate, and reactivity of the carbon precursor are the crucial variables determining the porous character of the activated carbons [22, 23, 30]. In some cases their careful control has allowed the production of highly porous activated carbons ($S_{BET} > 2500–3000$ m^2 g^{-1}) with narrow MPSDs.

Differently to H$_3$PO$_4$ and ZnCl$_2$, KOH and NaOH do not act as dehydrating agents on the precursor. These activating agents start to react above 700 °C and the hydroxide activation, as reported previously, is a solid–liquid reaction that seems to occur through the following stoichiometric reaction [30]:

$$6\,MOH + 2C \rightarrow 2M + 3H_2 + 2M_2CO_3$$

where M = Na or K.

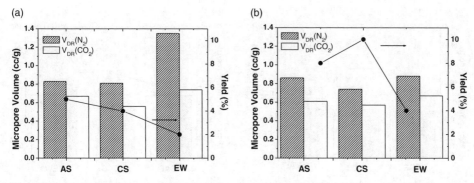

Figure 3.4 *Porosity characterization results (total MPV, $V_{DR}(N_2)$, and narrow MPV, $V_{DR}(CO_2)$) together with the activation yields, corresponding to the chemical activation of the pristine materials with NaOH (a) and KOH (b). (Values extracted with permission from [56]. © 2007 Elsevier.)*

In chemical activation with hydroxides, carbon atoms are removed from the carbon matrix and are transformed into an inorganic compound (i.e., carbonate), and a subsidiary process occurs in which the metal atoms formed through reduction of hydroxide are thought to be inserted/intercalated between the graphene layers of the residual carbon. This process can play a certain role in the global activation process [30].

In this section a series of examples of activated carbons prepared in our laboratory by KOH/NaOH activation of lignocellulosic materials have been selected [56] in order to illustrate how activated carbons with a highly developed homogeneous and controlled microporosity can be prepared from this type of precursors. The important effect that a pyrolysis step over the precursor previous to activation has on the porosity and the yield of the final activated carbon is remarked. These precursors include: eucalyptus wood (EW), coconut shell (CS), and almond shell (AS). These materials have been activated as such, and also the pyrolyzed samples (CCS, CEW and CAS) prepared by pyrolysis up to 850 °C of CS, EW, and AS precursors, respectively. The precursors were physically mixed either with NaOH or KOH using a hydroxide/precursor ratio of 3 : 1 (weight terms) and pyrolyzed up to 750 °C for 1 h (heating rate 5 °C min^{-1}).

Figure 3.4 presents the porosity characterization results (total MPV, $V_{DR}(N_2)$ (pore size smaller than 2 nm) and narrow MPV, $V_{DR}(CO_2)$ (pore size smaller than 0.7 nm) calculated applying the Dubinin–Radushkevich (DR) equation to the nitrogen adsorption isotherms at 77 K and to the CO_2 adsorption isotherms at 273 K, respectively) together with the activation yields, corresponding to the chemical activation of the pristine materials with KOH and NaOH. It can be seen that lignocellulosic precursors can be successfully activated by alkaline hydroxides, with porosity results comparable or even better to those from coals, achieving MPVs as high as 1.35 cm^3 g^{-1}. The result for the EW activated with NaOH should be noted, since it has the highest porosity development among all the materials studied (total MPV of 1.35 cm^3 g^{-1}) and a wide microporosity distribution, as it is shown by the difference between the $V_{DR}(N_2)$ (total MPV) and $V_{DR}(CO_2)$ (narrow MPV). The KOH activation for the same precursors shows MPVs of 0.88 cm^3 g^{-1}. In the activation of AS, similar porosities are achieved by NaOH and KOH activation, whereas NaOH shows higher performance than KOH for the EW and CS. This behavior can be explained considering the higher reactivity of EW and CS compared to AS. As concluded in our previous study,

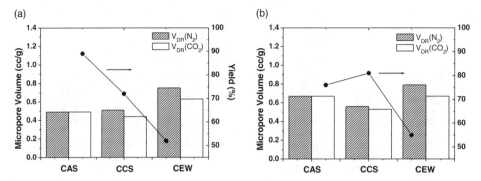

Figure 3.5 *Porosity characterization results (total MPV, $V_{DR}(N_2)$, and narrow MPV, $V_{DR}(CO_2)$) together with the activation yields, corresponding to the chemical activation of the pyrolyzed lignocellulosic materials with NaOH (a) and KOH (b). (Values extracted with permission from [56]. © 2007 Elsevier.)*

as a general trend it can be said that the higher the reactivity of the precursor, the better activation performance with NaOH in comparison to KOH, whereas those precursors with low reactivity can be more efficiently activated by KOH [56].

The results obtained for the carbonized lignocellulosic precursors (CEW, CCS and CAS) (see Figure 3.5) point out that pyrolysis previous to the activation causes a strong decrease in the activation of CAS and CCS, in comparison to CEW (note that CEW reactivity is higher than that for CAS and CCS). For the pyrolyzed lignocellulosic precursors, which present lower reactivity that pristine precursors, KOH manages to produce activated carbons with higher adsorption capacities than NaOH.

The yields values included in Figures 3.4 and 3.5 reveal that the obtained yields for pristine lignocellulosic materials are very low under the conditions studied (below 10 wt%). This is an important drawback for the use of these precursors for the preparation of activated carbons at large scales. However, when pyrolysis is performed prior to the activation, a strong increase in the activation yield is observed, which, as just mentioned, is joined to a decrease in the activation. For a better comparison, the carbonization yield should be also considered in the calculation of the global yield. Thus, for example, if the carbonization yield of sample CAS is taken into account in the calculation of the global yield for NaOH and KOH activation of sample CAS, these values are 20% and 17%, respectively. The comparison of these yields with those of NaOH and KOH activation of AS shows that direct activation leads to lower yield values, 5% and 8%, respectively. Thus, it can be stated that direct activation of a lignocellulosic material is not necessarily an advantage (even though it develops very high porosity), because the activation yield of the pristine lignocellulosic precursors is much lower than that for the carbonized ones.

3.3   Activated Carbons from Hydrothermally Carbonized Organic Materials and Biomass

As described in the previous sections, the typical approach to convert a lignocellulosic material into a char is a carbonization process. One of the main efforts in modern material research is the production of new and effective materials that can be synthesized from cheap natural precursors, and also through environmentally friendly and efficient processes.

In recent years, HTC has demonstrated its capability of converting biomass into carbon materials under very mild processing conditions [57, 58]. Even though this methodology was developed almost 100 years ago [59], as described in Chapter 1, its full potential as a synthetic route for carbon materials that have potential applications in several fields such as catalysis, energy storage, CO_2 sequestration, water purification, and agriculture, has only been fully understood in the last decade [60, 61].

As previously described, HTC consists of the heat treatment of an aqueous solution/dispersion of an organic material such as saccharides (glucose, cellulose, starch, or sucrose), simpler compounds such as furfural, or more complex substances such as biomass, at temperatures in the range 130–250 °C under autogeneous pressure [57]. This process has several advantages, such as: (i) the precursors are readily available, cheap, and renewable (i.e., saccharides or biomass), (ii) it is a "green" and simple process as it only involves water as the solvent and consists of a simple heat treatment in a closed autoclave, and (iii) the resulting solid carbon products exhibit attractive chemical and structural properties. The resulting solid carbon products (called sometime hydrochars) generally exhibit uniform chemical and structural properties as well as a very high content of oxygen-containing functional groups [62]. The amount and type of these groups can be tuned by modifying the operating conditions (i.e., temperature, solution concentration, reaction time, and precursor). Other functionalities (e.g., nitrogen based) can also be introduced into hydrochars by using dopant-containing carbon precursors or additives, as described in Chapter 5 [57, 63]. Unfortunately, the hydrochar material has the drawback of possessing almost no porosity, unless it is synthesized in the presence of a template as described in Chapter 2 [63, 64] or subjected to additional heat treatment at higher temperature [65].

However, the use of templates generates mainly microporous carbons, while the additional heat treatment produces a too narrow fraction of micropores exhibiting rather low pore volumes (maximum 0.3 ml g^{-1})

To make possible the use of porous materials based on hydrochar in emergent applications, such as hydrogen storage or electrical energy storage (supercapacitors), the synthesis of highly microporous carbon materials based on these materials is required. As explained above, and as reviewed in detail previously [30], chemical activation by KOH and NaOH is a way to successfully prepare highly microporous activated carbons from very different precursors (coals, lignocellulosic materials, carbon fibers, pitch, etc.). Thus, it can be envisaged that controlled chemical activation may also constitute an excellent method to activate these hydrochar products. Therefore, it is challenging to develop high MPV in this type of material.

Sevilla *et al.* were the first to report on the chemical activation of HTC materials as a way to generate highly porous materials [66]. They applied the procedure to HTC materials derived from glucose, starch, furfural, cellulose, and eucalyptus sawdust achieving large apparent surface areas, up to around 3000 m^2 g^{-1}, and pore volumes in the range of 0.6–1.4 cm^3 g^{-1}. Those materials are further characterized by narrow MPSDs in the supermicropore range (0.7–2 nm). Tuning of the MPSD was achieved through the modification of the activation temperature (600–850 °C) and the amount of KOH used (KOH/HTC weight ratio = 2 or 4). Applications of these microporous materials for supercapacitors and selective CO_2 adsorbents is exemplified in Chapter 7.

In the present chapter, we will focus on some results obtained in a broad study carried out in our laboratories with several types of hydrochar products and different activation conditions. The examples selected for this book correspond to three types of hydrochar products obtained from saccharides (glucose and cellulose) and biomass (rye straw), which have been hydrothermally treated at the same temperature (240 °C) and, afterwards, activated using KOH.

3.3.1 Hydrochar Materials: Synthesis, Structural, and Chemical Properties

Simple monosaccharide and oligosaccharides have been effectively employed as HTC starting materials [67]. Glucose especially, being one of the cheapest and most abundant carbohydrates, has been studied extensively. Recently, it has been shown that HTC represent an effective method to convert lignocellulosic biomass into high-value carbonaceous materials [68]. For that purpose a relatively low lignin content biomass was chosen: rye straw (composition in Table 3.4). Cellulose is a biopolymer made from glucose monomers and the main component of such raw biomass. This raw material is an agricultural waste product; therefore, it is available at costs well below petrochemical-based chemicals. Their use as carbon precursors is, however, definitely more complex than that of simple monosaccharides due to a higher degree of structural complexity [69], which has been demonstrated to affect the HTC mechanism [68].

For the preparation of the hydrochars the starting materials (glucose, cellulose, and biomass (rye straw)) were subjected to HTC. For each experiment 10 w/v of precursor in water was prepared and placed into 50-ml Teflon-lined stainless steel autoclaves. After sealing, the autoclaves were heated in a programmable oven for 24 h at 240 °C. The hydrothermal carbon obtained from each experiment (hydrochar) was abundantly washed with deionized water and then dried in a vacuum oven at 80 °C for 16 h. The yield for this process is around 40% for the three materials.

Hydrochars were characterized by elemental chemical analysis (C, N, O, S, H) and scanning electron microscopy (SEM). Elemental chemical analysis was performed on a Elementar Vario Micro Cube and SEM images were acquired on a LEO 1550 (LEO GmbH, Oberkochen) provided with a Everhard Thornley secondary electron and in-lens detectors.

From a chemical point of view, these materials possess a high concentration of oxygen groups, as evidenced by the elemental chemical analysis given in Table 3.5. SEM micrographs of hydrothermal carbons obtained from the three different precursors are included in Figure 3.6. It can be seen that spheres with a relatively large particle size and a homogeneous average size are obtained from pure glucose hydrothermally carbonized at 240 °C (Figure 3.6a). The morphology of cellulose-derived HTC carbons presented in Figure 3.6b points out that, similarly to glucose, spherical particles are also formed in the case of cellulose, but the overall morphology is not as homogeneous as for simple glucose and the particle sizes obtained are much smaller. In the case of rye straw, a detailed study

Table 3.4   *Rye straw composition (wt%).*

Cellulose	Hemicellulose	Lignin	Others
41.2	21.2	19.5	18.1

Table 3.5 *Elemental chemical analysis (%) and yield (%) for the three materials hydrothermally treated at 240 °C.*

Sample	N	C	H	S	O	Yield
Glucose	0.02	71.05	4.71	0.05	24.17	38
Cellulose	0.09	69.64	4.51	0.05	25.77	39
Rye straw	0.48	73.05	5.32	0.13	21.03	42

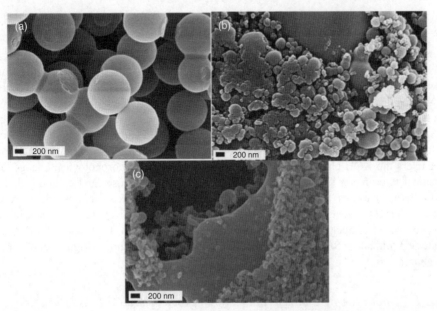

Figure 3.6 *SEM micrographs of HTC materials obtained from (a) glucose, (b) cellulose, and (c) rye straw at 240 °C.*

carried out by Falco *et al.*, and presented in more detail in Chapter 6 [68], showed that transformations of rye straw after hydrothermal treatment were very similar to those of cellulose. The lignocellulosic biomass does not undergo any structural disruption at low temperatures ($T = 180\,°C$) and the rye straw fibrous structure is maintained intact. When the biomass is hydrothermally treated at higher temperatures ($T = 240\,°C$), its fibrous network is disrupted and spherical particles start forming. A SEM image of the morphology of the rye straw after HTC at 240 °C is shown in Figure 3.6c, observing similar spherical particles to those previously shown for cellulose (Figure 3.6).

3.3.2   KOH Activation of Hydrochar Materials

A systematic study was done in our laboratories using several types of hydrochar products and different KOH activation conditions, with the aim of understanding the relationship between the chemical structure of hydrothermal carbons and their KOH chemical activation. As an example, in this section, results corresponding to KOH activation of hydrothermal carbons derived from glucose, cellulose, and rye straw synthesized at the same HTC temperatures (240 °C) are presented. Chemical activation with KOH of the three hydrochars was carried out following the procedure described elsewhere [56]. Physical mixtures of around 2 g of sample with corresponding amounts of KOH pellets were prepared using a hydroxide/precursor ratio of 3 : 1 (weight terms). The mixture was heated up to 750 °C (20 °C min^{-1}) during 2 h under nitrogen atmosphere (500 ml min^{-1} N$_2$ flow rate). After such heat treatment activation, the samples were washed sequentially with 5 M HCl and distilled water, and finally dried at 100 °C overnight.

3.3.2.1   Elemental Analysis and SEM of Activated Carbons

The elemental composition of the HTC carbon-derived activated carbons is characterized by very similar values regardless of the different precursors (Table 3.6). On the other hand, compared to hydrothermal carbons pyrolyzed at comparable temperatures (750 °C), their oxygen content is higher. This difference can be attributed to the higher degree of surface oxidation, which is typically observed after KOH chemical activation.

SEM was used to investigate the morphology of activated carbons derived from glucose, cellulose, and rye straw hydrothermal carbons (Figure 3.7). It is evident that for all three precursors the KOH chemical activation leads to a complete morphological change. The spherical micrometer-sized particles, characterizing both glucose- and cellulose-derived hydrothermal carbons (see Section 3.1), or the rye straw fiber-like structures, which are still present after HTC treatment, are not observed anymore. The activated carbon materials are now composed of macrometer-sized monolithic fragments with sharp edges. Furthermore higher magnification SEM micrographs (Figure 3.7b, d, and f) show a high level of surface roughness.

3.3.2.2   Porous Texture Characterization Results of Activated Carbons Produced from HTC-Derived Carbons

The nitrogen adsorption/desorption isotherms at 77 K for the activated carbons prepared by KOH activation of the three hydrothermal carbon precursors used (glucose, cellulose, and rye straw) using the same hydroxide/carbon ratio (3 : 1) are presented in Figure 3.8. In general, from the shapes of the obtained isotherms and in accordance with International Union of Pure and Applied Chemistry classification, it can be observed that all the samples present a marked microporous character. When the adsorption isotherms are analyzed in detail, some differences can be found among the activated materials. Thus, for glucose and

Table 3.6   *Elemental analysis (%) of HTC carbon-derived activated carbons and pyrolyzed HTC-derived carbon.*

Sample	C	H	O
Glucose-derived activated carbon	88.99	1.69	9.32
Cellulose-derived activated carbon	88.33	1.94	9.73
Rye straw-derived activated carbon	87.77	1.75	10.48
HTC-pyrolyzed at 750 °C	94.0	1.7	4.2

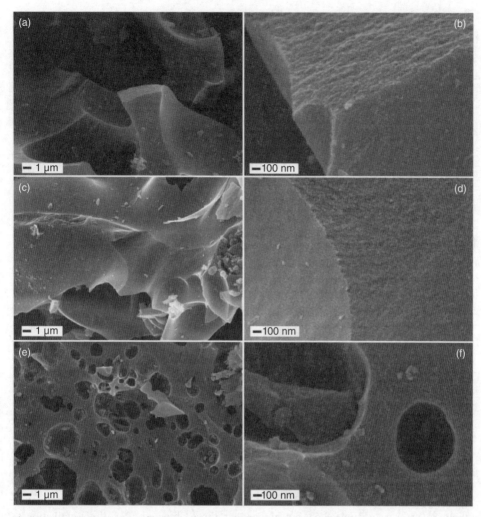

Figure 3.7 *SEM images of activated carbons derived from (a and b) glucose, (c and d) cellulose, and (e and f) rye straw hydrothermal carbons.*

cellulose, similar nitrogen adsorption isotherms are obtained, which show a slight slope at relative pressures higher than 0.2, indicating the presence of some mesopores. However, in the case of the activated rye straw, its isotherm shows a plateau at these relative pressures, reflecting a minimum presence of mesopores. Likewise, the knees of all the isotherms are quite wide, indicating a wide MPSD, although the knee for the rye straw is somewhat closer than the other two. In general terms, it can be concluded that the KOH activation performed for these samples provokes an important increase in the adsorption capacity.

Table 3.7 compiles the porous texture characterization results for the three activated carbons together with the activation yield (gram of activated carbon per 100 grams of hydrochar). It can be seen that, for the activated samples, the KOH activation process results

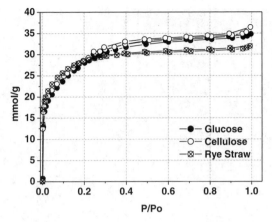

Figure 3.8   *Nitrogen adsorption/desorption isotherms at 77 K for the KOH HTC-derived activated carbons.*

in a high porosity development reaching, in all the cases, BET surface area values higher than 2000 m^2 g^{-1} and MPVs of 0.90 cm^3 g^{-1}. These high surface areas are over 4–5 times higher than that of porous carbons hydrothermally synthesized with the assistance of sacrificial templates reported previously [63]. From the MPVs calculated from CO$_2$ adsorption at 273 K (V_{DR}(CO$_2$)), which is related with narrow microporosity (i.e., micropore width less than 0.7 nm) [70, 71], it can be stated that (i) the three materials have narrow micropores, and (ii) the amount of narrow microporosity for glucose and cellulose is quite similar and lower than for the activated carbon obtained from the rye straw. In summary, these results indicate that the PSD, which is particularly important for many pore-size-dependent applications, is relatively narrow, and most of the pore volume and surface area arises from micropores.

Regarding the activation yields, as expected the values obtained for activation of HTC (obtained after HTC at 240 °C) are lower than those obtained after activation of other lignocellulosic precursors (EW, CS, and AS), previously carbonized at higher temperature (about 850 °C) (see Figure 3.5), but higher than those obtained after direct activation of the same precursors (no previous carbonization) (see Figure 3.4). For a better comparison, the HTC yield (or carbonization yield) should be also considered in the calculation of the global yield. Thus, for example, if the HTC yield of rye straw (42%) is taken into account in the calculation of the global yield for KOH activation, the value is 13%. This value

Table 3.7   *Porous texture characterization results of the KOH HTC-derived activated carbons.*

Sample	S_{BET} (m^2 g^{-1})	V_{DR}(N$_2$) (cm^3 g^{-1})	V_{DR}(CO$_2$) (cm^3 g^{-1})	Yield (%)
Glucose	2210	0.90	0.60	24
Cellulose	2250	0.90	0.56	21
Rye straw	2200	0.92	0.64	31

is similar to the global yield obtained for the almond shell (17%) if carbonization yield is considered.

3.3.2.3 Preliminary Results of CO_2 Adsorption and CH_4 Storage

To evaluate the effectiveness of the HTC-derived activated carbons as CO_2 adsorbents, CO_2 adsorption isotherms were carried out up to 3 MPa for two of the activated hydrochars (those obtained from glucose and rye straw) (see Figure 3.9). As described in our previous study [72], the shape of these isotherms provides information about the porosity of the samples. Both CO_2 adsorption isotherms are of type I, characteristic of microporous materials. The activated carbon prepared from rye straw presents a sharper knee than that prepared from glucose, suggesting that the former presents a narrower MPSD than the glucose-derived activated carbon. The CO_2 uptakes at 30 bar and 273 K for HTC-derived activated carbons are comparable to those obtained with superactivated carbons prepared by KOH activation of anthracites with similar MPV [72].

The same two HTC-derived activated carbons tested as CO_2 adsorbents were also studied as materials for high-pressure CH_4 storage. Figure 3.10 presents the methane adsorption isotherms on a gravimetric basis for both samples, which have very similar methane adsorption capacities, since, as just shown, the porosity development for both of them is quite similar. It is well established that CH_4 adsorption on porous materials at these conditions (room temperature and high pressures) takes place in micropores [73–75]. In this sense, the results obtained in a previous study with activated carbon fibers and powder activated carbons, covering a wide range of MPV (up to 1.5 cm^3 g^{-1}) [74] pointed out that a general trend exists for this type of materials: the higher the MPV, the higher the methane adsorption capacity. It was concluded that, for samples with relatively narrow MPSD, the total MPV ($V_{DR}(N_2)$) obtained from the DR equation could be a good indicator of the methane capacity. The preliminary study carried out with HTC-derived activated carbons has given CH_4 uptakes as high as 9 mmol CH_4 g^{-1} sample at 30 bar, being values comparable with

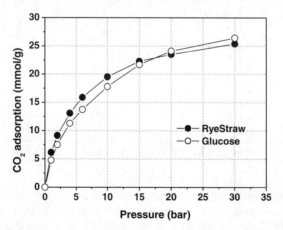

Figure 3.9 *CO_2 adsorption isotherms at 273 K and up to 30 bar for two activated hydrochars prepared from glucose and rye straw on a gravimetric basis.*

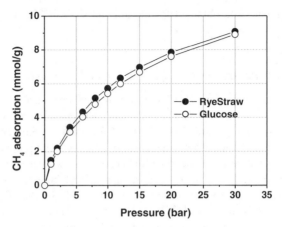

Figure 3.10 *CH₄ adsorption isotherms at 298 K and up to 30 bar for two activated hydrochars prepared from glucose and rye straw on gravimetric basis.*

activated carbons with similar $V_{DR}(N_2)$ obtained from other precursors [74], as shown in Figure 3.11.

In addition to MPV and MPSD, methane storage in volumetric basis also depends on another important parameter, namely the packing density. Table 3.8 contains the values corresponding to the "tap density" and the "packing density" for the three activated hydrochars. In our previous work carried out with activated carbons (obtained by KOH activation of an anthracite), it was clearly shown that, for both types of densities (tap and packing), a general trend was observed: the higher the porosity development of the materials, the lower the

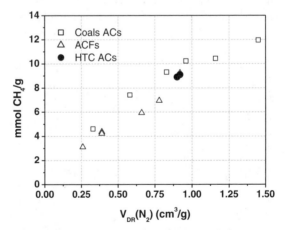

Figure 3.11 *Gravimetric methane adsorption capacity (at 298 K) corresponding to KOH anthracite-derived activated carbons (Coal ACs) (40 bar) [73], activated carbon fibers (ACFs) (40 bar) [74], and HTC-derived activated carbons (HTC ACs) obtained in the present study (30 bar) versus the MPV.*

Table 3.8   *Tap and packing densities of the three activated samples.*

Sample	Tap density (g/cm^3)	Packing density (g/cm^3)
Rye straw	0.13	0.43
Cellulose	0.12	0.39
Glucose	0.11	0.38

density, following a linear trend [76]. Comparing with those results, it can be said that the densities of the hydrochars prepared in the present study are lower than activated carbons from anthracite with similar MPV.

These excellent preliminary results could be further improved by tailoring the PSD of the activated carbons to the optimum pore width value for methane adsorption (around 0.8–1.1 nm) [73–75]. Such specific material pore size features could be achieved by tuning the KOH activation process parameters and the chemical structure of the parent HTC carbon.

3.4   Conclusions

This chapter deals with the production of activated carbons with a highly developed microporosity from lignocellulosic biomass. Two different synthetic routes are covered: (i) carbonization followed by (or together with) chemical activation with H_3PO_4, $ZnCl_2$, and KOH, and (ii) HTC of biomass, followed by chemical activation with alkaline hydroxides.

After presenting some examples of chemical activation (H_3PO_4, $ZnCl_2$, and KOH) of lignocellulosic chars obtained by the common high-temperature techniques, preliminary results have been included about the chemical activation (KOH) of hydrochars obtained using an alternative procedure (HTC), where very mild temperatures (180–250 °C) are needed. The results showed that HTC-derived carbons are excellent precursors for the synthesis of activated carbons via KOH chemical activation. Regardless of the parent biomass (i.e., glucose, cellulose, or rye straw), highly microporous activated carbons with a narrow MPSD could be prepared.

Preliminary testing of the synthesized activated carbons as adsorbents for either CO_2 capture or high-pressure CH_4 storage yielded very promising results. The measured uptakes of both adsorbates were comparable to top-performing and commercially available adsorbents usually employed for these end-applications. Further improvements of the synthesized activated carbon performance could certainly be achieved by optimizing the activation and HTC synthesis parameters, in such a way as to tailor their PSDs to the adsorbate.

Acknowledgments

The authors would like to thank the Spanish Ministerio de Economía y Competitividad and PLAN E funds (Projects CTQ2009-10813/PPQ, PLE2009-0021 and CTQ2012/31762) and GeneralitatValenciana and FEDER (PROMETEO/2009/047) for financial support.

References

(1) Bansal, R.C., Donnet, J.B., and Stoeckli, F. (1988) *Active Carbon*, Dekker, New York.
(2) Jankowska, H., Swiatkowski, A., and Choma, J. (1991) *Active Carbon*, Ellis Horwood, New York.
(3) Marsh, H., Heintz, E.A., and Rodríguez-Reinoso, F. (1997) *Introduction to Carbon Technologies*, 1st edn, Servicio de Publicaciones de la Universidad de Alicante, Alicante.
(4) Kyotani, T. (2003) Porous carbon, in *Carbon Alloys. Novel Concepts to Develop Carbon Science and Technology* (eds E. Yasuda, M. Inagaki, K. Kaneko *et al.*), Elsevier, Oxford.
(5) Burchell, T.D. (1999) *Carbon Materials for Advanced Technologies*, Elsevier, Oxford.
(6) Derbyshire, F., Jagtoyen, M., Andrews, R. *et al.* (2001) Activated carbon in environmental applications, in *Chemistry and Physics of Carbon*, vol. **27** (ed. L.R. Radovic), Dekker, New York.
(7) Radovic, L.R., Moreno-Castilla, C., and Rivera-Utrilla, J. (2001) Carbon materials as adsorbents in aqueous solutions, in *Chemistry and Physics of Carbon*, vol. **27** (ed. L.R. Radovic), Dekker, New York.
(8) Muñoz-Guillena, M.J., Illán-Gómez, M.J., Martín-Martínez, A., and Salinas-Martínez de Lecea, C. (1992) Activated carbons from Spanish coals. 1. Two-stage CO2 activation. *Energy and Fuels*, **6**, 9–15.
(9) Benaddi, H., Legras, D., Rouzaud, J.N., and Béguin, F. (1998) Influence of the atmosphere in the chemical activation of wood by phosphoric acid. *Carbon*, **36**, 306–309.
(10) Molina-Sabio, M., Rodríguez-Reinoso, F., Caturla, F., and Sellés, M.J. (1995) Porosity in granular carbons activated with phosphoric acid. *Carbon*, **33**, 1105–1113.
(11) Díaz-Díez, M.A., Gómez-Serrano, V., González, C.F. *et al.* (2004) Porous texture of activated carbons prepared by phosphoric acid activation of woods. *Applied Surface Science*, **238**, 309–313.
(12) Suárez-García, F., Martínez-Alonso, A., and Tascón, J.M.D. (2004) Nomex polyaramid as a precursor for activated carbon fibers by phosphoric acid activation. Temperature and time effects. *Microporous and Mesoporous Materials*, **75**, 73–80.
(13) Rodríguez-Reinoso, F. and Molina-Sabio, M. (1992) Activated carbons from lignocellulosic materials by chemical and/or physical activation: an overview. *Carbon*, **30**, 1111–1118.
(14) Ahmadpour, A. and Do, D.D. (1996) The preparation of active carbons from coal by chemical and physical activation. *Carbon*, **34**, 471–479.
(15) Caturla, F., Molina-Sabio, M., and Rodríguez-Reinoso, F. (1991) Preparation of activated carbon by chemical activation with ZnCl2. *Carbon*, **29**, 999–1007.
(16) Ibarra, J.V., Moliner, R., and Palacios, J.M. (1991) Catalytic effects of zinc chloride in the pyrolysis of Spanish high sulphur coals. *Fuel*, **70**, 727–732.
(17) Hayashi, J., Watkinson, A.P., Teo, K.C. *et al.* (1995) in *Proceedings of the Eighth International Conference on Coal Science* (eds J.A. Pajares and J.M.D. Tascón), Elsevier, Amsterdam.
(18) Carvalho, A.P., Gomes, M., Mestre, A.S. *et al.* (2004) Activated carbons from cork waste by chemical activation with K2CO3: application to adsorption of natural gas components. *Carbon*, **42**, 672–674.

(19) Hayashi, J., Uchibayashi, M., Horikawa, T. *et al.* (2002) Synthesizing activated carbons from resins by chemical activation with K2CO3. *Carbon*, **40**, 2747–2752.

(20) Illán-Gómez, M.J., García-García, A., Salinas-Martínez de Lecea, C., and Linares-Solano, A. (1996) Activated carbons from Spanish coals. 2. Chemical activation. *Energy Fuels*, **10**, 1108–1114.

(21) Evans, M.J.B., Halliop, E., and MacDonald, J.A.F. (1999) The production of chemically-activated carbon. *Carbon*, **37**, 269–274.

(22) Lozano-Castello, D., Lillo-Rodenas, M.A., Cazorla-Amoros, D., and Linares-Solano, A. (2001) Preparation of activated carbons from Spanish anthracite: I. Activation by KOH. *Carbon*, **39**, 741–749.

(23) Lillo-Ródenas, M.A., Lozano-Castelló, D., Cazorla-Amorós, D., and Linares-Solano, A. (2001) Preparation of activated carbons from Spanish anthracite: II. Activation by NaOH. *Carbon*, **39**, 751–759.

(24) Perrin, A., Celzard, A., Albiniak, A. *et al.* (2004) NaOH activation of anthracites: effect of temperature on pore textures and methane storage ability. *Carbon*, **42**, 2855–2866.

(25) Suárez-García, F., Martínez-Alonso, A., and Tascón, J.M.D. (2002) Pyrolysis of apple pulp: chemical activation with phosphoric acid. *Journal of Analytical and Applied Pyrolysis*, **63**, 283–301.

(26) Teng, H., Yeh, T.S., and Hsu, L.Y. (1998) Preparation of activated carbon from bituminous coal with phosphoric acid activation. *Carbon*, **36**, 1387–1395.

(27) Baker, F.S. (1995) US patent 5,416,056.

(28) Wennerberg, A.N. and O'Grady, T.M. (1978) US patent 4,082,694.

(29) Otowa, T., Tanibata, R., and Itoh, M. (1993) Production and adsorption characteristics of MAXSORB: high-surface-area active carbon. *Gas Separation & Purification*, **7**, 241–245.

(30) Linares-Solano, A., Lozano-Castelló, D., Lillo-Ródenas, M.A., and Cazorla-Amorós, D. (2008) in *Chemistry and Physics of Carbon*, vol. **30** (ed. L.R. Radovic), CRC Press, New York.

(31) Gonzalez, J.C., Gonzalez, M.T., Molina-Sabio, M. *et al.* (1995) Porosity of activated carbons prepared from different lignocellulosic materials. *Carbon*, **33**, 1175–1177.

(32) Rodriguez-Reinoso, F. and Linares-Solano, A. (1989) Microporous structure of activated carbons as revealed by adsorption methods, in *Chemistry and Physics of Carbon*, vol. **21** (ed. P.A. Thrower), Dekker, New York.

(33) Gonzalez, J.C., Sepulveda-Escribano, A., Molina-Sabio, M., and Rodriguez-Reinoso, F. (1997) Production, properties and applications of activated carbon, in *Characterization of Porous Solids IV* (eds B. McEnaney, T.J. Mays, F. Rouquerol *et al.*), Royal Society of Chemistry, Cambridge.

(34) Hu, Z., Srinivasan, M.O., and Ni, Y. (2001) Novel activation process for preparing highly microporous and mesoporous carbons. *Carbon*, **39**, 877–886.

(35) Marsh, H. and Rodríguez-Reinoso, F. (2006) *Activated Carbon*, Elsevier, Oxford.

(36) Laine, J., Calafat, A., and Labady, M. (1989) Preparation and characterization of activated carbons from coconut shell impregnated with phosphoric acid. *Carbon*, **27**, 191–205.

(37) Jagtoyen, M. and Derbyshire, F. (1993) Some considerations of the origins of porosity in carbons from chemically activated wood. *Carbon*, **31**, 1185–1192.

(38) Jagtoyen, M. and Derbyshire, F. (1998) Activated carbons from yellow poplar and white oak by H3PO4 activation. *Carbon*, **36**, 1085–1097.

(39) Suárez-García, F., Martínez-Alonso, A., and Tascón, J.M.D. (2002) Pyrolysis of apple pulp: effect of operation conditions and chemical additives. *Journal of Analytical and Applied Pyrolysis*, **62**, 93–109.

(40) Suárez-García, F., Martínez-Alonso, A., and Tascón, J.M.D. (2002) Pyrolysis of apple pulp: chemical activation with phosphoric acid. *Journal of Analytical and Applied Pyrolysis*, **63**, 283–301.

(41) Solum, M.S., Pugmire, R.J., Jagtoyen, M., and Derbyshire, F. (1995) Evolution of carbon structure in chemically activated wood. *Carbon*, **33**, 1247–1254.

(42) Castro, J.B., Bonelli, P.R., Cerrella, E.G., and Cukierman, A.L. (2000) Phosphoric acid activation of agricultural residues and bagasse from sugar cane: influence of the experimental conditions on adsorption characteristics of activated carbons. *Industrial & Engineering Chemistry Research*, **39**, 4166–4172.

(43) Rosas, J.M., Bedia, J., Rodríguez-Mirasol, J., and Cordero, T. (2009) HEMP-derived activated carbon fibers by chemical activation with phosphoric acid. *Fuel*, **88**, 19–26.

(44) Almansa, C., Molina-Sabio, M., and Rodriguez-Reinoso, F. (2004) Adsorption of methane into ZnCl$_2$-activated carbon derived discs. *Microporous and Mesoporous Materials*, **76**, 185–191.

(45) Jolly, R., Charcosset, H., Boudou, J.P., and Guet, J.M. (1988) Catalytic effect of ZnCl$_2$ during coal pyrolysis. *Fuel Processing Technology*, **20**, 51–60.

(46) Kirubakaran, C.J., Krishnaiah, K., and Seshadri, S.K. (1991) Experimental study of the production of activated carbon from coconut shells in a fluidized bed reactor. *Industrial & Engineering Chemistry Research*, **30**, 2411–2416.

(47) Haider-Usmani, T., Wahab-Ahmed, T., Zafar-Ahmed, S., and Yousufzai, A.H.K. (1996) Preparation and characterization of activated carbon from a low rank coal. *Carbon*, **34**, 77–82.

(48) Chong-Lua, A. and Yang, T. (2005) Characteristics of activated carbon prepared from pistachio-nut shell by zinc chloride activation under nitrogen and vacuum conditions. *Journal of Colloid and Interface Science*, **290**, 505–513.

(49) Encinar, J.M., Beltran, F.J., Ramiro, A., and Gonzalez, J.F. (1998) Pyrolysis/ gasification of agricultural residues by carbon dioxide in the presence of different additives: influence of variables. *Fuel Processing Technology*, **55**, 219–233.

(50) Tsai, W.T., Chang, C.Y., and Lee, S.L. (1998) A low cost adsorbent from agricultural waste corn cob by zinc chloride activation. *Bioresource Technology*, **64**, 211–217.

(51) Hayashi, J., Watkinson, A.P., Teo, K.C. *et al.* (1995) Production of activated carbon from Canadian coal by chemical activation. *Coal Science*, **1**, 1121–1124.

(52) Teng, H. and Hsu, L. (1999) High-porosity carbons prepared from bituminous coal with potassium hydroxide activation. *Industrial & Engineering Chemistry Research*, **38**, 2947–2953.

(53) Carrott, P.J.M., Ribeiro Carrott, M.M.L., and Mourao, P.A.M. (2006) Pore size control in activated carbons obtained by pyrolysis under different conditions of chemically impregnated cork. *Journal of Analytical and Applied Pyrolysis*, **75**, 120–127.

(54) Ubago-Perez, R., Carrasco-Marın, F., Fairen-Jimenez, D., and Moreno-Castilla, C. (2006) Granular and monolithic activated carbons from KOH-activation of olive stones. *Microporous and Mesoporous Materials*, **92**, 64–70.

(55) El-Hendawy, A.A. (2006) Variation in the FTIR spectra of a biomass under impregnation, carbonization and oxidation conditions. *Journal of Analytical and Applied
Pyrolysis*, **75**, 159–166.

(56) Lillo-Rodenas, M.A., Marco-Lozar, J.P., Cazorla-Amoros, D., and Linares-Solano,
A. (2007) Activated carbons prepared by pyrolysis of mixtures of carbon precur-
sor/alkaline hydroxide. *Journal of Analytical and Applied Pyrolysis*, **80**, 166–174.

(57) Titirici, M.M. and Antonietti, M. (2010) Chemistry and materials options of sustainable carbon materials made by hydrothermal carbonization. *Chemical Society
Reviews*, **39**, 103–116.

(58) Titirici, M.M., Thomas, A., and Antonietti, M. (2007) Back in the black: hydrothermal
carbonization of plant material as an efficient chemical process to treat the CO2
problem? *New Journal of Chemistry*, **31**, 787–789.

(59) Bergius, F. (1928) Beiträge zur Theorie der Kohleentstehung. *Die Naturwissenschaften*, **16**, 1–10.

(60) Kubo, S., Demir-Cakan, R., Zhao, L. *et al.* (2010) Porous carbohydrate-based materials via hard templating. *ChemSusChem*, **3**, 188–194.

(61) Hu, B., Wang, K., Wu, L. *et al.* (2010) Engineering carbon materials from the
hydrothermal carbonization process of biomass. *Advanced Materials*, **22**, 813–
828.

(62) White, R.J., Antonietti, M., and Titirici, M.M. (2009) Naturally inspired nitrogen
doped porous carbon. *Journal of Materials Chemistry*, **19**, 8645–8650.

(63) Joo, J.B., Kim, Y.J., Kim, W. *et al.* (2008) Simple synthesis of graphitic porous carbon
by hydrothermal method for use as a catalyst support in methanol electro-oxidation.
Catalysis Communications, **10**, 267–271.

(64) Titirici, M.M., Thomas, A., and Antonietti, M. (2007) Replication and coating of
silica templates by hydrothermal carbonization. *Advanced Functional Materials*, **17**,
1010–1018.

(65) Kim, P., Joo, J.B., Kim, W. *et al.* (2006) Graphitic spherical carbon as a support
for a PtRu-alloy catalyst in the methanol electro-oxidation. *Catalysis Letters*, **112**,
213–218.

(66) Sevilla, M., Fuertes, A.B., and Mokaya, R. (2011) High density hydrogen storage in
superactivated carbons from hydrothermally carbonized renewable organic materials.
Energy & Environmental Science, **4**, 1400–1410.

(67) Titirici, M.M., Antonietti, M., and Baccile, N. (2008) Hydrothermal carbon from
biomass: a comparison of the local structure from poly- to monosaccharides and
pentoses/hexoses. *Green Chemistry*, **10**, 1204–1212.

(68) Falco, C., Baccile, N., and Titirici, M.M. (2011) Morphological and structural differences between glucose, cellulose and lignocellulosic biomass derived hydrothermal
carbons. *Green Chemistry*, **13**, 3273–3281.

(69) O'Sullivan, A.C. (1997) Cellulose: the structure slowly unravels. *Cellulose*, **4**, 173–
207.

(70) Cazorla-Amoros, D., Alcañiz-Monge, J., and Linares-Solano, A. (1996) Characterization of activated carbon fibers by CO2 adsorption. *Langmuir*, **12**, 2820–2824.

(71) Cazorla-Amoros, D., Alcañiz-Monge, J., de la Casa-Lillo, M.A., and Linares-Solano,
A. (1998) CO2 as an adsorptive to characterize carbon molecular sieves and activated
carbons. *Langmuir*, **14**, 4589–4596.

(72) Lozano-Castello, D., Cazorla-Amoros, D., Linares-Solano, A., and Quinn, D.F. (2002) Micropore size distributions of activated carbons and carbon molecular sieves assessed by high-pressure methane and carbon dioxide adsorption isotherms. *Journal of Physical Chemistry B*, **106**, 9372–9379.

(73) Lozano-Castello, D., Cazorla-Amoros, D., Linares-Solano, A., and Quinn, D.F. (2002) Influence of pore size distribution on methane storage at relatively low pressure: preparation of activated carbon with optimum pore size. *Carbon*, **40**, 989–1002.

(74) Lozano-Castello, D., Alcañiz-Monge, J., de la Casa-Lillo, M.A. *et al.* (2002) Advances in the study of methane storage in porous carbonaceous materials. *Fuel*, **81**, 1777–1803.

(75) Alcañiz-Monge, J., Lozano-Castello, D., Cazorla-Amorós, D., and Linares-Solano, A. (2009) Fundamentals of methane adsorption in microporous carbons. *Microporous and Mesoporous Material*, **124**, 110–116.

(76) Jorda-Beneyto, M., Lozano-Castello, D., Suarez-García, F. *et al.* (2008) Advanced activated carbon monoliths and activated carbons for hydrogen storage. *Microporous and Mesoporous Materials*, **112**, 235–242.

4

Hydrothermally Synthesized Carbonaceous Nanocomposites

Bo Hu, Hai-Zhou Zhu, and Shu-Hong Yu

Division of Nanomaterials & Chemistry, Hefei National Laboratory for Physical Sciences at Microscale, Department of Chemistry, University of Science and Technology of China, China

4.1 Introduction

The hydrothermal carbonization (HTC) process, has recently become a powerful technique for the synthesis of unique and valuable carbonaceous nanomaterials and nanocomposites, which have shown promising applications as heavy metal sorption, biochemical, catalytic, CO_2 fixation, and energy storage materials [1]. As described in the previous chapters of this book, the advantages of this process include the low toxicological impact of materials and processes, environmentally friendly conditions, facile instruments and techniques, use of renewable resources, high energy and atom economy, high carbon efficiency, and abundant functional groups remaining on the product surface.

Carbon materials have shown a great impact on human daily life since the early stages of mankind. For example, carbon black, fabricated from fuel-rich partial combustion, has been widely used as ink, pigments, and tattoos for more than 3000 years [2]. Recently, with the discovery of fulerenes [3], carbon nanotubes (CNTs) [4], and graphene, carbon material science and technology has become a hot research area, and related invention and progress will greatly benefit human society. Carbonaceous materials, have occupied an important position in carbon materials and have shown promising applications [1]. Carbonaceous materials, made by the HTC process, not only have well-controlled morphology, composition, and structure, but also contain lots of functional groups, such as hydroxyl, aldehyde,

Sustainable Carbon Materials from Hydrothermal Processes, First Edition. Edited by Maria-Magdalena Titirici.
© 2013 John Wiley & Sons, Ltd. Published 2013 by John Wiley & Sons, Ltd.

and carboxyl groups, which can greatly improve their hydrophilicity and surface chemical reactivity. The advantages of these unique carbonaceous materials include the facile synthesis, well-controlled morphology, rich surface functional groups, high surface reactivity, good chemical stability, satisfactory biocomaptibility, promising carriers for biomolecular immobilization, the possibility of coating of most nanostructured surfaces, facile functionalization inorganic nanostructures, remarkable reactivity, and the capability for being hybridized with other nanomaterials.

In this chapter, we discuss the latest advances in the synthesis of such unique carbonaceous nanomaterials and nanocomposites via the HTC synthesis method. Some representative examples of the applications of carbonaceous nanomaterials and nanocomposites from the HTC process are briefly presented with a special focus on their environmental, catalytic energy, and biological applications. Finally, we provide a summary and outlook on this active research topic.

4.2   Hydrothermal Synthesis of Unique Carbonaceous Nanomaterials

The HTC process has provided remarkable developments in the synthesis of functional carbonaceous nanomaterials from waste biomass, carbohydrates, and organic molecules. Usually biomass precursors used for this process were directly obtained from agricultural residues, wood herbaceous energy crops, and so on. Carbohydrates normally included hexoses (glucose and fructose), disaccharides (sugar), polysaccharides (starch), hemicellulose, cellulose, and other dehydration products of glucose (2-furaldehyde, furfural, and hydroxymethylfurfural (HMF)). Organic molecules, such as glutaraldehyde, could greatly change the structures and surface functional groups of the final carbonaceous materials. In the following sections, we will review the advances in the area of controlled synthesis of various carbonaceous nanomaterials and nanocomposites via the HTC process.

4.2.1   Carbonaceous Nanomaterials

4.2.1.1   *Zero-Dimensional Carbonaceous Nanomaterials*

Semiconductor quantum dots that exhibit quantization effects with quantum confined in three dimensions are presently attracting a great deal of interest and are finding many important applications, such as in transistors, solar cells, light-emitting diodes, diode lasers, devices, and medical imaging [5]. Carbon quantum dots, as a new promising material, have been recognized as more benign substitutes for semiconductor quantum dots, due to their low toxicity, low cost, and high stability [6]. Currently, the methods of synthesis of carbon quantum dots (CQDs) mainly include direct thermal carbonization and ablation of graphite, graphene, or CNTs [7]. The as-obtained particles lack dispersibility in solvents and need a post-treatment. For example, a novel and simple hydrothermal approach for the cutting of graphene sheets into surface-functionalized graphene quantum dots (GQDs of around 9.6 nm average diameter) has been reported [8]. The functionalized GQDs were found to exhibit bright blue photoluminescence, which has never been observed in graphene sheets and graphene nanoribbons owing to their large lateral sizes. The blue luminescence and new ultraviolet (UV)/Vis absorption bands are directly induced by the large edge effect shown in the ultrafine GQDs.

The HTC process of carbohydrates has been used itself as an efficient technique for the one-step synthesis of hydrophilic CQDs. For example, a simple aqueous solution route was used by Travas-Sejdic *et al.* [9] to prepare luminescent carbogenic dots using carbohydrates as starting materials. In short, the carbohydrates were dehydrated using concentrated sulfuric acid, producing carbonaceous materials. The obtained carbonaceous materials were then broken down into individual carbogenic nanoparticles by treatment with nitric acid. Finally, the carbogenic nanoparticles were passivated using amine-terminated compounds, yielding luminescent carbogenic dots (Figure 4.1).

Similarly, Yang *et al.* reported a microwave approach to fluorescent carbon nanoparticles using poly(ethylene glycol) and saccharides (glucose, fructose) [10]. Kundi *et al.* has demonstrated that such glucose-derived CQDs are unique molecular transporters owing to their intrinsic fluorescence, surface functionalization, nontoxicity, and green synthesis method [11]. The affinity of carbon spheres for positively charged molecules such as polycations and negatively charged molecules such as DNA widens the scope of their use. Their immense potential in nuclear targeted entry has been harnessed by trafficking the cell-impermeable histone acetyltransferase (HAT) activator (*N*-(4-chloro-3-trifluoromethylphenyl)-2-ethoxybenzamide), which in turn could induce hyperacetylation

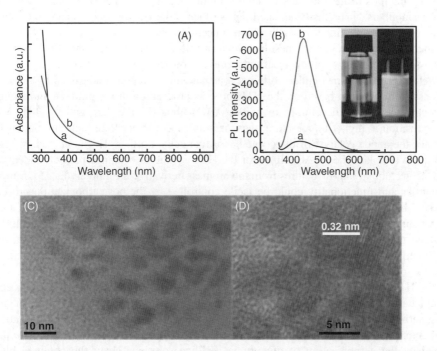

Figure 4.1 *Absorption (A) and emission (B) spectra of carbogenic dots prepared from glucose, before (a) and after (b) 4,7,10-trioxa-1,13-tridecanediamine (TTDDA) passivation. (Inset: carbogenic dots under ambient light (left) and UV light (365 nm) (right).) (C and D) TEM images of TTDDA-passivated carbogenic dots prepared from glucose. (Reproduced with permission from [9]. © 2009 American Chemical Society.)*

in vivo, leading to transcriptional activation and giving a positive impetus to the field of therapeutics involving direct HAT activation.

4.2.1.2   One-Dimensional Carbonaceous Nanomaterials

The hard or soft templates presented in Chapter 2 are powerful tools for controlling the synthesis of one-dimensional carbonaceous nanostructures. The presence of a template could effectively restrain the homogeneous nucleation and growth of carbonaceous spheres, and instead promote the heterogeneous deposition of carbonaceous products on the backbone of the template. For example, tellurium nanowires, of several nanometers in diameter and a high aspect ratio, have been successfully used as a template for the synthesis of uniform carbonaceous nanofibers by the HTC process from glucose [12]. Te@carbonaceous composite nanocables with a high aspect ratio were first synthesized. The diameter of nanocables could be controlled by facile adjustment of the HTC reaction time or the ratio of tellurium and glucose. The core structure of tellurium nanowires could be simply removed, forming well-defined ultralong and functionalized carbonaceous nanofibers. The typical as-synthesized carbonaceous nanofibers have an average diameter of 50 nm and are up to tens or hundreds of micrometers in length (Figure 4.2a and b). The diameter of carbonaceous nanofibers can be easily adjusted. The surface of these carbonaceous nanofibers had a polar, oxygen-containing layer with a high concentration of hydroxyl and carboxyl groups. Further, sodium dodecyl benzenesulfonate, as a soft template, could control the synthesis and assembly process to produce carbonaceous nanowires [13].

Well-aligned and open-ended carbonaceous nanotubes can be synthesized by the HTC process from glucose, based on the HTC process proceeding in the uniform and straight nanochannels of anodic aluminum oxide (AAO) films (Figure 4.2c and d) [14]. After carbonization at higher temperature, these nanotubes could decorate platinum nanoparticles for the anode catalyst of direct methanol fuel cells. Further, Titirici *et al.* have used this same template (i.e., AAO films) for the HTC synthesis of uniform, open-ended carbonaceous tubular nanostructures from the biomass derivative 2-furaldehyde [15]. Surface properties and functionality could be facile controlled by the postcarbonization process. These tubular nanostructures with a functional surface could selectively introduce some important chemical moieties. Thermoresponsive polymers (poly(*N*-isopropylacrylamide)) could be grafted on their surface, and the final products have shown temperature-dependent hydrophilic–hydrophobic altering behavior and consequent controlled dispersibility in the aqueous phase.

The HTC process can combine with other methods for the synthesis of carbonaceous nanomaterials. Based on the cooperative effect between self-assembly and the HTC process, carbonaceous submicrotubes have been successfully synthesized (Figure 4.2e and f) [16]. First, carbonaceous nanoparticles tethered by organic functional groups were produced by the carbonization of glucose in pyridine solution. Then, the obtained black solutions containing these carbonaceous nanoparticles were diluted with distilled water and the slow self-assembly process began. The self-assembly process was propelled by the different affinities between the functional groups and the solvents. Finally, these carbonaceous nanoparticles with hydrophobic surface groups assembled themselves, forming the submicrotubular structures.

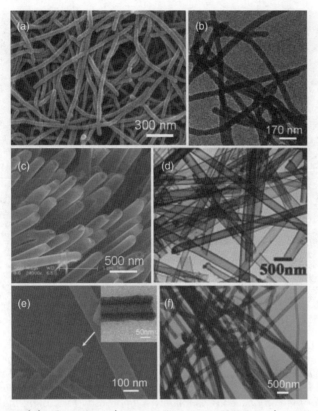

Figure 4.2 *(a and b) Scanning electron microscopy (SEM) and transmission electron microscopy (TEM) images of carbonaceous nanofibers. (Reproduced with permission from [12]. © 2006 American Chemical Society.) (c and d) SEM and TEM images of carbonaceous nanotubes. (Reproduced with permission from [14].) (e and f) SEM and TEM images of carbonaceous submicrotubules. (Reproduced with permission from [16]. © 2008 American Chemical Society.)*

4.2.1.3   Two-Dimensional Carbonaceous Nanomaterials

The carbonaceous nanofibers produced by the HTC process can easily form a free-standing carbonaceous membrane with a very narrow pore size distribution, which can be used as a filtration membrane for the precise separation of nanoparticles (Figure 4.3) [17]. The pore size distribution, filtration flux, and cut-off size of the membrane could be well-controlled by carefully changing the diameter of the composing carbonaceous nanofibers. Regulating the diameter of these carbonaceous nanofibers was easily done by a decrease of the ratio of the tellurium templates to glucose and prolonging the HTC reaction time. Based on these carbonaceous nanofibers, the solvent-evaporation-induced self-assembly process could effectively fabricate the free-standing carbonaceous membranes. After a casting and drying process, the flexible, intertwisted, mechanical robust membrane could be easily detached from the Teflon substrate without cracks or pinholes. The thickness of the membrane can be well-controlled via the concentration and the volume of the

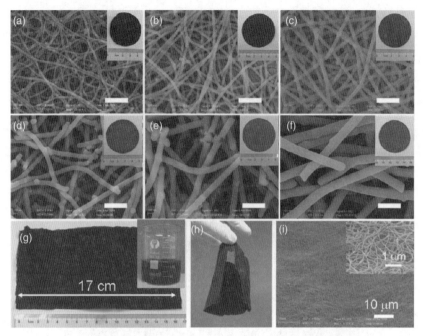

Figure 4.3 *(a–f) SEM images of carbonaceous nanofibers fabricated by a HTC process with different reaction times from 12 to 60 h. All images were collected at the same magnification. Scale bar: 400 nm. (Insets: photographs of the corresponding freestanding membranes.) (g and h) Optical images of the flexible carbonaceous membrane. The inset in (g) shows the optical image of the carbonaceous nanofibers solution used for casting the membrane. (i) Low- and (inset) high-magnification SEM images showing the surface morphology of the CNF-50 membrane. (Reproduced with permission from [17]. © 2010 WILEY-VCH Verlag GmbH & Co. KGaA, Weinheim.)*

suspension containing the carbonaceous nanofibers. The membrane was highly hydrophilic and porous – characteristics that are required for efficient filtrations from aqueous solutions.

4.2.1.4 Three-Dimensional Carbonaceous Nanomaterials

Colloidal carbonaceous micro- and nanospheres could be easily synthesized using the HTC process from carbohydrates, including sugar (Figure 4.4a) [18], glucose (Figure 4.4b) [19], cyclodextrins [20], fructose [21], sucrose [22], cellulose [23], and starch [22]. As will be described in detail in Chapter 6, the formation process of these materials usually included dehydration, condensation, polymerization, and aromatization [24]. The whole process can be regarded as "green," because water is the only reaction media involved without any organic solvents, initiators, or surfactants. The as-prepared spheres are nontoxic with rich surface functional groups, enabling the facile decoration of inorganic nanostructures and graft biomolecules.

Hollow carbonaceous spheres, with even porous walls, could be synthesized from glucose via the HTC process using the template of amino functionalization of silica particles or

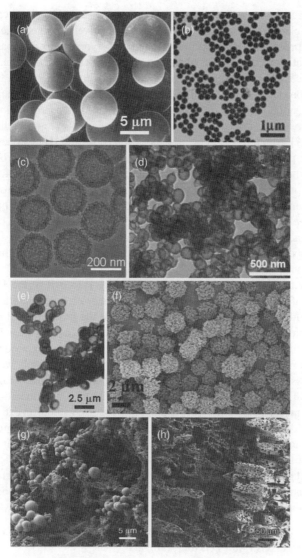

Figure 4.4 (a) SEM image of monodispersed hard carbon spherules. (Reproduced with permission from [18]. © 2001 Elsevier.) (b) TEM image of carbonaceous spheres. (Reproduced with permission from [19]. © 2004 WILEY-VCH Verlag GmbH & Co. KGaA, Weinheim.) (c) TEM image of hollow carbon materials. (Reproduced with permission from [25]. © 2007 American Chemical Society.) (d) TEM image of hollow carbonaceous nanospheres. (Reproduced with permission from [26]. © 2010 American Chemical Society.) (e) TEM image of hollow spheres. (Reproduced with permission from [27]. © 2006 WILEY-VCH Verlag GmbH & Co. KGaA, Weinheim.) (f) SEM images of carboxylate-rich carbonaceous materials. (Reproduced with permission from [28]. © 2009 American Chemical Society.) (g) SEM image of the coexistence of carbonaceous spheres and a microstructured carbonaceous species. (h) SEM image of carbonaceous scaffold replicating of the insoluble carbohydrates in rice grain. (Reproduced with permission from [27]. © 2006 WILEY-VCH Verlag GmbH & Co. KGaA, Weinheim.)

silica (core)–porous silica (shell) spheres (Figure 4.4c) [25]. The significant effect in this method was the electrostatic attraction between the positively charged silica and negatively charged carbon precursors. Hydroxyl-terminated polystyrene latex nanoparticles were also a suitable template for the synthesis of functional hollow carbon nanospheres from glucose via the HTC process, based on the hydrogen-bonding effect for the initial adsorption of HTC decomposition products (Figure 4.4d) [26]. Soft templates, such as the anionic surfactant sodium dodecyl sulfate (SDS), could also be used for the synthesis of hollow carbonaceous capsules with a reactive surface layer as well as tunable void size and shell thickness in the HTC process [29]. Iron ions could be used as a catalyst for the HTC process. For example, Fe^{2+} ions in the presence of $[Fe(NH_4)_2(SO_4)_2]$ can catalyze the HTC process and lead to the synthesis of hollow carbonaceous microspheres from starch (Figure 4.1e) [27].

Interestingly, industrial microorganism could be used as the carbon resources in the HTC process. For example, the budding yeast *Saccharomyces cerevisiae* has been successfully used for the fabrication of hollow carbonaceous microspheres with controlled meso- and macroporous shells [30]. These hollow carbonaceous microspheres have also shown amphiphilic properties, because their surface was functionalized with both hydrophilic and hydrophobic functional groups.

A small amount of organic monomers could easily take part in the HTC process for the production of carbonaceous materials with a variety of functionalities. For example, acrylic acid could take part in the HTC of glucose for the production of carboxylate-rich carbonaceous microspheres with rough raspberry-like surfaces (Figure 4.4f) [28].

Hierarchical three-dimensional carbonaceous nanostructures were produced by the HTC process from biomass. Usually, two kinds of carbonaceous materials were formed [31]. "Hard" plant tissues, with structural, crystalline cellulose scaffolds, can preserve their outer hierarchical shape and the large-scale structural features on the macro- and microscale, forming special mesoporous structures. "Soft" plant tissues, without an extended crystalline cellulose scaffold, could not keep their original hierarchical structure, forming globular carbonaceous nanoparticles with very small sizes and interstitial porosity. Recently, raw rice grains were treated by the HTC process with the catalyst of $[Fe(NH_4)_2(SO_4)_2]$ salts (Figure 4.4g and h) [27]. The insoluble parts kept their primary grain microstructures with larger, porous carbonaceous species, which were different along the body of the grain. Interestingly, the primary fibrillar nanostructures of the rice tissue were well preserved.

4.2.1.5   Carbonaceous Hydrogels and Aerogels

A new type of monolithic hydrogels/aerogels consisting of highly uniform carbonaceous nanofibers can be fabricated through a simple template-directed HTC process [32]. Compared with the conventional process for aerogel preparation, this synthetic method has some unique advantages: (i) directly scaling up from 30 mL to 12 L just by using a large autoclave without changing synthetic parameters, (ii) easy and precise control of the structural parameters and mechanical strength of gels over a wide range, and (iii) the extraordinary flexibility and high chemical reactivity of gels give them great application potential.

The main steps of the synthesis process are as proposed in Figure 4.5. First, ultrathin tellurium nanowires are used as templates and dispersed in glucose aqueous solution. After HTC at 80 °C for 12–48 h, a mechanically robust monolithic gel-like product was synthesized, which occupied the whole Teflon container. The as-prepared wet gel can be

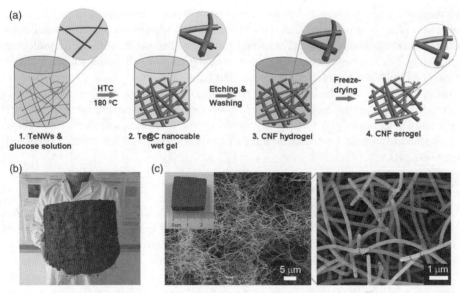

Figure 4.5 *(a) Schematic illustration of the synthetic steps (TeNWs = tellurium nanowires) and (b) photograph of an approximately 12-L monolithic wet gel. (Reproduced with permission from [12]. © 2006 American Chemical Society.) (c) SEM images of the aerogel at different magnifications showing the nanowire network structure. The inset in the right image of (c) shows a photograph of a aerogel obtained after freeze-drying. The dotted circles in the left image of (c) indicate the junctions between fibers. (Reproduced with permission from [32]. © 2012 WILEY-VCH Verlag GmbH & Co. KGaA, Weinheim.)*

easily cut into the desired shape for further application. Then, the hydrogel was formed by washing and chemical etching tellurium nanowires, and the aerogel was formed by removing water in the hydrogel via freeze-drying. SEM images of the aerogels also show the highly porous network structure consisting of disordered nanofibers of uniform size, and a homogeneous character without any apparent difference in nanofibers size and distribution over whole monolithic gel. Interestingly, these basic building blocks of nanofibers were interconnected with each other to a high degree through numerous junctions, which may be responsible for the outstanding mechanical properties of gels. Furthermore, this synthesis process is very versatile in scaling up and retaining the microstructures, chemical, and physical properties of gels. A nice case is the fabrication of a large monolithic gel with a volume of 12-L by using a 16-L autoclave and it is quite possible to satisfy industrial large-scale requirements by further enlarging the autoclave. On the other hand, this synthesis process can be well-controlled to prepare a series of gels with different nanofiber diameters, porosities, and mechanical strengths by easily regulating the synthetic parameters, mainly the amount of tellurium nanowire template and the HTC time. The as-prepared aerogel is a good complement to the low-density aerogels, because the density of the lightest aerogels with stable structure is only 3.3 mg cm^{-3}, belonging to the lowest among the reported values.

The unique intrinsic properties of the carbonaceous gels are very attractive. The robust mechanical behavior of gels includes the extraordinary flexibility, which allows for large deformations without fracture. The large amount of surface functional groups of the carbonaceous nanofibers in the gels has shown not only high adsorption capacity towards ionic pollutants, but also high chemical reactivity for further decoration with useful molecules or nanoparticles. The interpenetrating, open-pore network of gels permits rapid transport of gas- or liquid-phase molecular reactants and nanoscale objects into, through, and out of the gel. By exploiting the high surface reactivity of the carbonaceous nanofibers, extra-high porosity, and robust mechanical properties of gels, we have demonstrated the great application potential of gels for simple removal of dye pollutants, as selective adsorbents for oil-spill cleanup, and as versatile three-dimensional templates for creating functional composite gels.

4.2.2   Carbonaceous Nanocomposites

The controlled synthesis of carbonaceous nanocomposites has become a hot research area, and they have achieved many important results because of their combined and improved properties with highly potential value in many fields. Currently, there are three main methods for the production of well-controlled carbonaceous nanocomposites: (i) coating of preformed nanostructures, (ii) decorating inorganic nanostructures onto carbonaceous materials, and (iii) direct synthesis via a one-step method. Furthermore, the carbonaceous materials can themselves act as good sacrificial templates or as support shells for the synthesis of hollow and complex nanostructures, as well as free-standing membranes.

4.2.2.1   Coating of Preformed Nanostructures

For the HTC process, the most important and convenient advantage is that we can use it to rationally form the carbonaceous shell on the surface of preformed nanostructures (except for superhydrophobic surfaces). The coating process is easily controlled and the carbonaceous shell can form on the surface of most nanostructures, such as inorganic, organic, and polymer nanostructures. However, this strongly depends on the surface properties of these nanostructures. Such carbonaceous shells not only protect the respective nanostructures, but also produce additional unique chemical or physical properties. The final nanocomposites have shown important applications in many fields, such as catalysis, fuel cells, drug delivery, and bioimaging.

A particularly nice example is the synthesis of uniform Te@carbonaceous composite nanocables using the HTC process and ultralong tellurium nanowires with diameters of several nanometers. Tellurium nanowires can subsequently act as sacrificial templates and glucose as the starting reagent (Figure 4.6a and b) [12]. Tellurium nanowires can not only restrain the usual homogeneous nucleation of carbonaceous spheres, but also promote the heterogeneous deposition of a carbonaceous shell on the backbone of tellurium nanowires. The diameter of the final nanocables can be easily adjusted by changing the HTC time or the ratio of tellurium and glucose. Furthermore, the so-called silica/carbon-encapsulated core–shell spheres could also be synthesized by the HTC process with silica spheres as template and glucose as carbon precursor [33].

The carbonaceous shell can protect the nanostructures after some additional heat treatment and graphitization process, improving thus the physical properties of the newly resulting nanocomposites. For example, carbon-coated Fe_3O_4 nanospindles can be synthesized by the HTC process and subsequent heat treatment (Figure 4.6c) [34]. These composites

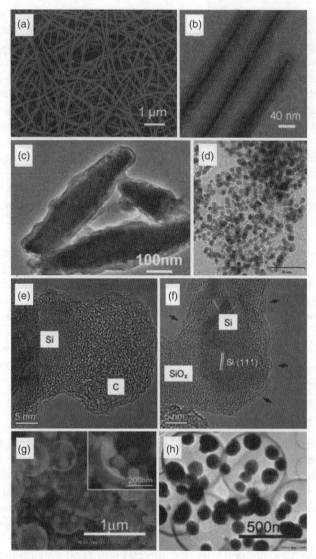

Figure 4.6 *(a and b) SEM and TEM images of Te@carbonaceous composite nanocables. (Reproduced with permission from [12]. © 2006 American Chemical Society.) (c) TEM images of carbon-coated Fe$_3$O$_4$ nanospindles. (Reproduced with permission from [34]. © 2008 WILEY-VCH Verlag GmbH & Co. KGaA, Weinheim.) (d) TEM images of carbon-decorated FePt nanoparticles. (Reproduced with permission from [38]. © 2007 WILEY-VCH Verlag GmbH & Co. KGaA, Weinheim.) (e and f) TEM images of carbon-coated silicon nanocomposites. (Reproduced with permission from [35]. © 2008 WILEY-VCH Verlag GmbH & Co. KGaA, Weinheim.) (g and h) SEM and TEM images of TNHCSs. (Reproduced with permission from [37]. © 2008 WILEY-VCH Verlag GmbH & Co. KGaA, Weinheim.)*

can be used as a superior anode material for Li-ion batteries with high coulombic efficiency in the first cycle, high reversible capacity, enhanced cycling performance, and high rate capability compared with bare hematite spindles and commercial magnetite particles. The improved performance is mainly attributed to the uniform and continuous carbon coating layers, the increase in the electrical conductivity, and the thin solid electrolyte interphase film on the surface. Meanwhile, carbon-coated silicon nanocomposites through the HTC process and heat treatment, have displayed significant improved lithium-storage performance because of the generation of a passivated layer of SiO_x and carbon (Figure 4.6e and f) [35].

The carbonaceous shell has been used as a powerful tool for the smart design synthesis process forming more complex nanocomposites. For instance, the synthesis of coaxial $SnO_2@C$ hollow nanospheres by removing silica from the original $SiO_2@SnO_2@C$ structures has been well designed [36]. The as-synthesized nanocomposites performed as high-energy anode materials for Li-ion batteries with exceptionally good life cycles over several hundred cycles and good rate capability.

Similarly, complex nanostructures of tin nanoparticles encapsulated in elastic hollow carbon spheres (TNHCSs) have been successfully synthesized (Figure 4.6g and h) [37]. SiO_2 spheres were used as the core for the deposition of the SnO_2 shell via the hydrolysis of Na_2SnO_3. After etching the core of SiO_2 by a sodium hydroxide solution, SnO_2 hollow spheres were coated with the carbonaceous shell via the HTC process. Finally, after additional heat treatment at 700 °C for 4 h under nitrogen atmosphere, complex TNHCS hybrid materials were synthesized. Such hybrid materials were promising anodes for high-performance Li-ion batteries with high specific capacity, high volume capacity, and good cycle performance.

4.2.2.2 Decorating Carbons with Inorganic Nanostructures

The important advantage of HTC-derived carbonaceous materials is their rich surface functional groups, such as hydroxyl, aldehyde, and carboxyl groups. These surface functional groups have shown remarkable reactivity and the capability of decorating the preformed materials with various functions. Based on the redox reaction, they could *in situ* reduce and stabilize noble metal ions, forming very fine noble metal nanostructures, such as silver, gold, platinum, and palladium [19, 39]. The whole process was "green" without using other reducing agents or stabilizing surfactants or polymers. The obtained noble metal nanostructures not only have uniform sizes, but also clean surfaces, which is beneficial for the further surface modification of biomolecules or drugs [11, 40].

We would like to illustrate in detail the synthesis of hybrid "golden fleece" nanocomposites. The so-called "golden fleece" nanocomposites are uniform carbonaceous nanofibers loaded with noble metal nanoparticles. A different diameter and distribution was noticed for different noble metals (Figure 4.7) [39a]. The carbonaceous nanofibers produced using the HTC process have rich surface functional groups that can *in situ* reduce the noble metal ions into noble metal nanoparticles, such as palladium, platinum, and gold. These nanocomposites have shown high permeation and stability, high surface areas, and enhanced chemical reactivity, which makes them ideal candidates for use in heterogeneous catalysis.

Further, carbonaceous nanofibers (CNF) can be decorated by other metal oxides nanoparticles, such as Fe_3O_4 and TiO_2, forming composite nanofibers that can be assembled into

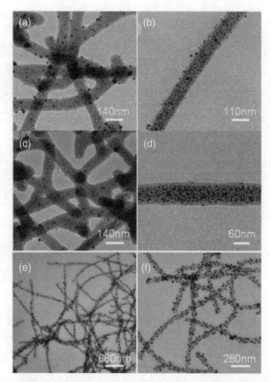

Figure 4.7 *TEM images of noble metal nanoparticles loaded on the surfaces of carbonaceous nanofibers. (a and b) Palladium nanoparticles with an average diameter of 6 nm. (c and d) Platinum nanoparticles with an average diameter of 7 nm. (e and f) Gold nanoparticles ranging from 7 to 15 nm in diameter. (Reproduced with permission from [39a]. © 2007 WILEY-VCH Verlag GmbH & Co. KGaA, Weinheim.)*

macroscopic free-standing membranes through a simple casting process (Figure 4.8) [41]. These multifunctional composite membranes have demonstrated their wide application properties in magnetic actuation, antibiofouling filtration, and continuous-flow catalysis.

It is necessary to illustrate the decoration of carbonaceous gels. The carbonaceous nanofiber gels can *in situ* reduce silver ions, forming the composite gels [32]. The abundant silver nanoparticles with the diameter of 20 nm are uniformly decorated on the carbonaceous nanofibers matrix, and the three-dimensional porous network of gels was not destroyed after functionalization. The mechanical behavior of this composite gel is similar to that of the pure gels.

Carbonaceous nanospheres synthesized by the HTC process have also successfully demonstrated their surface reactivity by loading silver and platinum nanoparticles on their surfaces without using other reduce and stabilize agents [19]. The redox surface reactions decrease the number of surface functional groups, as revealed by Fourier transform infrared spectroscopy.

Another concept making use of the reactivity of the HTC-derived materials is the self-assembly process, based on the electrostatic attraction of surface charge. Nanoparticles

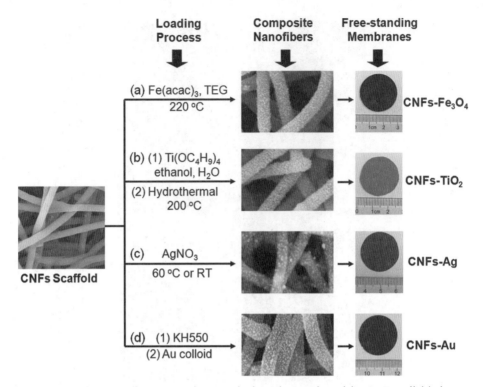

Figure 4.8 *Schematic illustrations showing the broad versatility of the CNF scaffolds for constructing free-standing multifunctional membranes, including (a) CNFs-Fe$_3$O$_4$, (b) CNFs-TiO$_2$, (c) CNFs-Ag, and (d) CNFs-Au. (Reproduced with permission from [41]. © 2011 American Chemical Society.)*

with different surface charge can be easily self-assembled on the surface of carbonaceous materials. For example, citrate-stabilized gold nanoparticles have been assembled on the surface of carbonaceous spheres to fabricate a core–shell hybrid material [40]. The resulting nanocomposites have shown satisfactory chemical stability and good biocompatibility, and are expected to be promising templates for bimolecular immobilization and biosensor fabrication.

4.2.2.3   One-Step HTC Synthetic Method

When noble metal ions were first added into the HTC process of carbohydrates or organic monomers, novel carbonaceous-encapsulated core–shell nanocomposites could be successfully produced by the one-step HTC process [42]. Controlling the reagent ratios and the reaction conditions could control the diameter and the length of the resulting composite materials and thickness of carbonaceous shell. The advantages of this one-step HTC process include: (i) the carbonization process, reduction and growth of noble metal ions could proceed simultaneously, (ii) the noble metal ions could catalyze the carbonization process while carbon reagents or intermediates could effectively reduce the noble metal

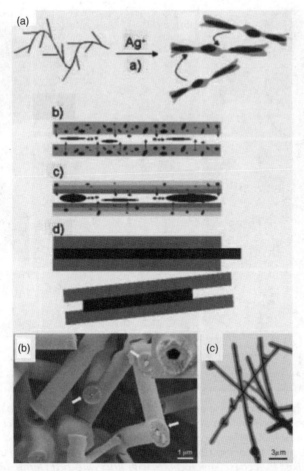

Figure 4.9 *(a) Schematic formation mechanism of the silver carbonaceous nanocables. (b and c) SEM and TEM images of silver carbonaceous nanocables. (Reproduced with permission from [43]. © 2004 WILEY-VCH Verlag GmbH & Co. KGaA, Weinheim.)*

ions, and (iii) this one-step HTC process could produce batches of novel and valuable nanocomposites on a large scale and at high quality.

A very nice case is also the one-step HTC synthesis of silver carbonaceous nanocables (Figure 4.9) [43]. These nanocables, with lengths up to 10 μm and diameters of 200–500 nm, have pentagonal-shaped silver nanowires as cores and tend to branch or fuse with each other. A possible mechanism for their formation has been proposed as follows. First, the silver ions are reduced by starch, forming silver nanostructures entrapped in a starch-gel matrix. Aggregation of these matrixes could enhance the carbonization process of starch. A partially hollow structure will be formed and the silver nanoparticles will move to this hollow interior and fuse with the already existing ones via an oriented attachment mechanism. Continuation of these steps could lead to the growth of the elongated silver nanostructures and carbonaceous shells forming the final complex structures. The

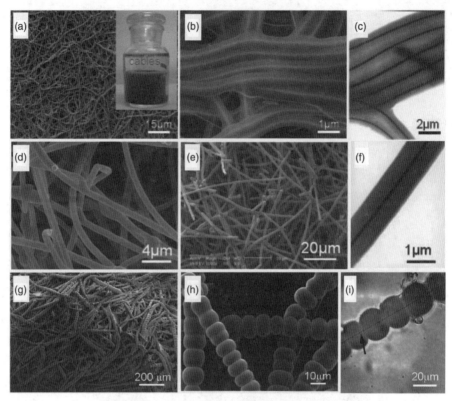

Figure 4.10 *(a–c) SEM and TEM images of flexible Ag@cross-linked PVA coaxial nanocables. (Reproduced with permission from [44a]. © 2005 American Chemical Society.) (d–f) SEM and TEM images of flexible Ag@C-rich composite submicrocables. (Reproduced with permission from [44b]. © 2006 Royal Society Chemistry.) (g–i) SEM and TEM images of unique necklace-like Cu@cross-linked PVA coaxial microcables with strict wire–bead forms. (Reproduced with permission from [44c]. © 2008 American Chemical Society.)*

as-synthesized individual silver carbonaceous nanocables have shown excellent conductivity, which makes them ideal materials for constructive interconnects in nanodevices.

A series of interesting work regarding the poly(vinyl alcohol) (PVA)-assisted synthesis of coaxial nanocables has gained much attention in HTC technology [44]. First, flexible Ag@cross-linked PVA coaxial nanocables have been fabricated by the HTC process (Figure 4.10a–c) [44a]. These nanocables, normally with a diameter of 0.7–4 μm and length up to 100 μm, are comprised of a smooth core about 150–200 nm in diameter and a surrounding sheath about 0.5–1 μm in thickness. These nanocables can pack in a parallel fashion, forming a bundle structure. A synergistic soft–hard template mechanism has been proposed. The synergistic effects of both the stabilization of PVA and the binding interaction of cross-linked PVA with the silver nanowires are responsible for the formation of cables and bundles. Further, some glucose-based saccharides, including glucose, β-cyclodextrin, starch, and maltose, can take part in this PVA-assisted synthesis process, forming flexible Ag@C-rich composite submicrocables (Figure 4.10d–f) [44b].

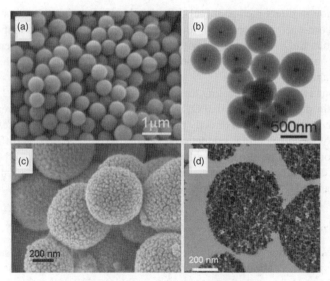

Figure 4.11 *(a and b) SEM and TEM images of monodisperse Ag/phenol formaldehyde resin core–shell nanospheres. (Reproduced with permission from [45a]. © 2008 WILEY-VCH Verlag GmbH & Co. KGaA, Weinheim.) (c and d) SEM and TEM images of mesoporous SnO₂ microspheres with nanoparticle assembly. (Reproduced with permission from [46]. © 2008 American Chemical Society.)*

This PVA-assisted synthesis process can be extended to the synthesis of Cu@cross-linked PVA coaxial microcables [44c, 44d]. Unique necklace-like Cu@cross-linked PVA coaxial microcables with strict wire–bead forms have been successfully synthesized (Figure 4.8g–i) [44c]. The necklace-like microcables have a length of several hundreds of micrometers to millimeters and a diameter in the range from several to 20 μm. The synergistic soft–hard template mechanism could explain the formation process of these cable-like structures. Further, Cu@carbonaceous submicrocables can be synthesized in the presence of some glucose-based saccharides, including glucose, β-cyclodextrin, starch, and maltose [44e].

Core–shell nanoparticles can be synthesized directly by the one-step HTC process [45]. For example, biocompatible and green luminescent monodisperse Ag/phenol formaldehyde resin core–shell nanospheres have been successfully produced by the one-step HTC process (Figure 4.11a and b) [45a]. These nanospheres have some interesting architectures including centric, eccentric, and coenocyte core–shell spheres with diameters of 180–1000 nm, which can be used potentially as *in vivo* bioimaging agents. Further, Pd@C core–shell nanoparticles have also been successfully synthesized by the one-step HTC process [45b], which have shown selectively catalytic capability for the batch partial hydrogenation of hydroxyl aromatic derivatives.

The one-step HTC process can synthesize carbonaceous nanocomposites with more complex structures and specific properties [46, 47]. For instance, core–shell Pt@C nanoparticles can embed in the porous channels of mesoporous silica (SBA-15) by the one-step HTC process [47a]. After high-temperature carbonization and removal of SBA-15, the final complex nanocomposites have shown high catalytic capability for methanol-tolerant oxygen electroreduction in direct methanol fuel cells. Furthermore, using the one-step HTC process, hollow core–shell mesospheres with aggregated crystalline SnO_2 nanoparticles [47b] and mesoporous SnO_2 microspheres with nanoparticle assembly have been synthesized (Figure 4.11c and d) [46]. Both of these SnO_2 nanocomposites have showed high capability of Li-ion storage.

As an alternative to the postcoating technologies reported in Section 4.2.2.1, a one-step synthesis for the production of a $LiFePO_4$ mesocrystals coated with a thin layer of carbon or nitrogen-doped carbon has been directly produced by simply adding glucose/glucosamine as a carbon to the solvothermal synthetic procedure [48]. There was no difference in the morphology or crystalline structure with or without the addition of the carbon precursor. However, the composite made in the presence of the carbon precursor showed an increased conductivity and an enhanced performance when utilized as a cathode in Li-ion batteries.

4.2.2.4   Carbonaceous Sacrificial Template Method

The as-synthesized carbonaceous nano- or micromaterials have been used as sacrificial templates for the synthesis of hollow and complex nanostructures, which have shown important potential applications in catalysis, sensing, chemical biological separation, and Li-ion batteries [18, 19, 49]. The advantages of this carbonaceous sacrificial template method include: (i) the well-controlled heat treatment method could easily remove/graphitize the carbonaceous materials depending on the atmosphere under which is conducted (O_2/N_2), (ii) a series of metal or bimetal oxide hollow structures have been produced, and (iii) it is an economic and facile technique, and can produce complex metal oxide hollow structures on a large scale and at high quality. The general synthesis process is based on the large surface negative charge of carbonaceous materials (Figure 4.12a–c) [50]. First, metal ions are adsorbed from solution to the surface layer of carbonaceous materials. Then heat treatment process followed by the controlled removal of the carbonaceous core or shell is accomplished via calcination.

This carbonaceous sacrificial template method has been widely used to synthesize metal or dimetal oxide hollow spheres, such as Ga_2O_3 [50b, 51], GaN [51], WO_3 [52], SiC [53], ZnO [54], manganese-doped ZnO [37], SnO_2 [55], NiO [56], CoO [50b], In_2O_3 [57], TiO_2 [58], SiO_2 [33], CuO [50c], MgO [50a], CeO_2 [50a], MnO_2 [59], Mn_3O_4 [50b], Al_2O_3 [50b], ZrO_2 [60], Cr_2O_3 [50b], layered double hydroxide [61], Y_2O_3:Eu [62], La_2O_3 [50b], Y_2O_3 [50b], Lu_2O_3 [50b], MFe_2O_4 (M = Zn, Co, Ni, and Cd) [63], Fe_2O_3 [64], and Bi_2WO_6 [65]. The size and structures of these hollow spheres can be well controlled by the templates. These metal oxide hollow spheres have shown high potential applications in gas sensitivity or catalysis. Further, metal oxide hollow nanofibers or nanotubes have been successfully synthesized using carbonaceous nanofibers as sacrificial templates (Figure 4.12d and e) [63, 66].

More complex nanocomposites can be synthesized by this carbonaceous sacrificial template route, which can be extended to synthesize various metal oxide shells [67]. For example, a series of jingle-bell-shaped hollow structured nanomaterials marked as Ag@MFe_2O_4 (M = Ni, Co, Mg, and Zn), consisting of ferrite hollow shells and silver nanoparticle cores, has been successfully produced by this sacrificial template method (Figure 4.12f–h) [67a]. The synthesis process included first coating the silver nanoparticle core with a glucose-derived carbonaceous layer via the HTC process, directly adsorbing metal cations Fe^{3+} and M^{2+} on the surface of the carbonaceous layers, and finally calcinating the sample to

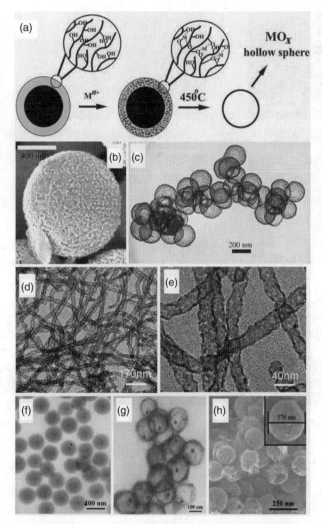

Figure 4.12   *(a) Schematic formation mechanism of metal oxide hollow spheres by using carbonaceous spheres as templates. (b and c) SEM and TEM images of SnO$_2$ hollow nanospheres. (Reproduced with permission from [50b]. © 2006 WILEY-VCH Verlag GmbH & Co. KGaA, Weinheim.) (d and e) SEM and TEM images of TiO$_2$ nanotubes. (Reproduced with permission from [66b]. © 2009 Royal Society of Chemistry.) (f) TEM image of Ag@C core–shell microspheres. (g and h) TEM and SEM images of jingle-bell-shaped hollow structured nanomaterials. (Reproduced with permission from [67a]. © 2008 Elsevier.)*

remove the middle carbonaceous shell and transform the metal ions into pure phase ferrites. These nanocomposites have shown multifunctional properties, including magnetic, optical, and antibacterial properties. Further, using a quite similar procedure, Fe$_3$O$_4$@TiO$_2$ core–shell microspheres were synthesized, which have shown highly specific capture ability of phosphopeptides for the application of direct matrix-assisted laser desorption ionization time-of-flight mass spectrometry analysis [67b].

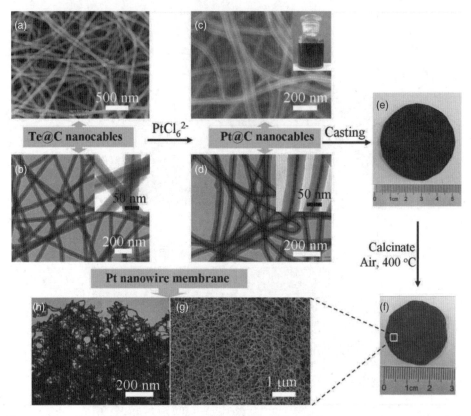

Figure 4.13   *Schematic formation process of free-standing platinum nanowire membranes by using the carbonaceous sacrificial template method. (Reproduced with permission from [68]. © 2011 WILEY-VCH Verlag GmbH & Co. KGaA, Weinheim.)*

Interestingly, this carbonaceous sacrificial template method can be applied for the synthesis of two-dimensional free-standing macroscopic membranes with high porosity, good flexibility, large area per unit volume, and an interconnected open pore structure. For instance, a free-standing platinum nanowire membrane has been prepared via this multistep sacrificial template method (Figure 4.13) [68]. (i) Tellurium nanowires were used as template for the synthesis of well-defined Te@carbonaceous nanocables via the HTC process. (ii) These Te@C nanocables were transformed into Pt@C nanocables with the galvanic replacement reaction. (iii) These Pt@C nanocables could be assembled into a two-dimensional free-standing membrane by a simple casting process. (iv) After calcination of the Pt@C membrane at high temperature, the free-standing platinum nanowire membrane was formed. SEM and TEM images have shown that the platinum nanowire membrane was composed of very fine wire-like nanostructures with lengths of several tens of micrometers and these platinum nanowires interconnected with each other to form a highly porous nanowire network structure. This unique porous network structures could greatly facilitate electron transport and gas diffusion. The platinum nanowire membrane was a good electrocatalyst for the oxygen reduction reaction, and has higher catalytic activity and stability than commercial Pt/C and platinum black catalysts.

4.3 Conclusion and Outlook

In summary, a palette of carbonaceous nanomaterials with focus on the nanocomposites derived from biomass, carbohydrates, and organic molecules via the HTC process have been discussed. We have learned that some promising strategies have been designed for the synthesis of these materials, which we have briefly presented here while leading the interested reader to specific references. Until now, the morphology of carbonaceous nanomaterials has been well controlled, expanding from zero to one, two, and three dimensions, and hydrogels/aerogels. Carbonaceous nanocomposites with combined and improved properties have been synthesized by three main methods: coating of preformed nanostructures, decorating the preformed HTC materials with inorganic nanostructures, and direct synthesis by a one-step method. Furthermore, carbonaceous materials can act as good sacrificial templates or support shells for the production of hollow and complex nanostructures, as well as free-standing membranes.

Importantly, the as-synthesized carbonaceous nanomaterials and nanocomposites produced via the HTC process have already demonstrated significant value in many practical application fields, such as environmental, catalytic, energy, and biological applications that we have pointed out while describing various synthetic approaches. However, more emphasis on the multitude of applications of HTC material is presented in detail in Chapter 7.

Although carbonaceous nanomaterials and nanocomposites produced via the HTC process have been achieved with so far excellent success in both synthesis and applications, there are still many challenges and questions that should be faced and solved. Further exploration will facilitate the rational design of a variety of carbonaceous nanomaterials and nanocomposites with hierarchical porosity, interesting functionalities, and significant practical value.

Acknowledgments

We acknowledge funding support from the Ministry of Science and Technology of China (grant 2012BAD32B05-4), the National Basic Research Program of China (Grant 2010CB934700), the Chinese Academy of Sciences (grant KJZD-EW-M01-1), the National Natural Science Foundation of China (grants 91022032, 21061160492, and 912271032), and the Principal Investigator Award from the National Synchrotron Radiation Laboratory at the University of Science and Technology of China.

References

(1) (a) Hu, B., Yu, S.H., Wang, K. *et al.* (2008) *Dalton Transactions*, 5414–5423; (b) Hu, B., Wang, K., Wu, L.H. *et al.* (2010) *Advanced Materials*, **22**, 813–828.

(2) Suh, W.H., Suslick, K.S., Stucky, G.D., and Suh, Y.-H. (2009) *Progress in Neurobiology*, **87**, 133–170.

(3) Kroto, H.W., Heath, J.R., Brien, S.C. *et al.* (1985) *Nature*, **318**, 162–163.

(4) Iijima, S. (1991) *Nature*, **354**, 56–58.

(5) (a) Nozik, A.J., Beard, M.C., Luther, J.M. *et al.* (2010) *Chemical Reviews*, **110**, 6873–6890; (b) De Franceschi, S., Kouwenhoven, L., Schonenberger, C., and Wernsdorfer,

W. (2010) *Nature Nanotechnology*, **5**, 703–711; (c) Pinaud, F., Clarke, S., Sittner, A., and Dahan, M. (2010) *Nature Methods*, **7**, 275–285.

(6) Bourlinos, A.B., Stassinopoulos, A., Anglos, D. *et al.* (2008) *Small*, **4**, 455–458.

(7) (a) Zong, J., Zhu, Y.H., Yang, X.L. *et al.* (2011) *Chemical Communications*, **47**, 764–766; (b) Sun, Y.P., Zhou, B., Lin, Y. *et al.* (2006) *Journal of the American Chemical Society*, **128**, 7756–7757; (c) Weiss, S., Rashba, E.I., Kuemmeth, F. *et al.* (2010) *Physical Review B*, **82**, 165427; (d) Pan, D.Y., Zhang, J.C., Li, Z., and Wu, M.H. (2010) *Advanced Materials*, **22**, 734–738.

(8) Pan, D., Zhang, J., Li, Z., and Wu, M. (2010) *Advanced Materials*, **22**, 734–738.

(9) Peng, H. and Travas-Sejdic, J. (2009) *Chemistry of Materials*, **21**, 5563–5565.

(10) Zhu, H., Wang, X., Li, Y. *et al.* (2009) *Chemical Communications*, 5118–5120.

(11) Selvi, B.R., Jagadeesan, D., Suma, B.S. *et al.* (2008) *Nano Letters*, **8**, 3182–3188.

(12) Qian, H.S., Yu, S.H., Luo, L.B. *et al.* (2006) *Chemistry of Materials*, **18**, 2102–2108.

(13) Yan, Y., Yang, H.F., Zhang, F.Q. *et al.* (2006) *Small*, **2**, 517–521.

(14) Wen, Z.H., Wang, Q., and Li, J.H. (2008) *Advanced Functional Materials*, **18**, 959–964.

(15) Kubo, S., Tan, I., White, R.J. *et al.* (2010) *Chemistry of Materials*, **22**, 6590–6597.

(16) Zhan, Y.J. and Yu, S.H. (2008) *Journal of Physical Chemistry C*, **112**, 4024–4028.

(17) Liang, H.W., Wang, L., Chen, P.Y. *et al.* (2010) *Advanced Materials*, **22**, 4691–4695.

(18) Wang, Q., Li, H., Chen, L.Q., and Huang, X.J. (2001) *Carbon*, **39**, 2211–2214.

(19) Sun, X.M. and Li, Y.D. (2004) *Angewandte Chemie International Edition*, **43**, 597–601.

(20) Shin, Y., Wang, L.Q., Bae, I.T. *et al.* (2008) *Journal of Physical Chemistry C*, **112**, 14236–14240.

(21) Yao, C., Shin, Y., Wang, L.Q. *et al.* (2007) *Journal of Physical Chemistry C*, **111**, 15141–15145.

(22) Sevilla, M., Lota, G., and Fuertes, A.B. (2007) *Journal of Power Sources*, **171**, 546–551.

(23) Sevilla, M. and Fuertes, A.B. (2009) *Carbon*, **47**, 2281–2289.

(24) Sevilla, M. and Fuertes, A.B. (2009) *European Journal of Chemistry*, **15**, 4195–4203.

(25) Ikeda, S., Tachi, K., Ng, Y.H. *et al.* (2007) *Chemistry of Materials*, **19**, 4335–4340.

(26) White, R.J., Tauer, K., Antonietti, M., and Titirici, M.M. (2010) *Journal of the American Chemical Society*, **132**, 17360–17363.

(27) Cui, X.J., Antonietti, M., and Yu, S.H. (2006) *Small*, **2**, 756–759.

(28) Demir-Cakan, R., Baccile, N., Antonietti, M., and Titirici, M.M. (2009) *Chemistry of Materials*, **21**, 484–490.

(29) (a) Sun, X.M. and Li, Y.D. (2005) *Journal of Colloid and Interface Science*, **291**, 7–12; (b) Wen, Z.H., Wang, Q., Zhang, Q., and Li, J.H. (2007) *Electrochemistry Communications*, **9**, 1867–1872.

(30) Ni, D.Z., Wang, L., Sun, Y.H. *et al.* (2010) *Angewandte Chemie International Edition*, **49**, 4223–4227.

(31) Titirici, M.M., Thomas, A., Yu, S.H. *et al.* (2007) *Chemistry of Materials*, **19**, 4205–4212.

(32) Liang, H.-W., Guan, Q.-F., Chen, L.-F. *et al.* (2012) *Angewandte Chemie International Edition*, **51**, 5101–5105.

(33) Wan, Y., Min, Y.L., and Yu, S.H. (2008) *Langmuir*, **24**, 5024–5028.
(34) Zhang, W.M., Wu, X.L., Hu, J.S. *et al.* (2008) *Advanced Functional Materials*, **18**, 3941–3946.
(35) Cakan, R.D., Titirici, M.M., Antonietti, M. *et al.* (2008) *Chemical Communications*, 3759–3761.
(36) Lou, X.W., Li, C.M., and Archer, L.A. (2009) *Advanced Materials*, **21**, 2536–2541.
(37) Zhang, W.M., Hu, J.S., Guo, Y.G. *et al.* (2008) *Advanced Materials*, **20**, 1160–1165.
(38) Caiulo, N., Yu, C.H., Yu, K.M.K. *et al.* (2007) *Advanced Functional Materials*, **17**, 1392–1396.
(39) (a) Qian, H.S., Antonietti, M., and Yu, S.H. (2007) *Advanced Functional Materials*, **17**, 637–643; (b) Tang, S.C., Vongehr, S., and Meng, X.K. (2010) *Journal of Physical Chemistry C*, **114**, 977–982; (c) Hu, B., Zhao, Y., Zhu, H.Z., and Yu, S.H. (2011) *ACS Nano*, **5**, 3166–3171; (d) Lu, Y.M., Zhu, H.Z., Li, W.G. *et al.* (2013) *Journal of Materials Chemistry A*, **1**, 3783–3788.
(40) Cui, R.J., Liu, C., Shen, J.M. *et al.* (2008) *Advanced Functional Materials*, **18**, 2197–2204.
(41) Liang, H.W., Zhang, W.J., Ma, Y.N. *et al.* (2011) *ACS Nano*, **5**, 8148–8161.
(42) (a) Sun, X.M. and Li, Y.D. (2005) *Advanced Materials*, **17**, 2626–2630; (b) Song, X.C., Zhao, Y., Zheng, Y.F. *et al.* (2008) *Crystal Growth & Design*, **8**, 1823–1826; (c) Wang, W.Z., Xiong, S.L., Chen, L.Y. *et al.* (2006) *Crystal Growth & Design*, **6**, 2422–2426; (d) Fang, Z., Tang, K.B., Lei, S.J., and Li, T.W. (2006) *Nanotechnology*, **17**, 3008–3011; (e) Barone, P.W., Baik, S., Heller, D.A., and Strano, M.S. (2005) *Nature Materials*, **4**, 86–92; (f) Wang, W.Z., Qiu, S., Xi, B.J. *et al.* (2008) *Chemistry – An Asian Journal*, **3**, 834–840; (g) Hu, X.L., Yu, J.C., and Gong, J.M. (2007) *Journal of Physical Chemistry C*, **111**, 5830–5834; (h) Chen, G.Y., Deng, B., Cai, G.B. *et al.* (2008) *Crystal Growth & Design*, **8**, 2137–2143; (i) Wang, W.Z., Sun, L., Fang, Z. *et al.* (2009) *Crystal Growth & Design*, **9**, 2117–2123; (j) Deng, B., Xu, A.W., Chen, G.Y. *et al.* (2006) *Journal of Physical Chemistry B*, **110**, 11711–11716.
(43) Yu, S.H., Cui, X.J., Li, L.L. *et al.* (2004) *Advanced Materials*, **16**, 1636–1640.
(44) (a) Luo, L.B., Yu, S.H., Qian, H.S., and Zhou, T. (2005) *Journal of the American Chemical Society*, **127**, 2822–2823; (b) Luo, L.B., Yu, S.H., Qian, H.S., and Gong, J.Y. (2006) *Chemical Communications*, 793–795; (c) Zhan, Y.J. and Yu, S.H. (2008) *Journal of the American Chemical Society*, **130**, 5650–5651; (d) Gong, J.Y., Luo, L.B., Yu, S.H. *et al.* (2006) *Journal of Materials Chemistry*, **16**, 101–105; (e) Gong, J.Y., Yu, S.H., Qian, H.S. *et al.* (2007) *Journal of Physical Chemistry C*, **111**, 2490–2496.
(45) (a) Guo, S.R., Gong, J.Y., Jiang, P. *et al.* (2008) *Advanced Functional Materials*, **18**, 872–879; (b) Makowski, P., Cakan, R.D., Antonietti, M. *et al.* (2008) *Chemical Communications*, 999–1001.
(46) Demir-Cakan, R., Hu, Y.S., Antonietti, M. *et al.* (2008) *Chemistry of Materials*, **20**, 1227–1229.
(47) (a) Wen, Z.H., Liu, J., and Li, J.H. (2008) *Advanced Materials*, **20**, 743–747; (b) Deng, D. and Lee, J.Y. (2008) *Chemistry of Materials*, **20**, 1841–1846.
(48) Popovic, J., Demir-Cakan, R., Tornow, J. *et al.* (2011) *Small*, **7**, 1127–1135.

(49) Titirici, M.M., Antonietti, M., and Baccile, N. (2008) *Green Chemistry*, **10**, 1204–1212.

(50) (a) Titirici, M.M., Antonietti, M., and Thomas, A. (2006) *Chemistry of Materials*, **18**, 3808–3812; (b) Sun, X.M., Liu, J.F., and Li, Y.D. (2006) *European Journal of Chemistry*, **12**, 2039–2047; (c) Qian, H.S., Lin, G.F., Zhang, Y.X. *et al.* (2007) *Nanotechnology*, **18**, 355602.

(51) Sun, X.M. and Li, Y.D. (2004) *Angewandte Chemie International Edition*, **43**, 3827–3831.

(52) Li, X.L., Lou, T.J., Sun, X.M., and Li, Y.D. (2004) *Inorganic Chemistry*, **43**, 5442–5449.

(53) Zhang, Y., Shi, E.W., Chen, Z.Z. *et al.* (2006) *Journal of Materials Chemistry*, **16**, 4141–4145.

(54) (a) Wang, X., Hu, P., Yuan, F.L., and Yu, L.J. (2007) *Journal of Physical Chemistry C*, **111**, 6706–6712; (b) Zhang, Y., Shi, E.W., Chen, Z.Z., and Xiao, B. (2008) *Materials Letters*, **62**, 1435–1437.

(55) Wang, C.H., Chu, X.F., and Wu, M.M. (2007) *Sensors and Actuators B – Chemical*, **120**, 508–513.

(56) (a) Wang, X., Yu, L.J., Hu, P., and Yuan, F.L. (2007) *Crystal Growth & Design*, **7**, 2415–2418; (b) Li, C.C., Liu, Y.L., Li, L.M. *et al.* (2008) *Talanta*, **77**, 455–459.

(57) Guo, Z., Liu, J.Y., Jia, Y. *et al.* (2008), *Nanotechnology*, **19**, 345–704.

(58) (a) Ao, Y.H., Xu, J.J., Fu, D.G., and Yuan, C.W. (2008) *Electrochemistry Communications*, **10**, 1812–1814; (b) Ao, Y.H., Xu, J.J., Fu, D.G., and Yuan, C.W. (2008) *Catalysis Communications*, **9**, 2574–2577.

(59) Wang, N., Gao, Y., Gong, J. *et al.* (2008) *European Journal of Inorganic Chemistry*, 3827–3832.

(60) Guo, C.Y., Hu, P., Yu, L.J., and Yuan, F.L. (2009) *Materials Letters*, **63**, 1013–1015.

(61) Gunawan, P. and Xu, R. (2009) *Chemistry of Materials*, **21**, 781–783.

(62) Jia, G., Yang, M., Song, Y.H. *et al.* (2009) *Crystal Growth & Design*, **9**, 301–307.

(63) Murugan, A.V., Muraliganth, T., Ferreira, P.J., and Manthiram, A. (2009) *Inorganic Chemistry*, **48**, 946–952.

(64) Yu, J.G., Yu, X.X., Huang, B.B. *et al.* (2009) *Crystal Growth & Design*, **9**, 1474–1480.

(65) Shang, M., Wang, W.Z., and Xu, H.L. (2009) *Crystal Growth & Design*, **9**, 991–996.

(66) (a) Yuan, R.S., Fu, X.Z., Wang, X.C. *et al.* (2006) *Chemistry of Materials*, **18**, 4700–4705; (b) Gong, J.Y., Guo, S.R., Qian, H.S. *et al.* (2009) *Journal of Materials Chemistry*, **19**, 1037–1042.

(67) (a) Li, S.H., Wang, E.B., Tian, C.G. *et al.* (2008) *Journal of Solid State Chemistry*, **181**, 1650–1658; (b) Li, Y., Wu, J.S., Qi, D.W. *et al.* (2008) *Chemical Communications*, 564–566.

(68) Liang, H.W., Cao, X., Zhou, F. *et al.* (2011) *Advanced Materials*, **23**, 1467–1471.

5

Chemical Modification of Hydrothermal Carbon Materials

Stephanie Wohlgemuth[1], Hiromitsu Urakami[1], Li Zhao[2] and Maria-Magdalena Titirici[3]

[1]*Max Planck Institute of Colloids and Interfaces, Colloid Chemistry, Germany*
[2]*Chinese Academy of Sciences, National Center for Nanoscience and Technology, China*
[3]*School of Engineering and Materials Science, Queen Mary, University of London, UK*

5.1 Introduction

Carbon-based materials have immense potential when it comes to finding alternative, greener technologies in the present and in the future. Their wide availability along with their diverse physicochemical properties, such as chemical and thermal stability, electrical conductivity, and the ability to adopt a wide range of morphologies, makes them attractive candidates for high-end applications. Non-fossil-fuel-based carbon feedstocks such as biomass present promising renewable energy sources and building blocks for more advanced applications than simple heat and energy generation via combustion. Carbon-based materials find use in almost all aspects of life, ranging from basic applications such as water purification [1] and fertilizers [2] through to sorption [3, 4], pigments [5], chromatography [6–8], energy storage and conversion [9, 10], as well as catalysis (for more details, see Chapter 7) [11–13].

For the development of nanostructured materials for high-end applications, some of the most important factors concerning material properties are:

- Conductivity (e.g., electronic applications).
- Surface functionality (e.g., adsorption, acid/base catalysis).
- Surface area (virtually all applications).

Sustainable Carbon Materials from Hydrothermal Processes, First Edition. Edited by Maria-Magdalena Titirici.
© 2013 John Wiley & Sons, Ltd. Published 2013 by John Wiley & Sons, Ltd.

Pure carbon materials can fulfill these criteria to a certain extent, but to broaden the range of possible applications, heteroatom modification ("doping") is a useful tool to tune material properties. The introduction of, for example, sulfur or nitrogen into a carbon material can result in improved electrical conductivity, material stability, and catalytic performance due to an increased number of active sites [14–18]. Remarkable progress has been made in recent years regarding the use of such carbon-based materials and it can be expected that this trend will continue in the future [19–21]. Current synthetic procedures for carbon materials most commonly involve high-temperature heat treatment under inert conditions (pyrolysis) [22, 23], chemical vapor deposition [21, 24], or arc-discharge techniques [25, 26]. In addition to being very energy-consuming, these techniques also often require the use of harmful precursors that are in turn produced from fossil fuels. It is therefore a point of great interest to not only focus on the sustainability of the final application (e.g., electric vehicles, hydrogen fuel cells, solar cells, etc.), but equally on the sustainability of the entire process, namely from choosing the precursor, the synthesis conditions through to the final application. Hydrothermal carbonization (HTC) can offer the basis for such technologies. The advantages of HTC are the possibility of using cheap, readily available starting materials in a simple, low-cost synthesis that requires only water as solvent [27, 28]. Also, the abundance of heteroatoms in natural molecules such as amino acids and proteins allows for the synthesis of doped carbon materials in one step directly from biowaste, such as HTC of prawn shells [29]. The main part of this chapter will deal with such *in situ* doping procedures. Alternatively, there is the possibility to postmodify hydrothermal carbons and thereby introduce heteroatoms onto the surface. Since the term "doping" tends to be used when referring to the bulk material rather than surface-modified materials, we shall refer to the latter as "postfunctionalized" hydrothermal carbons.

5.2   *In Situ* Doping of Hydrothermal Carbons

5.2.1   Nitrogen

Nitrogen is by far the most commonly used nonmetal dopant for carbon-based materials.

As described in detail in Chapter 7, nitrogen-doped carbon materials are currently applied in several technological fields, especially in that of renewable energy. Thus, the electrochemical performance of carbons can be further enhanced by nitrogen doping providing supercapacitance (in supercapacitors), inherent catalytic activity in fuel cells as well as extra lithium storage sites.

For example, Li-ion battery anodes employing nitrogen-doped graphene demonstrate capacities as high as 1043 and 1549 mAh g^{-1} [30, 31]. It was proposed that the nitrogen atom could drastically alter the electronic performance, offer more active sites, and enhance the interaction between the carbon structure formed and lithium. Thus, it is expected to improve the kinetics of lithium diffusion and transfer – both beneficial for electrochemical performance [32, 33].

The nitrogen functionalities of (micro) porous carbons are valuable for supercapacitor applications as well as they are able to undergo redox reactions with the electrolyte, thus increasing their capacity [34–38]. In addition, nitrogen-doped carbons showed intrinsic properties as metal-free catalysts in the oxygen reduction reaction (ORR) at the cathodic site of a fuel cell [39–43]. Density functional theory calculations suggest that a possible explanation would be the high electronegativity of the nitrogen atom that then can polarize the C—N bond, resulting in a reduced energy barrier of the adjacent carbon atom towards the ORR [44].

The synthesis of nitrogen-doped carbons has been achieved via a variety of pathways, such as post-treatment of carbon with ammonia [45], amines, or urea [46] and also more direct approaches using acetonitrile [47], pyrrole [48], polyacetonitrile [49], or polyaniline [50] as starting products.

In the context of availability and sustainability, nature offers various nitrogen-containing precursors, such as amino acids, peptides/glycopeptides, proteins, and aminated/amidated saccharides and polysaccharides (e.g., glucosamine, *N*-acetyl glucosamine). It has been repeatedly reported that these biomonomers and biopolymers can give rise to nitrogen-doped carbon materials. The first report on nitrogen-doped hydrothermal carbon was published in 2009 by White *et al.* who used prawn shell waste as the precursor in HTC [51]. Prawn shells are natural hybrid composite materials formed from chitin (a polysaccharide containing *N*-acetyl glucosamine units) and calcium carbonate. HTC of chitin results in the formation of nitrogen-containing carbonaceous material around the inorganic scaffold (see Figure 2.11). Following HTC treatment, the scaffold was removed by washing with acid, leaving behind a porous, nitrogen-doped material (as discussed in Chapter 2). The nitrogen content of their material was 5.8 wt% according to elemental analysis and X-ray photoelectron spectroscopy (XPS) revealed the presence of various nitrogen species (pyridinic, pyrrolic, and quartenary as seen in Figure 5.1).

In 2010, Baccile *et al.* reported on the hydrothermal treatment of glucose in the presence of the glycoprotein ovalbumin [52]. They showed that the nitrogen contained in the protein was incorporated into the final hydrothermal carbon framework while the presence of the protein also had an effect on the material morphology. Nitrogen contents of up to 8.2 wt% could be obtained. The carbon material consisted of smaller particles than when pure glucose was used as a precursor, presumably due to a surface stabilizing effect of the protein in the hydrothermal reaction mixture. Through further investigation, White *et al.* later published the formation of nitrogen-doped organic aerogels (i.e., interconnected small particles arranged in a three-dimensional porous structure) obtained by addition of ovalbumin to glucose (see Figure 2.14) [53]. Other examples include the use of glucosamine or chitosan directly as carbon and nitrogen precursors [54, 55], as well as the addition of glycine to the hydrothermal treatment of glucose [56]. Interestingly, the nitrogen content from pure nitrogen-containing precursors (i.e., using only glucosamine or chitosan) ranged from 6.5 to 9.4 wt% (depending on the temperature used), which is similar to the approximately 8.5 wt% obtained by mixing 0.4 g of glycine with 1 g of glucose. Zhang *et al.* prepared nitrogen-doped carbons from sucrose in the presence of ammonia [57], while Liu *et al.* produced nitrogen-doped fluorescent nanoparticles upon the HTC of grass [58].

Overall, it seems to be the case that the level of nitrogen doping via the aforementioned methods is limited, but at the same time it is possible to achieve relatively high doping levels by using only small amounts of nitrogen source together with a readily available carbon

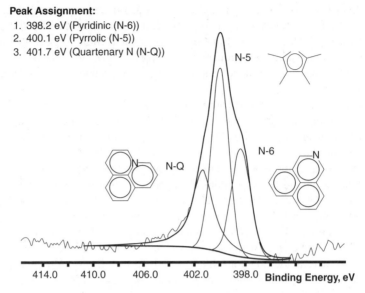

Peak Assignment:
1. 398.2 eV (Pyridinic (N-6))
2. 400.1 eV (Pyrrolic (N-5))
3. 401.7 eV (Quartenary N (N-Q))

N-5

N-6

N-Q

Binding Energy, eV

414.0   410.0   406.0   402.0   398.0

Figure 5.1   *High-resolution XPS spectra of the N 1(s) photoelectron region for acid-washed nitrogen-doped carbon material (peaks assigned using charge correction). (Reproduced with permission from [51]. © 2009 Royal Society of Chemistry.)*

source such as glucose. This is an advantage of HTC over other methods such as simple pyrolysis because it allows for more efficient use of heteroatom-containing precursors. During hydrothermal treatment at mild temperatures stable (aromatic) nitrogen species are formed that then remain in the carbonaceous scaffold during further pyrolysis at higher temperatures. How does the nitrogen incorporation during HTC take place? This question is under debate; however, it is generally accepted that Maillard chemistry plays a crucial role in the process [52, 56, 59].

The Maillard reaction is not one specific reaction, but rather refers to a group of reactions occurring between reducing sugars and amino acids. In the process, hundreds of different compounds can be created that can themselves react further in various ways. It is therefore impossible to provide a definite mechanistic pathway for the formation of nitrogen-doped HTC-derived materials derived from saccharides and various nitrogen-containing dopant molecules [60]. Some of the relevant steps in the Maillard reaction cascades are briefly presented here in order to demonstrate some of the ways in which heteroatoms can be incorporated into hydrothermal carbon from glucose or other reducing sugars.

The process starts with the nucleophilic attack of an amine (e.g., in amino acids) on the aldehyde of the sugar to produce glycosylamines (Figure 5.2, **I**), which subsequently loses water to give a Schiff base (Figure 5.2, **II**). The α-hydroxy aldehyde motif allows for the rearrangement of the Schiff bases to aminoketoses or so-called Amadori compounds (Figure 5.2, **III** and **IV**) [61]. Compound **III** can form α-dicarbonyl species (Figure 5.2, **V**) which could go through successive dehydration to form hydroxymethylfurfural (HMF) – the main reactive intermediate in HTC, as will be explained in Chapter 6).

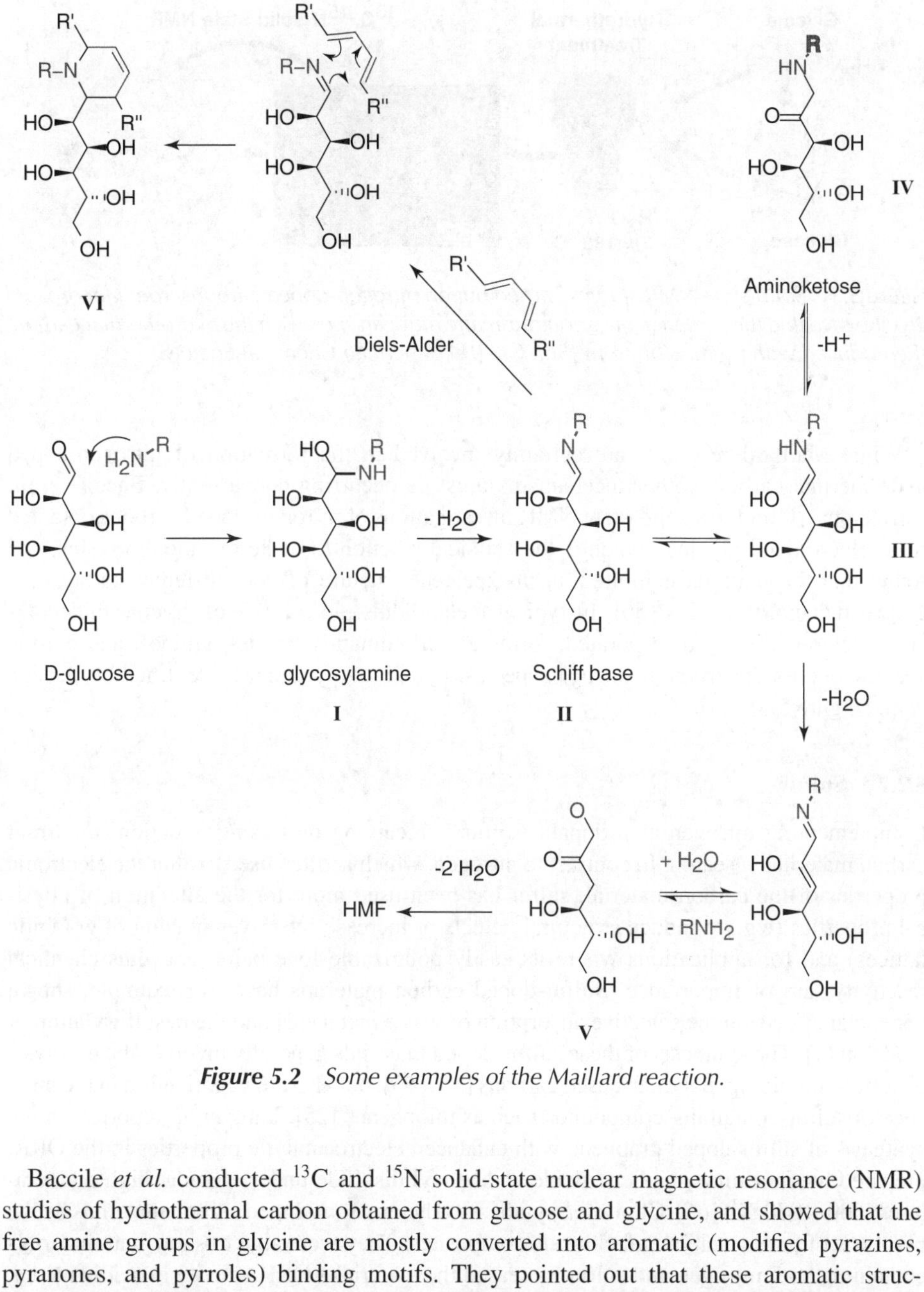

Figure 5.2 *Some examples of the Maillard reaction.*

Baccile *et al.* conducted ^{13}C and ^{15}N solid-state nuclear magnetic resonance (NMR) studies of hydrothermal carbon obtained from glucose and glycine, and showed that the free amine groups in glycine are mostly converted into aromatic (modified pyrazines, pyranones, and pyrroles) binding motifs. They pointed out that these aromatic structures are most likely formed as a consequence of degradation reactions such as the Strecker degradation [56, 62, 63]. Network-forming Maillard reactions, on the other hand, are responsible for the formation of cyclic amines and other nonaromatic nitrogen species.

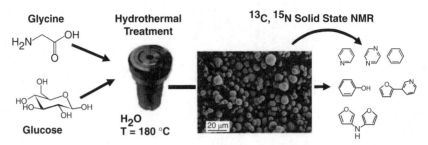

Figure 5.3 *Solid-state NMR studies carried out on nitrogen-doped carbons from glucose and glycine revealed the presence of various aromatic nitrogen species in the hydrothermal carbon. (Reproduced with permission from [56]. © 2011 American Chemical Society.)*

While Maillard reactions are certainly involved in the formation of nitrogen-doped hydrothermal carbons, other mechanisms must be occurring concurrently. Baccile *et al.* carried out ^{13}C and ^{15}N solid-state NMR investigations of nitrogen-doped carbons obtained from glucose and glycine, which is the classical reaction to make melanoidin resins, and found that the material exhibited an unexpected structure of very different nature than typical melanoidin resins [56]. In typical melanoidins, up to 60% of glycine molecules are stoichiometrically incorporated, forming predominantly amides, amines, and pyrolic species, but not the aromatic nitrogen species (pyrazines, pyranones) identified by Baccile *et al.* (Figure 5.3) [64].

5.2.2   Sulfur

Complementing nitrogen as a dopant, sulfur is receiving increasing attention in current carbon materials research. In contrast to nitrogen, which is often used to alter the electronic properties of the carbon material, sulfur has been used more for the alteration of physical properties (e.g., to induce structural defects or increase interlayer spacing of graphitic lattices) and for applications where its easily polarizable lone pairs (and thus chemical reactivity) are of importance. Sulfur-doped carbon materials have, for example, shown beneficial effects on the selective adsorption of waste metals [3] and the desulfurylation of crude oil [4]. The synthesis of these sulfur-doped materials generally involves the pyrolysis of sulfur-containing, polymer-based carbons [3, 4, 65], but also arc vaporization in the presence of sulfur-containing compounds such as thiophenes [25]. Yang *et al.* reported on the synthesis of sulfur-doped graphene with enhanced electrocatalytic properties in the ORR, prepared by annealing graphene oxide and benzyl disulfide under argon at high temperatures (600–1050 °C) [66]. Choi *et al.* synthesized heteroatom-doped carbon materials by the pyrolysis of amino acid/metal chloride composites. They were able to show that materials containing both nitrogen and sulfur increased the material's ORR activity in acidic media, relative to undoped or purely nitrogen-doped carbons [22]. Tsubota *et al.* prepared sulfur and nitrogen dual-doped carbons via the pyrolysis of thiourea and formaldehyde [23]. Their material exhibited superior performance as supercapacitor electrodes than commercial activated carbons. The combined incorporation of nitrogen and sulfur may therefore also yield interesting properties.

In the context of HTC, analogous experiments to the aforementioned nitrogen-doped systems have been carried out by Wohlgemuth *et al.* who used small molecules (cysteine (Cys), thienyl-cysteine (TCys)) as comonomers in the HTC of glucose [67]. Considering that both Cys and TCys are amino acids, incorporation of Cys and TCys is thought to involve the Maillard reaction pathway. However, both molecules contain an additional nucleophile (thiol or thienyl moiety to the free amine). As sulfur nucleophiles cannot undergo classical Maillard reaction cascades (they cannot form Schiff bases), alternative reactions must take place. The authors could show that in the case of sulfur, the heteroatom binding state in the final hydrothermal carbon may be controlled during HTC. If an aliphatic sulfur source like Cys is used as an additive, nitrogen and sulfur dual-doped carbon materials with aliphatic (thio-ether or thiol) sulfur-binding motifs are obtained (after HTC at 180 °C). Note that HMF also contains an α,β-unsaturated aldehyde motif, allowing Michael addition to take place, representing one way of forming thio-ethers [68, 69]. In contrast, the use of aromatic sulfur precursors such as TCys results in nitrogen and sulfur dual-doped materials with aromatic (thiophene) binding motifs. This was confirmed by XPS analysis (Figure 5.4) [67].

In another example, Wohlgemuth *et al.* synthesized sulfur (and nitrogen)-doped carbon aerogels using thiophene carboxaldehyde as sulfur source [18]. Again, the thiophene motif

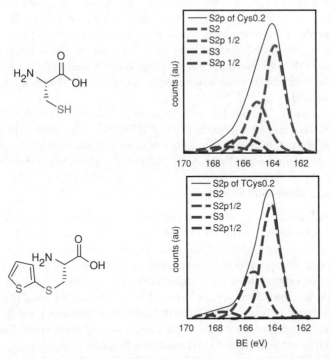

Figure 5.4 *Deconvoluted S 2(p) photoelectron envelopes for hydrothermal carbon obtained from glucose and Cys or glucose and TCys after HTC at 180 °C. The different dashed lines correspond to aliphatic (thiol, thio ether), aromatic (thiophene), and oxidized (sulfone, sulfonic acid, sulfoxide) sulfur species, respectively.*

was retained in the as-synthesized hydrothermal carbons, implying that these types of monomers are incorporated either via an electrophilic aromatic substitution or chemistry involving the aldehyde. However, detailed mechanistic investigations are still required.

5.2.3   Boron

Boron is another highly desired dopant in carbon materials. Boron-doped materials have shown to be useful in important applications in hydrogen storage due to the modification of the energy landscape of adsorbing surfaces and strong surface heterogeneity [70]. Boron doping is also considered to be one of the very few promising candidates for chemical protection of C/C composite materials against oxidation [71]. This was thoroughly studied by Radovic *et al.* [72]. In addition, boron-doped carbons recently demonstrated superior electrochemical properties [73–76]. As in the case of other heteroatom-doped carbons, boron doping was mainly achieved in carbon nanotubes (CNTs) and recently graphene, using high-temperature procedures [77].

Up to now there has been no pure hydrothermal process to produce boron-doped carbon materials from biomass precursors. However, there are a few reports on the fabrication of two-dimensional ordered mesoporous carbons (OMCs) via a one-pot organic–organic aqueous self-assembly approach, using resorcinol/formaldehyde as the carbon precursor and triblock copolymer Pluronic® F127 as the mesoporous structure template. In addition, Zhang *et al.* used boric acid and/or phosphoric acid, where the F127 soft template underwent a self-assembly process under strong acidic conditions to form a polymer with ordered mesostructure, which was then carbonized at 800 °C in a nitrogen atmosphere to form boron-incorporated, phosphorous-incorporated, or boron/phosphorous-coincorporated OMCs. Such resulting heteroatom-incorporated OMCs exhibited superior electrochemical performances to nonincorporated counterparts when used as electrodes of supercapacitors [78].

In a very recent work, Su together with the previous authors showed that the capacitance of such materials can be increased up to $177\,\mathrm{F\,g^{-1}}$ [79]. This is assumed to be due to the fact that boron and phosphorous elements were introduced in the form of oxygen-containing groups, and may interact mutually as the hydrothermal temperature is increased.

5.2.4   Organic Monomers Sources

Along with Maillard chemistry, non-Maillard reactions are also thought to simultaneously take place in hydrothermal environments. HMF and other furanic intermediates produced during HTC process (see Chapter 5) are considered to play a major role in non-Maillard pathway to form nitrogen-doped hydrothermal carbons. One of the possible pathways for nitrogen incorporation is via Diels–Alder (DA) cycloaddition. Imines (Figure 5.2, **II**) formed through Maillard chemistry could most likely go through imino-DA reactions with reactive dienes found in furans resulting in the formation of heterocycles (Figure 5.2, **VI**) [80]. The reactivity of furans towards dienophiles was confirmed upon addition of vinyl monomers such as acrylic acid and vinyl imidazole [81, 82].

Demir-Cakan *et al.* determined that the addition of acrylic acid and vinyl imidazole resulted in hydrothermal carbons rich in carboxylate groups or doped with up to 8.2 wt% nitrogen (in the form of imidazoles), respectively. Solid-state carbon NMR studies of acrylic acid-functionalized hydrothermal carbons revealed the decreased content in sp^2

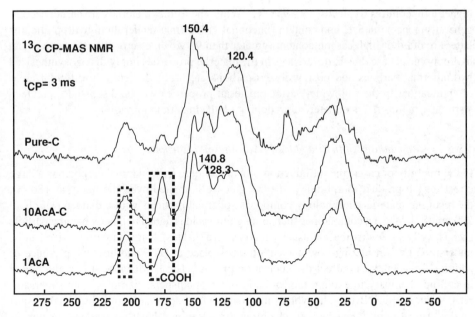

Figure 5.5 *^{13}C solid-state NMR of acrylic acid-functionalized carbonaceous materials. Top: no acrylic acid added; middle: 10 wt% acrylic acid added as a dopant; bottom: 1 wt% acrylic acid added as a dopant. (Reproduced with permission from [81]. © 2009 American Chemical Society.)*

carbons (and increased signal from the sp^3 region) from modified carbons compared to the standard unfunctionalized hydrothermal carbon. Furthermore, the signals from the furanic moiety become smaller whereas increases in aromatic peaks are observed as more acrylic acid is added to the reaction mixture (Figure 5.5).

The aforementioned possible mechanisms are especially relevant to hydrothermal carbon formation with dopants other than nitrogen, where Maillard chemistry does not apply.

In summary, there are various reaction pathways that can take place when heteroatom-containing molecules are added to a HTC mixture of pure carbohydrate (i.e., glucose). The additives may react directly with the sugar, but also with HTC decomposition products, of which HMF is expected to be the most abundant. For amino acids and proteins, Maillard reactions are possible with nucleophilic amines, which in turn give rise to many more intermediates that can react further. The reactivity of the heteroatom source is important concerning the ways in which the heteroatoms can be incorporated into the final structure. Heterocycles can, in principle, react via electrophilic aromatic substitution, which leaves the aromatic heteroatom-binding motif intact after incorporation into the hydrothermal carbon.

5.2.5 Properties of Heteroatom-Doped Carbon Materials

When considering the physicochemical properties of hydrothermal carbons, it is always important to distinguish between as-synthesized materials after HTC at usually 180–200 °C and hydrothermal carbons that have been subjected to an additional pyrolysis step at higher temperatures (typically 550–950 °C). While the former materials are carbonaceous, generally nonconductive, and contain functional groups that render them hydrophilic, the latter are often of high carbon content (greater than 90 wt%), electrically conductive, and hydrophobic. These drastic differences in the material properties imply that as-synthesized hydrothermal carbons and postpyrolyzed carbons are useful for entirely different areas of application. In the following, various material properties will be discussed separately, starting with the effects of heteroatom doping on the material morphology.

5.2.5.1   Morphology of In Situ Doped Hydrothermal Carbons

In the majority of cases, the introduction of a dopant into hydrothermal carbon has a large effect on the product morphology. If pure glucose is treated hydrothermally at 180 °C, the resulting material morphology comprises spherical particles with a diameter of about 200 nm [83]. When a comonomer such as an amino acid is added to this recipe, the particle size as well as size distribution usually increases [67]. This change in particle size can be understood by considering how nucleation takes place in HTC and how the presence of additional reactants can affect the nucleation process. The LaMer model is widely cited in colloid chemistry to explain the formation of monodisperse particles from supersaturated solutions [84]. Briefly, it involves the increase in concentration of a dissolved species until a critical point is reached, at which rapid nucleation ("burst nucleation") occurs. A growth phase follows, during which the remaining monomer in solution diffuses to the nuclei formed in the seeding stage. The first nucleation stage can be described as homogeneous nucleation, whereas the growth phase is basically heterogeneous nucleation, because monomers add onto pre-existing nuclei [85]. Once the first species has formed nuclei via homogeneous nucleation, the next species may preferentially nucleate heterogeneously, and therefore determine the final number of particles and their size. The difference in solubility of formed intermediates in water and the changing solvent properties of water itself as it is heated to 180 °C results in variations in the critical concentrations needed to drive nucleation. The more complex the reaction mixture, the more different products are formed at different points in time [86, 87]. Having additional comonomers such as amino acids in the hydrothermal reaction mixture is therefore expected to result in an increased particle size and overall size distribution. However, when HTC is carried out with pure nitrogen-containing precursors such as glucosamine or chitosan, smaller particle sizes than those from pure glucose have been observed (Figure 5.6) [54]. This may be due to the fast rates at which Maillard reactions occur, resulting in early nucleation and hence smaller final particle size.

In some cases the addition of a comonomer may give rise to smaller particles than obtained from pure glucose via some surface stabilizing effect. One example is the previously mentioned HTC of glucose in the presence of albumin, but also HTC of glucose in the presence of ethylene diamine (EDA), which dramatically alters the product morphology as shown in Figure 5.7 [52, 59, 88].

To study the formation mechanism of the unique net-cross structure of nitrogen-doped carbons, Sun *et al.* performed a series of control experiments. First, to determine the effect of reactant concentration on structure, EDA amounts were tuned from 0.2, 0.5, 1.0 up to 2.0 ml with a fixed glucose concentration of 1 M at 180 °C for 8 h (Figure 5.7). The

Figure 5.6 *Scanning (a–c) and transmission (d–f) electron micrographs of the nitrogen-doped carbons obtained upon HTC of (a and d) chitosan (HC-CH), (b and e) glucosamine (HC-GA), and (c and f) glucose (HC-G).*

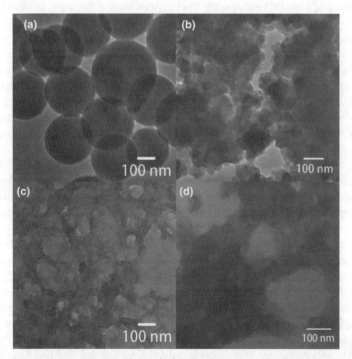

Figure 5.7 *Transmission electron microscopy (TEM) images of nitrogen-doped carbons prepared in different ratios of glucose/EDA: (a) 1 M/0.2 ml, (b) 1 M/0.5 ml, (c) 1 M/1 ml, and (d) 1 M/2 ml. (Reproduced with permission from [88]. © 2012 Royal Society of Chemistry.)*

morphologies of the carbon materials changed significantly with increasing EDA amount. Only spherical material particles were obtained when the EDA amount was less than 0.2 ml (Figure 5.7a). The product after addition of 0.5 ml EDA seemed to be composed of many aggregated spheres with an average diameter of 80 nm (Figure 5.7b), obviously smaller than the previous sample. When the EDA amounts were more than 1 ml (Figure 5.7c and d), the carbonaceous materials were completely converted into a similar mesoporous structure, in which the links between carbon spheres became clearer. The results confirmed that under mild conditions, the addition of EDA depressed the size of these spherical units and promoted their links to form a net-cross structure.

In the same paper, the authors also study the effect of glucose concentration and time on the final morphology of the material.

5.2.5.2   Surface Chemistry

In some cases, the function of the added small molecules can be transferred over to the final *in situ* functionalized HTC-derived material. In another words, HTC-derived material becomes the solid support for the added molecule, where its functionalities are presented on the material surface.

Demir-Cakan *et al.* reported two examples of HTC-derived carbons with specific surface chemistry such as carboxylic acids [81] or imidazole groups [82]. These carboxylic acid- and imidazole-bearing hydrothermal carbons were prepared by simply adding the corresponding vinyl monomers that are assumed to be incorporated through cycloaddition during the hydrothermal carbon formation process, as explained in the previous section. In both cases, the original characteristics of the functional groups of the additives are maintained upon introduction to the carbon surface. As seen with free carboxylates, the carboxylate-rich HTC-derived materials exhibited metal-chelating characteristic with high capacities of 351.4 mg g^{-1} for Pb(II) and 88.8 mg g^{-1} for Cd(II), which demonstrates the material's high potential for water purification applications [81] (see more details in Chapter 7).

Synthesis of imidazole-bearing hydrothermal carbons was carried out by addition of vinyl imidazole to the HTC reaction mixture. Zeta potential and Fourier transform infrared spectroscopy (FTIR) experiments of the resulting carbonaceous material implied the presence of unaltered imidazole groups. To further verify the presence of imidazole groups on the hydrothermal carbon surface, these materials were successfully used as catalysts for various transesterification, Knoevenagel, and Aldol reactions. To improve performance, the imidazole surface groups were converted to imidazolium groups by alkylation using butyl bromide (sample name HC-10Bu2ImBr, Figure 5.8) [82].

Zeta potential measurements on the nitrogen-doped carbon materials prepared by Zhao and Titirici from glucosamine and chitosan clearly showed the existence of positive zeta potentials below pH around 5 and 6, indicating the existence of positively charged, accessible amino groups at the particle's surface with values up to 30 mV at pH 2.5 (Figure 5.9). These data are coherent with XPS analyses, which have showed that roughly 6% of nitrogen sites. On the contrary, pure hydrothermal carbon from glucose shows no positive zeta potential value over the whole range.

These are clear examples on how "*in situ*" functionalization affects the surface properties of HTC-derived materials.

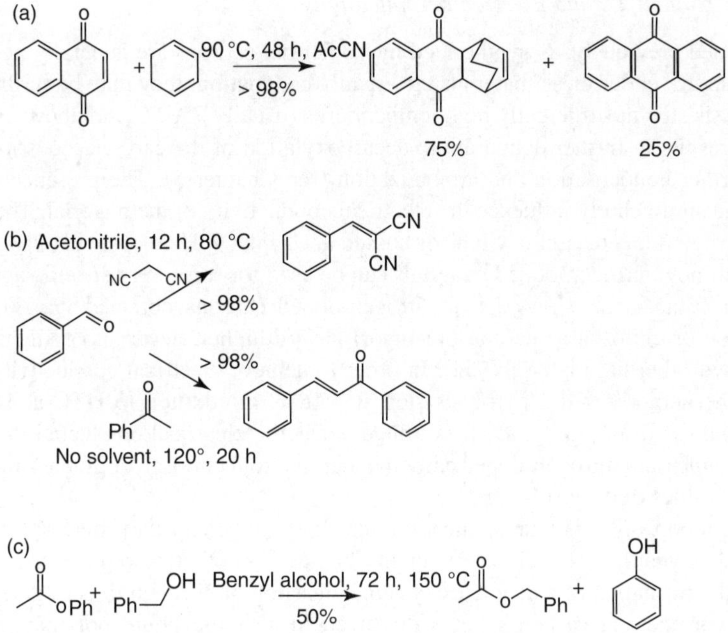

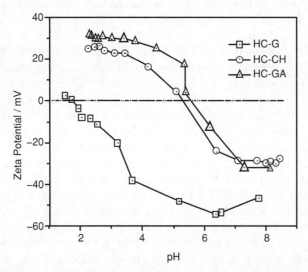

Figure 5.8 *Reactions catalyzed by imidazole-functionalized hydrothermal carbon (HC-10Bu2ImBr): (a) DA, (b) Knoevenagel and Aldol condensation, and (c) transesterification.*

Figure 5.9 *Zeta potential experiments of hydrothermal carbons from glucose, chitosan, and glucosamine. (Reproduced with permission from [54]. © 2010 Elsevier.)*

5.2.5.3   *Conductivity and Electronic Applications*

As mentioned previously, as-synthesized hydrothermal carbons are generally nonconductive. In order to render these materials electrically conducting, they may be further treated in a pyrolysis step at sufficiently high temperatures (usually 750 °C and above). This heat treatment results in further dehydration, decarboxylation of the carbonaceous material as well as further condensation and aromatization (see Chapter 5). The presence of *in situ* dopants can significantly influence the electrical conductivity of the material. The applications of nitrogen-doped carbon will be discussed in Chapter 7. Here, we will only give a few examples of how nitrogen-doped materials can be used for some very specific application.

In 2010, Zhao *et al.* showed that nitrogen-doped carbons obtained via HTC (using glucosamine or chitosan as carbon precursor) have a higher electrical conductivity than undoped hydrothermal materials [54]. In order to achieve electrical conductivity of their samples, the authors used a pyrolysis step at 750 °C in addition to HTC at 180 °C. A sample obtained from pure glucose exhibited a specific electrical conductivity of around 80 S m^{-1} while both nitrogen-doped carbon materials from chitosan or glucosamine exhibited higher values of over 100 S m^{-1}.

The first report on a specific application, namely as supercapacitor materials, appeared later that same year [38]. Details are given in Chapter 7. Another example of specific applications is the formation of a nanolatex system consisting of nitrogen-doped (glucosamine-derived) hydrothermal carbon spheres dispersed in an ionic liquid polymer [55]. Film formation of the nanolatex gave rise to cohesive films with high electrical conductivity (when pyrolyzed carbons were used) and promising thermal insulation properties.

Another example where nitrogen-doped carbons demonstrated their superior properties is provided by Liu *et al.* [88]. They loaded platinum nanoparticles onto nitrogen-doped mesoporous carbon materials obtained from glucose and EDA as described previously (see Figure 5.7; however, annealed at a comparatively low temperature of 400 °C). Their material was used to catalyze *p*-nitrophenol reduction by NaBH$_4$, and compared to supports without nitrogen doping as well as doped and undoped supports without platinum nanoparticles. The better catalytic performance of Pt@N-doped hydrothermal carbon was mainly attributed to a uniform dispersion of active-site platinum nanoparticles on the surface of the mesoporous carbons, as revealed by the TEM analysis. It was clearly observed that platinum nanoparticles loaded on nitrogen-doped hydrothermal carbon were smaller and exhibited a quite discrete dispersion while those on nondoped hydrothermal carbon tended to agglomerate. From the size distributions of Pt@N-doped hydrothermal carbon it was confirmed that platinum loaded on nitrogen-doped hydrothermal carbon produced more uniform nanoparticles, with a smaller average diameter of 4.6 nm than those on nondoped hydrothermal carbon (6.7 nm). The reason why nitrogen doping could help platinum nanoparticles assemble uniformly on the carbon surface was explained by the interactions between the empty orbital of a platinum atom and the free electron pair of a nitrogen atom, which could limit the mobility of the platinum nanoparticles and prevent them from agglomerating (Figure 5.10) [88]. The material doped with nitrogen showed significantly higher conversion than the undoped material or the supports without platinum nanoparticles.

Concerning sulfur-doped hydrothermal carbon materials, the list of reports is less extensive. It was shown that sulfur and nitrogen dual-doped carbon spheres exhibited higher electrical conductivity than undoped carbon materials after pyrolysis at 900 °C [67]. It is,

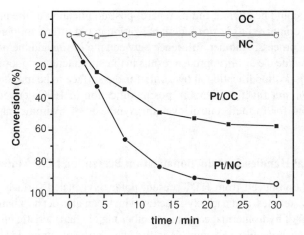

Figure 5.10 *Conversions of p-nitrophenol obtained with nondoped hydrothermal carbon (OC), platinum-doped hydrothermal carbon (PT/OC), nitrogen-doped hydrothermal carbon (NC), and Pt@N-doped hydrothermal carbon catalysts (Pt/NC) (catalyst: 5 mg, p-nitrophenol: 0.15 mM, NaBH4: 47 mM, room temperature and atmospheric pressure).*

however, unclear whether this effect comes from the sulfur or the nitrogen dopant. The presence of sulfur atoms in the carbon scaffold was shown to increase the graphitic inter-layer spacing due to the large size of the sulfur atom. The authors were unable to test the materials for specific applications due to the low surface area of the carbon microspheres.

In their later report on sulfur/nitrogen dual-doped carbon aerogels they could show that the presence of sulfur in addition to nitrogen greatly improves the electrocatalytic performance towards oxygen reduction of the pyrolyzed (900 °C) materials in both acidic and alkaline media (see Chapter 7) [18]. This was postulated to be due to a synergistic effect between nitrogen (which activated oxygen dissociation) and sulfur (which aids proton transport), although detailed mechanistic investigations are still required.

Overall, it is clear that *in situ* doping of carbon materials via HTC can greatly influence the final material's physicochemical properties. *In situ* doping has the great advantage of being simple to carry out, usually requiring only a one-pot reaction and, if needed, a postpyrolysis step. This approach will alter the bulk material properties, although certain variations (e.g., in the type of heteroatom species present) between batches are to be expected since the method allows for very little chemical control.

For a more controlled modification of hydrothermal carbons, postmodification for the introduction of heteroatoms may be a more appropriate approach. These postmodified carbons will be discussed in the next section.

5.3   Postmodification of Carbonaceous Materials

The potential downside of the *in situ* functionalization method is the HTC condition itself (greater than 180 °C, water as a solvent, pressure), where structures of many conventional small molecules could be altered. In this regard, postmodification of the preformed HTC-derived material compared to *in situ* functionalization could offer a milder and controlled functionalization process. Another difference between the *in situ* and the postmodification approach is that while the *in situ* approach results in the introduction of additives throughout the bulk material, postmodification allows only for the surface to be modified. Regardless, along with the *in situ* functionalization, postmodification of HTC-derived materials has potential to allow for further control over the synthesis of hydrothermal carbons with desired properties.

5.3.1   Chemical Handles for Functionalization Present on HTC Materials

Typical materials produced from HTC present polar oxygenated surface groups such as hydroxyl groups, ketones, and carboxylic acids, which can also act as a handle for the surface modification of hydrothermal carbons. Another major characteristic of a hydrothermal carbon (discussed in detail in Chapter 6) is the abundance of furans [81]. Postpyrolysis of the hydrothermal carbon leads to a decreased content of the polar functional groups as well as the furanic structure, rearranging to more condensed arene structure. The presence/absence of the polar functional groups as well as the furanic/arene core depends on the pyrolysis temperature. In this regard, the functionalization strategy may alter, based on the preparation conditions of the hydrothermal carbon. Herein, a short list of postmodifications of hydrothermal carbons as well as alternative potential postmodification strategies is introduced and discussed.

5.3.1.1   Nucleophilic Substitutions

Etherification of carbonaceous materials has been achieved by addition of 3-chloropropyl amine to ordered hydrothermal carbons (OHC) prepared at 180 °C (Figure 5.11a) [89]. The modification was confirmed from elemental analysis (around 4.5 N%), FTIR, and thermogravimetric analysis. Furthermore, while amino-modified OHC exhibited a positive zeta potential in acidic media, the unmodified OHC was found to have a negative zeta potential even in the acidic media, further supporting the success of the substitution. Chemical

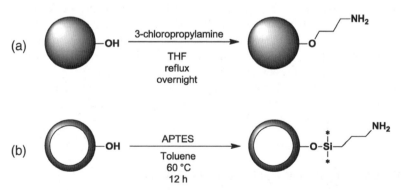

Figure 5.11   *(a) Amination of carbonaceous spheres using 3-chloropropylamine. (b) Modification of hollow spheres with APTES.*

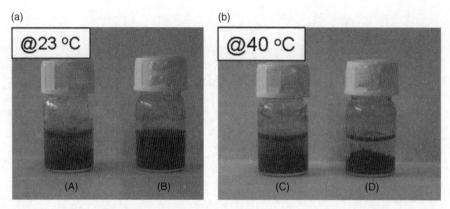

Figure 5.12 (a) Hydrothermal carbon (A) and HTC-pNIPAM (0.3 wt%) (B) in water at 23 °C. (b) Hydrothermal carbon (C) and HTC-pNIPAM (0.3 wt%) (D) in water at 40 °C.

modification was found to have little influence of the structure of the OHC, as observed from small-angle X-ray scattering (SAXS) patterns and nitrogen adsorption isotherms. As an alternative strategy to modify the surface hydroxyl groups, White *et al.* silylated the hydroxyl groups of the carbon hollow spheres prepared at 550 °C using (3-aminopropyl) triethoxysilane (APTES) (Figure 5.11b) [90]. FTIR of the APTES functionalized carbon hollow spheres showed the appearance of symmetric N—H and C—H stretching vibration bands in the 3000–2800 cm^{-1} region, confirming the successful introduction of surface amino groups. The introduction of the amine was also confirmed by elemental analysis (4.1 N%) as well as the pH-sensitive water dispersion behavior.

Surface amino groups of APTES-modified hydrothermal carbons have been further used as a handle for subsequent surface functionalization. Amidation of *N*-hydroxysuccinimide-activated poly(*N*-isopropylacrylamide) (pNIPAM, around 30 kDa) has been carried out to prepare polymer-grafted carbons [91]. Basing on the results found from elemental analysis, surface grafting density of pNIPAM on the hydrothermal carbon surface was determined to be 1.9 mg m^{-2}. The thermoresponsive behavior of pNIPAM-grafted carbons was tested by cycling the temperature of the carbon/aqueous dispersion between 15 and 40 °C. Clear differences in dispersion behavior were observed at temperatures below and above the lower critical solution temperature of pNIPAM as shown in Figure 5.12. Such stimuli-responsive carbon capsules with porous walls have a great potential to be used as drug delivery platforms where the release could be triggered by a temperature change. In this respect we need to mention that the lowest critical solution temperature where the polymer changes its properties from hydrophilic to hydrophobic (from expanded to coiled structure respectively) could be modulated to body temperature. In the future other stimuli responses could be added to such carbon structures, such as light and pH.

5.3.1.2 Cycloadditions

Despite the popularity of DA cycloaddition-mediated functionalization of carbon materials such as CNTs [92–94], graphene [95, 96], and fullerene [97, 98], the application of DA for the preparation of surface-modified carbonaceous material has yet to be fully explored.

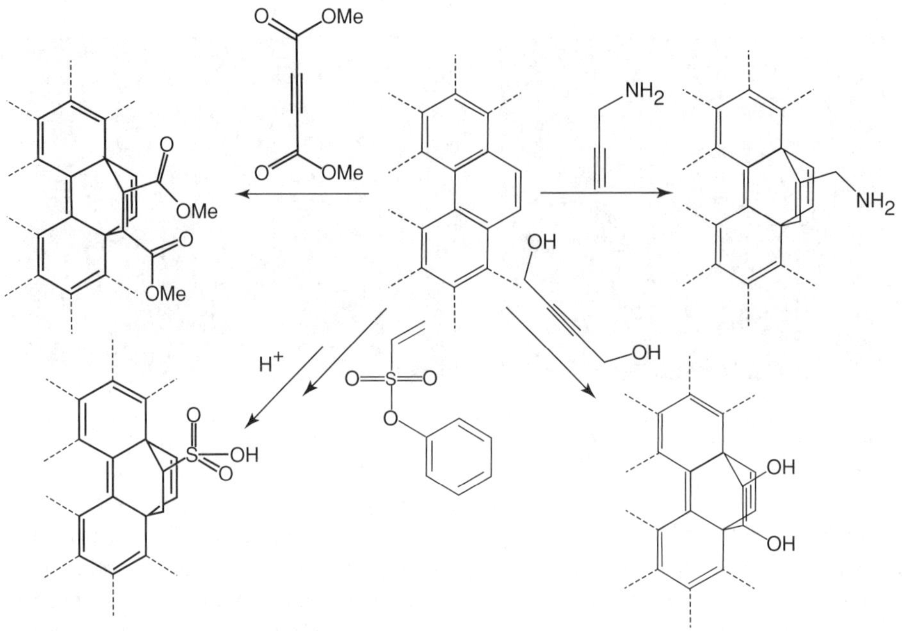

Figure 5.13   *Suggested reaction scheme for the functionalization of carbonaceous materials by a DA reaction.*

To our knowledge, the first example of the DA addition-mediated surface modification of preformed carbonaceous materials was carried out by Kaper *et al.* [99]. In their study, five types of carbon/carbonaceous materials were modified using potential dienophiles as shown in Figure 5.13. The addition was confirmed along with the decreased surface area, pore volume, and pore size of the carbon material.

Urakami *et al.* also attempted to surface functionalize hydrothermal carbon spheres using dienophiles [100]. HTC-derived material prepared at 180 °C and carbon spheres postpyrolyzed at 550 °C were modified using maleimide, tetracyanoethylene, and 4,5-dicyano-1,3-dithiol-2-one (DCDTO). DCDTO-functionalized HTC-derived samples were then hydrolyzed in an attempt to obtain thiol-functionalized HTC@180 and @550 (Figure 5.14a). As shown in Figure 5.14b, FTIR showed the disappearance of the dithiocarbonyl band at 1730 cm^{-1} upon acid-catalyzed alcoholysis. SEM images showed no apparent morphological differences between the unfunctionalized, DCDTO-functionalized, and the hydrolyzed HTC-derived samples. To prove the accessibility of surface-bound thiols, poly(ethylene glycol) (PEG480) acrylate and anthracene acrylate were successfully attached onto the carbon surface via conjugate addition.

5.3.1.3   Other Approaches

Macia-Agullo *et al.* prepared sulfonated HTC microspheres by treating the HTC-derived material prepared from glucose/225 °C with concentrated H$_2$SO$_4$ for 15 h at 150 °C [101]. Despite the harsh reaction conditions, the sulfonation process did not alter the

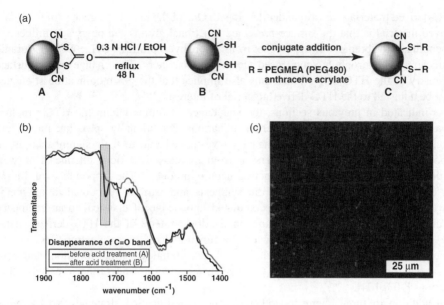

Figure 5.14 (a) Proposed reaction scheme of the alcoholysis and the subsequent conjugate addition. (b) FTIR before (A) and after (B) the acid catalyzed alcoholysis. (c) Fluorescent image of anthracene-functionalized HTC@180.

spherical morphology, particle size, no the surface area of the HTC-derived material. The sulfur content was determined by elemental analysis (less than 2%) and the S 2(p) peak from the XPS spectrum corresponded to that of an oxidized sulfur. The sulfonated HTC-derived material was applied as a solid acid catalysis for esterification of oleic acid for the production of biofuels and exhibited higher catalytic activity than commercial solid acids, such as Amberlyst-15 (see Chapter 7).

Shoujian Li *et al.* reported that a large number of oxygen-containing functional groups could be additionally created on the surface of HTC-derived material by simple thermal treatment at low temperature (i.e., 300 °C) under air [102]. Furthermore, surface character (e.g., polarity) and functional group type can be tuned by employing simple postcarbonization under an inert atmosphere. By heat treatment at 350 °C, —OH groups disappear and the surface groups such as "—C=O" and —C(O)OH become more dominant. Additional treatment at 500 °C eliminates some "carbonyl" features, while most of the carboxylic groups remain. Further treatment at higher temperatures (i.e., above 550 °C) eliminates the majority of the residual oxygenated groups and converts the carbon structure into a turbostratic-like disordered carbon structure with an increasingly more classical aromatic character and hydrophobic properties.

5.3.2 Perspectives on HTC Postmodification Strategies

Compared to other carbon materials such as CNTs, fullerene, graphene/graphene oxides, and porous carbon materials (i.e., CMK series), the postmodification strategy of

HTC-derived materials are still under developed. One of the unique characters of the HTC-derived material is that the surface-presented functional groups can be greatly influenced depending on the preparation approaches from a more hydrophilic (—OH, —COOH)/furanic character to a hydrophobic/condensed arene-like characters. The change in the surface chemistry of the HTC-derived materials also implies that the postmodification strategies must be adjusted to the HTC-derived material of interest.

As indicated in previous sections, the abundance of furans within the HTC structure makes an attractive target for further modifications. Traditionally speaking, furans and furan-related molecules have been heavily investigated due to their significance as an important intermediate in the conversion from biomass to various chemicals as well as their presence in numerous bioactive natural products. The importance of furans has caused chemists to investigate the synthesis and various functionalization methods of furans. In this regard, the accumulated knowledge of classical furan chemistry should be exploited towards new functionalization strategy of the HTC-derived material. As an example, furans are known to be functionalized using electrophiles, and go through transformations such as acylation, alkylation, halogenation, nitration, and condensations. Furthermore, furans can go through metallations and further cross-coupling reactions [103, 104].

In addition to furan chemistry, other chemical modification strategies incorporating oxygenated polar functional groups should be further investigated. Other than etherification and silylation, surface-present hydroxyl groups may also be used to react with electrophilic carbonyls (activated esters, isocyanates). Carbonyl groups on the HTC surface may also be utilized as a handle to introduce heteroatoms through imine formation, conjugate addition, as well as the introduction of sulfur atoms via the use of Lawesson's reagent.

For the modification of the HTC-derived material with a more condensed arene structure, functionalization knowledge from other established carbon materials (fullerene, CNTs, graphene/graphene oxide, various porous carbon materials) could be used as an example [105–109]. Other than [4 + 2], other cycloadditions such as 1,3-dipolar cycloaddition or [2 + 1] carbene, nitrene chemistry may be adopted [107, 110, 111]. Swager *et al.* have recently showed that other pericyclic reactions such as Claisen rearrangements could be applied for the functionalization of graphene oxides [112]. Electrophilic aromatic substitution such as Friedel–Crafts arylation or aryl diazonium chemistry have also been a utilized as a modification strategy [113–116]. Other often used modification approaches include halogenation and addition of carbon radicals [117–120].

One of platform modification strategies has been the oxidation of carbon materials. Classically, carboxylic acid functionality has been introduced to these materials via the use of strong acids such as HNO_3/H_2SO_4 or in basic conditions such as ammonium peroxide/hydrogen peroxide [121–123]. More recently, milder alternative oxidation conditions have been investigated [124, 125]. The carboxylic groups are often activated, and further modified to esters and amides. Carbon materials can also be altered through ozonolysis, where reductive or oxidative work-up could be applied to the molozonide intermediate to form various oxygenated functional groups [126, 127]. Epoxidation is another attractive modification approach for carbon materials [128], due to its facile installation and the possibility for subsequent nucleophilic ring-opening for further modification [129].

References

(1) Shannon, M.A., Bohn, P.W., Elimelech, M. *et al.* (2008) *Nature*, **452**, 301.
(2) Sohi, S.P., Krull, E., Lopez-Capel, E., and Bol, R. (2010) *Advances in Agronomy*, **105**, 47.
(3) Petit, C., Peterson, G.W., Mahle, J., and Bandosz, T.J. (2010) *Carbon*, **48**, 1779.
(4) Seredych, M., Khine, M., and Bandosz, T.J. (2011) *ChemSusChem*, **4**, 139.
(5) Kordas, K., Mustonen, T., Toth, G. *et al.* (2006) *Small*, **2**, 1021.
(6) Leboda, R., Lodyga, A., and Gierak, A. (1997) *Materials Chemistry and Physics*, **51**, 216.
(7) Leboda, R., Lodyga, A., and Charmas, B. (1998) *Materials Chemistry and Physics*, **55**, 1.
(8) Hanai, T. (2003) *Journal of Chromatography A*, **989**, 183.
(9) Zhai, Y., Dou, Y., Zhao, D. *et al.* (2011) *Advanced Materials*, **23**, 4828.
(10) Su, D.S. and Schlögl, R. (2010) *ChemSusChem*, **3**, 136.
(11) Serp, P., Corrias, M., and Kalck, P. (2003) *Applied Catalysis A – General*, **253**, 337.
(12) Jasinski, R. (1964) *Nature*, **201**, 1212.
(13) Guilminot, E., Fischer, F., Chatenet, M. *et al.* (2007) *Journal of Power Sources*, **166**, 104.
(14) Dai, H.J. (2002) *Accounts of Chemical Research*, **35**, 1035.
(15) Gong, K.P., Du, F., Xia, Z.H. *et al.* (2009) *Science*, **323**, 760.
(16) Terrones, M., Ajayan, P.M., Banhart, F. *et al.* (2002) *Applied Physics A – Materials Science & Processing*, **74**, 355.
(17) Marchand, A. and Zanchetta, J.V. (1966) *Carbon*, **3**, 483.
(18) Wohlgemuth, S.-A., White, R.J., Willinger, M.-G. *et al.* (2012) *Green Chemistry*, **14**, 1515.
(19) Thostenson, E.T., Ren, Z.F., and Chou, T.W. (2001) *Composites Science and Technology*, **61**, 1899.
(20) Avouris, P., Chen, Z.H., and Perebeinos, V. (2007) *Nature Nanotechnology*, **2**, 605.
(21) Dai, H.J. (2002) *Surface Science*, **500**, 218.
(22) Choi, C.H., Park, S.H., and Woo, S.I. (2011) *Green Chemistry*, **13**, 406.
(23) Tsubota, T., Takenaka, K., Murakami, N., and Ohno, T. (2011) *Journal of Power Sources*, **196**, 10455.
(24) Xia, Y. and Mokaya, R. (2004) *Advanced Materials*, **16**, 1553.
(25) Glenis, S., Nelson, A.J., and Labes, M.M. (1999) *Journal of Applied Physics*, **86**, 4464.
(26) Ebbesen, T.W. and Ajayan, P.M. (1992) *Nature*, **358**, 220.
(27) Hu, B., Wang, K., Wu, L. *et al.* (2010) *Advanced Materials*, **22**, 813.
(28) Funke, A. and Ziegler, F. (2010) *Biofuels, Bioproducts and Biorefining*, **4**, 160.
(29) White, R.J., Antonietti, M., and Titirici, M.M. (2009) *Journal of Materials Chemistry*, **19**, 8645.
(30) Wang, H., Zhang, C., Liu, Z. *et al.* (2011) *Journal of Materials Chemistry*, **21**, 5430.
(31) Wu, Z.-S., Ren, W., Xu, L. *et al.* (2011) *ACS Nano*, **5**, 5463.
(32) Ma, X. and Wang, E.G. (2001) *Applied Physics Letters*, **78**, 978.
(33) Ma, X., Wang, E.G., Tilley, R.D. *et al.* (2000) *Applied Physics Letters*, **77**, 4136.

(34) Ania, C.O., Khomenko, V., Raymundo-Pinero, E. *et al.* (2007) *Advanced Functional Materials*, **17**, 1828.

(35) Jeong, H.M., Lee, J.W., Shin, W.H. *et al.* (2011) *Nano Letters*, **11**, 2472.

(36) Kim, N.D., Kim, W., Joo, J.B. *et al.* (2008) *Journal of Power Sources*, **180**, 671.

(37) Li, L., Liu, E., Li, J. *et al.* (2010) *Journal of Power Sources*, **195**, 1516.

(38) Zhao, L., Fan, L.-Z., Zhou, M.-Q. *et al.* (2010) *Advanced Materials*, **22**, 5202.

(39) Gong, K., Du, F., Xia, Z. *et al.* (2009) *Science*, **323**, 760.

(40) Kundu, S., Nagaiah, T.C., Xia, W. *et al.* (2009) *Journal of Physical Chemistry C*, **113**, 14302.

(41) Maldonado, S. and Stevenson, K.J. (2005) *Journal of Physical Chemistry B*, **109**, 4707.

(42) Niwa, H., Horiba, K., Harada, Y. *et al.* (2009) *Journal of Power Sources*, **187**, 93.

(43) Qu, L., Liu, Y., Baek, J.-B., and Dai, L. (2010) *ACS Nano*, **4**, 1321.

(44) Sidik, R.A., Anderson, A.B., Subramanian, N.P. *et al.* (2006) *Journal of Physical Chemistry B*, **110**, 1787.

(45) Jaouen, F., Lefèvre, M., Dodelet, J.-P., and Cai, M. (2006) *Journal of Physical Chemistry B*, **110**, 5553.

(46) Pietrzak, R., Wachowska, H., and Nowicki, P. (2006) *Energy & Fuels*, **20**, 1275.

(47) Matter, P.H., Wang, E., Arias, M. *et al.* (2006) *Journal of Physical Chemistry B*, **110**, 18374.

(48) Glenis, S., Nelson, A.J., and Labes, M.M. (1996) *Journal of Applied Physics*, **80**, 5404.

(49) Iijima, T., Suzuki, K., and Matsuda, Y. (1995) *Synthetic Metals*, **73**, 9.

(50) Li, L., Liu, E., Yang, Y. *et al.* (2010) *Materials Letters*, **64**, 2115.

(51) White, R.J., Antonietti, M., and Titirici, M.-M. (2009) *Journal of Materials Chemistry*, **19**, 8645.

(52) Baccile, N., Antonietti, M., and Titirici, M.-M. (2010) *ChemSusChem*, **3**, 246.

(53) White, R.J., Yoshizawa, N., Antonietti, M., and Titirici, M.-M. (2011) *Green Chemistry*, **13**, 2428.

(54) Li, Z., Baccile, N., Gross, S. *et al.* (2010) *Carbon*, **48**, 3778.

(55) Zhao, L., Crombez, R., Caballero, F.P. *et al.* (2010) *Polymer*, **51**, 4540.

(56) Baccile, N., Laurent, G., Coelho, C. *et al.* (2011) *Journal of Physical Chemistry C*, **115**, 8976.

(57) Zhang, D.Y., Hao, Y., Ma, Y., and Feng, H.X. (2012) *Applied Surface Science*, **258**, 2510.

(58) Liu, S., Tian, J., Wang, L. *et al.* (2012) *Advanced Materials*, **24**, 2037.

(59) White, R.J., Yoshizawa, N., Antonietti, M., and Titirici, M.-M. (2011) *Green Chemistry*, **13**, 2428.

(60) Steinhart, H. (2005) *Angewandte Chemie International Edition*, **44**, 7503.

(61) Ledl, F. and Schleicher, E. (1990) *Angewandte Chemie International Edition in English*, **29**, 565.

(62) Capuano, E. and Fogliano, V. (2011) *LWT – Food Science and Technology*, **44**, 793.

(63) Koch, J., Pischetsrieder, M., Polborn, K., and Severin, T. (1998) *Carbohydrate Research*, **313**, 117.

(64) Fang, X. and Schmidt-Rohr, K. (2009) *Journal of Agricultural and Food Chemistry*, **57**, 10701.

(65) Paraknowitsch, J.P., Thomas, A., and Schmidt, J. (2011) *Chemical Communications*, **47**, 8283.

(66) Yang, Z., Yao, Z., Li, G. *et al.* (2012) *ACS Nano*, **6**, 205.

(67) Wohlgemuth, S.-A., Vilela, F., Titirici, M.-M., and Antonietti, M. (2012) *Green Chemistry*, **14**, 741.

(68) Yaylayan, V.A. and Locas, C.P. (2007) *Molecular Nutrition & Food Research*, **51**, 437.

(69) Wondrak, G.T., Tressl, R., and Rewicki, D. (1997) *Journal of Agricultural and Food Chemistry*, **45**, 321.

(70) Kuchta, B., Firlej, L., Roszak, S., and Pfeifer, P. (2010) *Adsorption – Journal of the International Adsorption Society*, **16**, 413.

(71) McKee, D.W. (1991) in *Chemistry and Physics of Carbon*, vol. **23** (ed. P.A. Thrower), Dekker, New York, p. 173.

(72) Radovic, L.R., Karra, M., Skokova, K., and Thrower, P.A. (1998) *Carbon*, **36**, 1841.

(73) Frackowiak, E., Kierzek, K., Lota, G., and Machnikowski, J. (2008) *Journal of Physics and Chemistry of Solids*, **69**, 1179.

(74) Jin, Z., Wei, X., Shuping, Z., and Yi, Z. (2011) *Solid State Sciences*, **13**, 2000.

(75) Tanaka, U., Sogabe, T., Sakagoshi, H. *et al.* (2001) *Carbon*, **39**, 931.

(76) Zhai, X.-l., Song, Y., Zhi, L.-j. *et al.* (2011) *New Carbon Materials*, **26**, 211.

(77) Panchakarla, L.S., Govindaraj, A., and Rao, C.N.R. (2010) *Inorganica Chimica Acta*, **363**, 4163.

(78) Zhao, X., Wang, A., Yan, J. *et al.* (2010) *Chemistry of Materials*, **22**, 5463.

(79) Zhao, X.C., Zhang, Q., Zhang, B.S. *et al.* (2012) *Journal of Materials Chemistry*, **22**, 4963.

(80) Buonora, P., Olsen, J.-C., and Oh, T. (2001) *Tetrahedron*, **57**, 6099.

(81) Demir-Cakan, R., Baccile, N., Antonietti, M., and Titirici, M.-M. (2009) *Chemistry of Materials*, **21**, 484.

(82) Demir-Cakan, R., Makowski, P., Antonietti, M. *et al.* (2010) *Catalysis Today*, **150**, 115.

(83) Titirici, M.M. and Antonietti, M. (2010) *Chemical Society Reviews*, **39**, 103.

(84) Lamer, V.K. and Dinegar, R.H. (1950) *Journal of the American Chemical Society*, **72**, 4847.

(85) Laaksonen, A., Talanquer, V., and Oxtoby, D.W. (1995) *Annual Review of Physical Chemistry*, **46**, 489.

(86) Billaud, C., Maraschin, C., Peyrat-Maillard, M.N., and Nicolas, J. (2005) *Annals of the New York Academy of Sciences*, **1043**, 876.

(87) Billaud, C., Maraschin, C., Chow, Y.N. *et al.* (2005) *Molecular Nutrition & Food Research*, **49**, 656.

(88) Liu, Z., Zhang, C., Luo, L. *et al.* (2012) *Journal of Materials Chemistry*, **22**, 12149.

(89) Titirici, M.-M., Thomas, A., and Antonietti, M. (2007) *Journal of Materials Chemistry*, **17**, 3412.

(90) White, R.J., Tauer, K., Antonietti, M., and Titirici, M.-M. (2010) *Journal of the American Chemical Society*, **132**, 17360.

(91) Kubo, S., Tan, I., White, R.J. *et al.* (2010) *Chemistry of Materials*, **22**, 6590.

(92) Munirasu, S., Albuerne, J., Boschetti-de-Fierro, A., and Abetz, V. (2010) *Macromolecular Rapid Communications*, **31**, 574.

(93) Delgado, J.L., de la Cruz, P., Langa, F. *et al.* (2004) *Chemical Communications*, 1734.

(94) Chang, C.-M. and Liu, Y.-L. (2009) *Carbon*, **47**, 3041.

(95) Bekyarova, E., Sarkar, S., Niyogi, S. *et al.* (2012) *Journal of Physics D – Applied Physics*, **45**, 154009.

(96) Sarkar, S., Bekyarova, E., Niyogi, S., and Haddon, R.C. (2011) *Journal of the American Chemical Society*, **133**, 3324.

(97) Diederich, F. and Thilgen, C. (1996) *Science*, **271**, 317.

(98) Krautler, B. and Puchberger, M. (1993) *Helvetica Chimica Acta*, **76**, 1626.

(99) Kaper, H., Grandjean, A., Weidenthaler, C. *et al.* (2012) *Chemistry – A European Journal*, **18**, 4099.

(100) Urakami, H., Antonietti, M., and Vilela, F. (2012) *Chemical Communications*, **48**, 10984.

(101) Macia-Agullo, J.A., Sevilla, M., Diez, M.A., and Fuertes, A.B. (2010) *ChemSusChem*, **3**, 1352.

(102) Chen, Z., Ma, L., Li, S. *et al.* (2011) *Applied Surface Science*, **257**, 8686.

(103) Dean, F.M. (1982) *Advances in Heterocyclic Chemistry*, **30**, 167.

(104) Dean, F.M. (1982) *Advances in Heterocyclic Chemistry*, **31**, 237.

(105) Stein, A., Wang, Z. and Fierke, M.A. (2009) *Advanced Materials*, **21**, 265.

(106) Liang, C., Li, Z. and Dai, S. (2008) *Angewandte Chemie International Edition*, **47**, 3696.

(107) Balasubramanian, K. and Burghard, M. (2005) *Small*, **1**, 180.

(108) Huang, X., Yin, Z.Y., Wu, S.X. *et al.* (2011) *Small*, **7**, 1876.

(109) Kuila, T., Bose, S., Mishra, A.K. *et al.* (2012) *Progress in Materials Science*, **57**, 1061.

(110) Yao, Z.L., Braidy, N., Botton, G.A., and Adronov, A. (2003) *Journal of the American Chemical Society*, **125**, 16015.

(111) Holzinger, M., Vostrowsky, O., Hirsch, A. *et al.* (2001) *Angewandte Chemie International Edition*, **40**, 4002.

(112) Collins, W.R., Lewandowski, W., Schmois, E. *et al.* (2011) *Angewandte Chemie International Edition*, **50**, 8848.

(113) Bahr, J.L. and Tour, J.M. (2001) *Chemistry of Materials*, **13**, 3823.

(114) Strano, M.S., Dyke, C.A., Usrey, M.L. *et al.* (2003) *Science*, **301**, 1519.

(115) Balaban, T.S., Balaban, M.C., Malik, S. *et al.* (2006) *Advanced Materials*, **18**, 2763.

(116) Liu, B., Bunker, C.E., and Sun, Y.P. (1996) *Chemical Communications*, 1241.

(117) Olah, G.A., Bucsi, I., Lambert, C. *et al.* (1991) *Journal of the American Chemical Society*, **113**, 9385.

(118) Hamilton, C.E., Lomeda, J.R., Sun, Z. *et al.* (2010) *Nano Research*, **3**, 138.

(119) Umek, P., Seo, J.W., Hernadi, K. *et al.* (2003) *Chemistry of Materials*, **15**, 4751.

(120) Ying, Y.M., Saini, R.K., Liang, F. *et al.* (2003) *Organic Letters*, **5**, 1471.

(121) Rosca, I.D., Watari, F., Uo, M., and Akaska, T. (2005) *Carbon*, **43**, 3124.

(122) Datsyuk, V., Kalyva, M., Papagelis, K. *et al.* (2008) *Carbon*, **46**, 833.

(123) Zhang, J., Zou, H.L., Qing, Q. *et al.* (2003) *Journal of Physical Chemistry B*, **107**, 3712.

(124) Zhang, L., Ni, Q.Q., Fu, Y.Q., and Natsuki, T. (2009) *Applied Surface Science*, **255**, 7095.

(125) Zhang, L., Hashimoto, Y., Taishi, T., and Ni, Q.Q. (2011) *Applied Surface Science*, **257**, 1845.
(126) Banerjee, S. and Wong, S.S. (2002) *Journal of Physical Chemistry B*, **106**, 12144.
(127) Kakade, B., Mehta, R., Durge, A. *et al.* (2008) *Nano Letters*, **8**, 2693.
(128) Ogrin, D., Chattopadhyay, J., Sadana, A.K. *et al.* (2006) *Journal of the American Chemical Society*, **128**, 11322.
(129) Collins, W.R., Schmois, E., and Swager, T.M. (2011) *Chemical Communications*, **47**, 8790.

6

Characterization of Hydrothermal Carbonization Materials

Niki Baccile[1], Jens Weber[2], Camillo Falco[3], and Maria-Magdalena Titirici[4]
[1] Laboratoire de Chimie de la Matière Condensée de Paris, Collège de France, France
[2] Max Planck Institute of Colloids and Interfaces, Colloid Chemistry, Germany
[3] Institute for Advanced Sustainability Studies, Earth, Energy and Environment Cluster, Germany
[4] School of Engineering and Materials Science, Queen Mary, University of London, UK

6.1 Introduction

The chemical structure elucidation, elemental composition, type of surface functionalities, and porosity of carbon, carbonaceous, and coal-like materials have been intensively investigated topics over the last 50 years [1]. Several characterization techniques (e.g., Fourier transform infrared spectroscopy (FTIR), X-ray photoelectron spectroscopy (XPS), Raman) and theoretical models have been employed to accomplish such investigations.

In this chapter, we provide the reader with a comprehensive array of techniques used for the characterization of hydrothermal carbonization (HTC) materials with the final aim of a better understanding of the special characteristics of these materials, which are different from the classical amorphous carbons. Where possible, the HTC materials will always be compared with other well-known carbon counterparts.

The chapter opens by analyzing the morphology of HTC materials starting with the model system glucose precursor, and then moving further to other carbohydrates and, finally, to cellulose and biomass. This is followed by a very brief analysis of the yields and elemental composition of HTC materials from different precursors synthesized at

Sustainable Carbon Materials from Hydrothermal Processes, First Edition. Edited by Maria-Magdalena Titirici.
© 2013 John Wiley & Sons, Ltd. Published 2013 by John Wiley & Sons, Ltd.

different temperatures, then by FTIR experiments on HTC carbons, XPS for the analysis of the surface functional groups, zeta potential for determining the surface charge of HTC materials, and, finally, X-ray diffraction (XRD) for understanding the degree of crystallinity in HTC materials. The thermal behavior of a glucose-derived HTC material is also described in detail. This model is, in general, valid for the other carbohydrate-derived materials.

However, the interpretation of results based on FTIR, XPS, and XRD has always shown a certain degree of ambiguity in the case of HTC due to the high degree of chemical heterogeneity characterized by an amorphous structure [2]. Therefore, a combination of techniques is necessary to fully understand the material characteristics.

Although HTC materials have been intensively characterized by FTIR, XPS, Raman, zeta potential, and electron microscopy techniques, which will be briefly presented [3], we choose in this chapter to focus on two main and very important techniques for the characterization of HTC and carbon materials in terms of general chemical structure: solid-state nuclear magnetic resonance (NMR) and porosity analysis.

Solid-state NMR has long been employed in the study of carbon-based materials, but only recently have theoretical and technical developments allowed us to fully exploit its analytical potential. For instance, magic angle spinning (MAS) and cross-polarization (CP) are routinely employed NMR techniques that improve spectra resolution and sensitivity. However, in some cases they might not be sufficient. For this reason, further advances in NMR probe engineering (e.g., higher rotational speed, reliable microelectronics), stronger field spectrometers (up 1 GHz), and spin manipulation engineering have been necessary to perform experiments with a sensitivity and spectral resolution that were inconceivable no longer than 20 years ago. Nonetheless, the analysis and chemical interpretation of highly heterogeneous carbonaceous materials can still represent a major challenge. Dr. Niki Baccile, the author of the Section 6.9 dedicated to the use of solid-state NMR, provides a comprehensive overview of the characterization of various crystalline and amorphous (mainly biomass-derived) carbon materials with a strong focus on the characterization of HTC materials.

We also choose to briefly describe various techniques providing a very important understanding of the carbon material's porous structure. Section 6.10 dedicated to porosity analysis starts with a short overview of the classification of porosity and the different methods that can be used to characterize them. Within Section 6.10, the author, Dr. Jens Weber, introduces the reader to the main techniques of porosity analysis. Some of the fundamental techniques are discussed in detail, mainly gas adsorption, which is probably the method most frequently used, but also mercury intrusion and X-ray scattering. Other less frequently used techniques will be discussed only briefly. Whenever possible, the main points will be emphasized using examples based on research on hydrothermal carbons or carbons in general.

6.2   Morphology of Hydrothermal Carbon Materials

Before going into the deep characterization of HTC materials we need to briefly remind the reader how these materials can be formed.

The two processes, biomass to biofuels and biomass to hydrothermal carbons, have as a common path the hydrolysis of cellulose. Bergius, the discoverer of HTC [4], was one of the first to perform studies on cellulose hydrolysis in the 1930s [5]. In his process, the hydrolysis of cellulose was carried out in 40 wt% HCl at room temperature. Under these conditions, cellulose and hemicellulose are solubilized in the reaction medium, whereas lignin remains insoluble. Cellulose breaks downs into oligosaccharides and glucose within a few hours with the formation of large amounts of dehydration products, such as hydroxymethylfurfural (HMF) and levulinic acid. Hemicellulose is also hydrolyzed, producing mannose, xylose, galactose, glucose, and fructose, mainly as oligomers.

Since this pioneering work, many other processes for cellulose hydrolysis, based on the same principle, have been reported [6–9]. Many of these processes employ hydrothermal conditions. This allows adjusting the ionic product and dielectric constant of water over a wide range by means of pressure and temperature. The critical parameters for water are 374.2 °C and 22.1 MPa, which signifies that, depending on the conditions, water is in the sub- or supercritical regime [10]. Water in such an "overheated" state is a very acidic and reactive substance, permitting chemical reactions such as cellulose hydrolysis and biomass refining in general without the addition of strong acids [11].

The transformation of cellulose into its glucose monomers and of biomass, in general, into easily hydrolyzable carbohydrates is a crucial prerequisite for both biofuels and HTC formation. Under hydrothermal conditions, the dehydration of glucose and other sugars is generally favored [12]. HMF and furfural are the resulting dehydration products of sugars [13]. Furfural is generated from pentoses, such as xylose and ribose, whereas HMF is built from glucose, such as fructose and mannose. With increasing persistence time in aqueous medium, HMF successively decomposes into levulinic acid and formic acid, on the one hand, and into polymeric carbonaceous material, on the other hand [14].

While investigators tried to avoid the formation of a solid during such processes by polymerization of HMF, our group and others rediscovered these processes in the sense of Bergius for the production of green and valuable carbon and carbon hybrid materials [4, 15–18]. Mostly depending on the applied conditions (i.e., the grade of carbonization) the products are called humins [19], hydrothermal carbons [20], or hydrochars [21].

Here, we focus on the morphology of the as-formed HTC materials produced from various precursors. We start with the model system glucose, which is the main constituent "monomer" of cellulose. We then compare the HTC glucose morphology with other monocarbohydrate-derived hydrothermal carbons and with disaccharide- and polysaccharide-derived hydrothermal carbons. We also discuss the morphology of cellulose- and biomass-derived hydrothermal carbons. The morphological changes occurring upon heteroatom doping were discussed in detail in Chapter 5, while the shaping of HTC materials in the presence of various templates/additives was described in Chapter 2 dedicated to porous materials. Therefore, we focus here only on the as-formed HTC and the products resulting upon additional calcination.

6.2.1 Morphology of Glucose-Derived Hydrothermal Carbons

Under hydrothermal conditions (180–280 °C; self-generated pressure) the isomerization of glucose into fructose appears according to the Lobry de Bruyn–Alberta van Ekstein rearrangement [22]. It is considered that HMF is derived from the dehydration of fructose and not from glucose directly [23]. Once HMF is formed, it is *in situ* "polymerized" to form hydrothermal carbon [19]. What we want to understand in this study is why do we obtain micrometer-sized spherical particles, how does the HMF polymerization occur, and what is the final chemical structure of the as-formed material. A correlation between HTC chemical structure obtained using ^{13}C solid-state NMR with other data acquired using a variety of characterization techniques will be presented later in the chapter.

In order to understand the nucleation in HTC, we started investigating the morphology of the glucose-derived hydrothermal carbon spheres at different temperatures, residence times, and concentrations. Figure 6.1 shows the morphology of the hydrothermal carbon spheres derived from HTC of 10 wt% glucose solution for 12 h at different temperatures. It can be observed that the temperature affects both the carbonaceous particle average diameter as well as the size distribution: higher temperatures (280 °C) lead to larger particles and a more homogeneous average size. This observation was also confirmed by dynamic light scattering (DLS) measurements. In the case of the material prepared at 160 °C, an average particle hydrodynamic radius of 474 nm was measured, while for the 280 °C samples, the mean value was 685 nm. For samples prepared at higher temperature, the correlation between DLS and SEM measurements is reasonable. On the other hand, for the materials prepared at 160 °C, the measured values using DLS are higher than those observed in the SEM images. This might be due to the wider size distribution of this sample and therefore to the presence of very large particles, which could affect the overall diameter obtained from light scattering and therefore increase the average error. Another factor that may possibly contribute to the value mismatch might be the presence of particle aggregates in the 160 °C sample, which could increase the value of the measured hydrodynamic particle radius. However, although the DLS measurements do not allow us to determine a

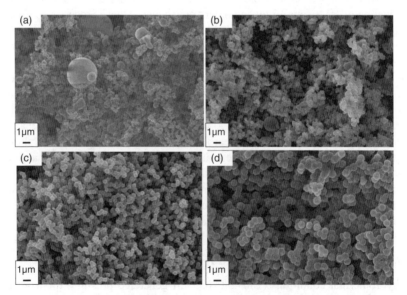

Figure 6.1   *Scanning electron microscopy (SEM) micrographs of HTC materials obtained from 10 wt% glucose for 12 h at (a) 160, (b) 180, (c) 200, and (d) 280 °C.*

Figure 6.2   *Changes in 10 wt% glucose solutions upon HTC at 180 °C.*

precise estimate of the particle size for the sample processed at lower temperature, it is important to underline that they quantitatively confirm the increase in the average particle diameter as the HTC processing temperature becomes higher, which could be observed from SEM measurements.

The next parameter we changed was the residence time while keeping constant the concentration of glucose solution (10 wt%) and temperature (180 °C, Figure 6.2). Nothing happens after a hydrothermal treatment of 2 h. After 4 h, the color of the solution becomes dark orange suggesting that a polymerization/aromatization occurred. Past 5 h, the first solid precipitate is observed. A closer look at this precipitate using SEM (Figure 6.3a) reveals the tendency to aggregate and form spherically shaped nuclei of about 150 nm. After an additional 8 h of hydrothermal treatment, a brown colloidal dispersion is formed. Examining the newly formed solid with SEM, we clearly see a homogenous burst of nucleation with monodisperse particles of around 200 nm being formed. Leaving the reaction proceed further leads to the formation of a final black-brown precipitate after 12 h (Figure 6.2).

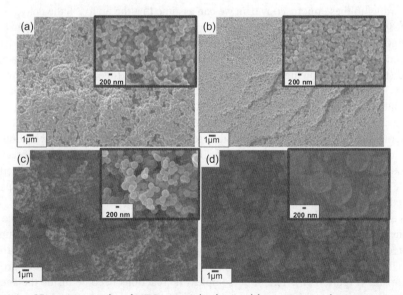

Figure 6.3   *SEM micrographs of HTC materials obtained from 10 wt% glucose at 180 °C after (a) 5, (b) 8, (c) 12, and (d) 20 h.*

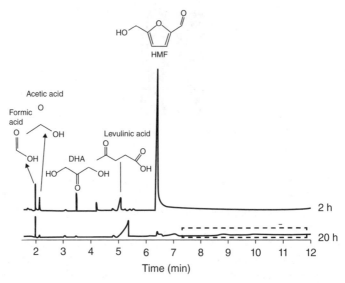

Figure 6.4   *GC experiments on residual liquors from hydrothermal treatment of glucose.*

The solid formed after 12 h consist of spherically shaped particles of around 500 nm, aggregated together (Figure 6.3c). The size of the particles increases even further (around 1.5 μm) if the hydrothermal treatment is continued for longer times (Figure 6.3d). We need to mention that these experiments were performed in the absence of any catalyst known to accelerate the HTC process such as acids or iron ions [24].

According to these microscopy data we can assume that the up to 2 h the dehydration of glucose into HMF occurs via the previously mentioned Lobry de Bruyn–Alberta van Ekstein rearrangement [22]. This has been also confirmed by gas chromatography (GC)-mass spectrometry (MS) experiments of the 10 wt% solution after 2 h hydrothermal treatment at 180 °C (Figure 6.4) showing indeed the formation of HMF, together with the rehydration products levulinic acid, dihydrohyacetone, acetic acid, and formic acid. The inevitable rehydration of HMF to levulinic and formic acid lowers the pH of the solution to a value of around 3. Thus, the *in situ* formed organic acids catalyze the dehydration and further polymerization of HMF. GC-MS of the same residual solution obtained upon hydrothermal treatment of 10 wt% glucose at 180 °C for 20 h, after the filtration of the black precipitate, shows that at this stage there is no residual HMF left. HMF has been transformed via a cascade of chemical reactions [19] into hydrothermal carbon with the spherical morphology depicted in Figure 6.3d.

It is extremely difficult to determine exactly which chemical reactions take place in the autoclave during the HTC [25]. Once such polymeric species are formed (after around 5 h) they precipitate out of the aqueous solution leading to the morphology shown in Figure 6.3a. Upon chemical reactions occurring *in situ* in the autoclave these resulting nuclei grow to a uniform size (Figure 6.3b). The more complex the reactions become inside the autoclave upon increasing the residence time, the growth of the particles occurs heterogeneously, resulting in rather polydisperse and agglomerated particles. The growth process continues

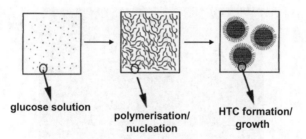

glucose solution

polymerisation/
nucleation

HTC formation/
growth

Figure 6.5 *Schematic representation of the nucleation growth process for the glucose hydrothermal carbon spheres. (Adapted with permission from [15]. © 2004 WILEY-VCH Verlag GmbH & Co. KGaA, Weinheim.)*

until all the monomer (HMF) has reacted (Figure 6.4) and the final particles attain a final particles size of around 1.5 μm after 20 h [25]. According to our experimental results, and others [15], the growth of carbon spheres seems to conform to the LaMer model [26] as shown schematically in Figure 6.5. No carbon spheres were formed when a 10 wt% glucose solution was hydrothermally treated below 160 °C. However, the orange or red color and increased viscosity of the resulting solutions indicate that some aromatic compounds and oligosaccharides are formed [27], in what has been denoted the "polymerization" step. When the solution reached a critical supersaturation (e.g., 10 wt%, 160 °C, 5 h), a short single burst of nucleation resulted.

Unfortunately it is very difficult to monitor *in situ* the nucleation growth process during the HTC process due to the relatively high temperatures and pressures inside the autoclaves. Yet, in 2009 Iversen *et al.* reported a high-temperature high-pressure study on the hydrothermal formation of zirconia nanoparticles by time-resolved *in situ* synchrotron XRD [28]. We are currently in contact with the authors for a future collaboration on HTC of glucose using the high-temperature and -pressure setup and small-angle X-ray scattering (SAXS) instead, since our materials are not crystalline. However, until these interesting experiments are conducted, we can only assume nucleation growth in HTC as that described above.

The effects of further heat treatment on the structural texture of glucose-derived HTC carbons were also investigated using high-resolution TEM (Figure 6.6). The surface of the 600 °C postpyrolyzed sample appears rougher in comparison to that of standard HTC-Glucose-180 °C. This could be associated with the formation of short and highly curved aromatic pregraphinic domains leading to micropore formation within the carbon structure, as will be shown in detail in Section 6.10.2.3. An additional increase in the pyrolysis temperature (900 °C) leads to cross-linking between the intermediate aromatic structures (Figure 6.6c). A highly curved profile and isotropic orientation is still observed related to a partial closure of the structural voids developed during the thermal degradation of the HTC carbon.

6.2.2 Morphology of Other Carbohydrate-Derived Hydrothermal Carbons

After determining the final structure of the glucose-based hydrothermal carbons we embarked on a comparative structural study of hydrothermally synthesized carbon materials obtained from different saccharides classified according to their number of carbons

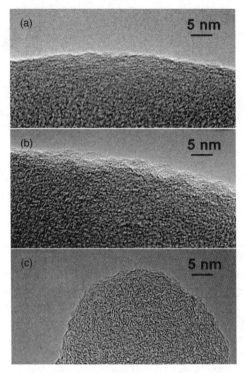

Figure 6.6 *High-resolution transmission electron microscopy (TEM) of glucose-derived HTC synthesized at 180 °C (a) before and after pyrolysis at (b) 600 and (c) 900 °C.*

(pentoses versus hexoses) and growing complexity (monosaccharides versus disaccharides versus. polysaccharides). Details about these morphological differences are provided elsewhere [3, 29].

Looking at the Figure 6.7a we can observe that while the morphology of glucose-derived HTC glucose-based carbon is characterized by a mixture of interconnected 1-μm spheres, the HTC-derived spheres from xylose (Figure 6.7j) are less agglomerated and interconnected. They are mainly two particle size distributions around 1 μm and 500 nm. Although both are monosaccharides, glucose is a hexose and forms hydrothermal carbon via HMF, while xylose is a pentose and forms hydrothermal carbon via furfural. This morphological difference is correlated with their different final chemical structure, which is more "aromatic" in the case of the xylose-derived hydrothermal carbon (see Section 6.9.4.1). Presumably because of this increased aromaticity and hence hydrophobicity, the xylose-derived hydrothermal carbon particles tend to aggregate less. Moving from monosaccharides to disaccharides (sucrose, Figure 6.7d) and to easily hydrolyzable polysaccharides (i.e., starch, Figure 6.7g) we observe that the heterogeneity of the samples is higher as we notice larger particles (5–7 μm) together with smaller particles (1 μm). This is presumably due to the multitude of reactions taking place simultaneously in the autoclaves. The hydrolysis to the constituent monosaccharides and their dehydration to HMF take place at various reaction rates, resulting in different particle sizes. What is also important to notice

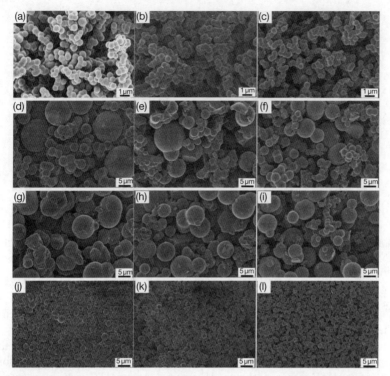

Figure 6.7 *SEM images of HTC and postcalcined carbons: (a) HTC-Glucose, (b) HTC-Glucose-550, (c) HTC-Glucose-950, (d) HTC-Sucrose, (e) HTC-Sucrose-550, (f) HTC-Sucrose-950, (g) HTC-Starch, (h) HTC-Starch-550, (i) HTC-Starch-950, (j) HTC-Xylose, (k) HTC-Xylose-550, and (l) HTC-Xylose-950.*

is that the morphology of any of the samples does not change upon further heat treatment at higher temperatures. This further heat treatment removes the functional groups from the hydrothermal carbon surface (see Sections 6.4, 6.5, and 6.9.4), increases their electric conductivity [30], and creates some micropores upon the elimination of small organic molecules (i.e., levulinic acid; see Section 6.10.2.3).

6.2.3 Morphology of Cellulose- and Biomass-Derived Hydrothermal Carbons

Thus far we have presented fundamental characterization data regarding the morphology of HTC materials derived from simple carbohydrates. The next challenging step in HTC is the effective exploitation of lignocellulosic biomass as carbon precursors, which would guarantee a readily accessible and carbon-negative feedstock supply. Lignocellulosic biomass is composed of three main components: cellulose, hemicellulose, and lignin. The relative proportion of cellulose varies according to the biomass origin from 20% to 45% for agricultural wastes [31].

For the purpose of this investigation, a relatively low lignin content biomass was chosen: rye straw (cellulose/hemicellulose/lignin = 41.2 : 21.2 : 19.5). This raw material is an

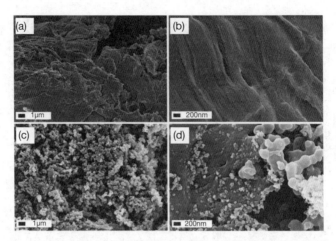

Figure 6.8 *SEM micrographs of hydrothermal carbons obtained from cellulose at 160 (a and b) and 220 °C (c and d). (Reproduced with permission from [32]. © 2011 Royal Society of Chemistry.)*

agricultural waste product; therefore, it is available at costs well below petrochemical-based chemicals.

The morphology of hydrothermal carbon spheres using glucose as a precursor at different carbonization temperatures has been already shown in Figure 6.1. The morphology of cellulose-derived HTC carbons is shown in Figure 6.8. At low HTC processing temperatures (Figure 6.8a and b), cellulose is resistant to hydrothermal treatment. Its fibers are still intact and arranged in the characteristic cellulose network [33]. On the other hand, as observed in Figure 6.8c and d, upon increasing the HTC temperature, spherical particles start forming, which is similar to the HTC materials obtained from glucose. However, the overall morphology is not as homogeneous as for simple sugars (Figures 6.1, 6.2, and 6.7).

Combining these observations with the results shown in Figure 6.9, a remarkable difference in the mechanism of particle formation can be noticed between glucose and cellulose. In the former case, particles form through a nucleation step from a homogeneous solution upon formation and polymerization of HMF [15]. Initially, their diameter is in the range of approximately 100–200 nm (Figure 6.5a). As the residence time increases, they

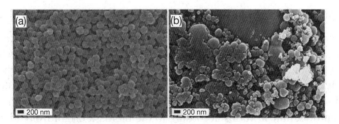

Figure 6.9 *SEM micrographs of hydrothermal carbons obtained from (a) glucose at 180 °C and (b) cellulose at 240 °C for 6 h. (Reproduced with permission from [32]. © 2011 Royal Society of Chemistry.)*

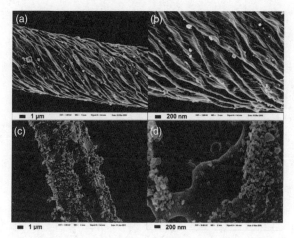

Figure 6.10 *SEM micrographs of HTC materials obtained from HTC of rye straw at 160 (a and b) and 240 °C (c and d). (Reproduced with permission from [32]. © 2011 Royal Society of Chemistry.)*

keep on growing until they reach their final size, which depends upon the HTC processing temperature (Figure 6.1).

On the other hand, in the case of cellulose, upon increasing the temperature, the fibrous network starts to be disrupted at several points (Figure 6.9b), leading to the formation of nano/microsized cellulose fragments, which, not being soluble in water, adopt a spherical shape to minimize their contacting interface with the surroundings.

After investigating the hydrothermal transformation of cellulose and comparing it to the reference glucose-derived hydrothermal carbon, we proceeded with HTC of rye straw – an agricultural biowaste. The SEM micrographs in Figure 6.10 show that the morphological transformations of rye straw after hydrothermal treatment are very similar to those of cellulose. The lignocellulosic biomass does not undergo any structural disruption at low ($T = 180$ °C) temperatures (Figure 6.10a and b) and the rye straw fibrous structure is maintained intact. When the biomass is hydrothermally treated at higher temperatures ($T = 240$ °C), its fibrous network is disrupted and spherical particles start forming, similar to what has been previously described for cellulose (Figure 6.10c and d).

In addition, Figure 6.10c clearly shows that part of the biomass natural macrostructure persists even after the spherical particles start forming. It can be seen that the particle formation takes place on the surface of the rye straw fibers, which overall still maintain their original structural scaffold. This is due to the lignin fraction, which cannot be converted during HTC (see Sections 6.9.3.1 and 6.9.4.2) [32].

6.3 Elemental Composition and Yields

Figure 6.11a compares the yields obtained upon HTC of glucose, cellulose, lignin, and rye straw at different temperatures. Here, we need to point out that, contrary to water-soluble glucose where the yield is zero at low processing temperatures, for all the other

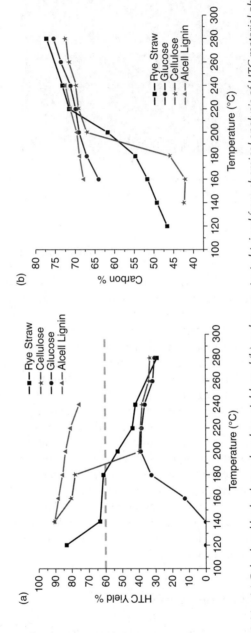

Figure 6.11 (a) Calculated hydrothermal carbon yields and (b) carbon content obtained from chemical analysis of HTC material obtained from glucose, cellulose, Alcell© lignin, and rye straw at different processing temperatures. Carbon (wt%) values for the raw materials are: glucose = 40%, cellulose = 41.75%, Alcell lignin = 66.10%, and rye straw = 45.86%. (Reproduced with permission from [32]. © 2011 Royal Society of Chemistry.)

raw materials (cellulose, lignin, and rye straw), the HTC yield gradually decreases from an initial value of 100% due to the increasing extents of liquefaction and gasification of the biomass at higher temperatures, which are parallel reactions to the process of hydrothermal carbon formation [34].

Yields were calculated according to:

$$\text{HTC yield } (\%) = \frac{\text{Amount of recovered solid after HTC (g)}}{\text{Initial amount of biomass (g)}} \tag{6.1}$$

Figure 6.11 underlines a very interesting feature of cellulose HTC behavior. Both the HTC yield and carbon content can be observed to undergo only minor changes within the low-temperature range (140–180 °C). On the other hand, between 180 and 200 °C, a sharp transition takes place. The HTC yield and the carbon content, respectively, become approximately 37% and 65%. After reaching this critical point, the changes are no longer so drastic and only small variations occur. The presence of this sharp transition supports the idea of the existence of a well-defined temperature threshold, above which cellulose structure is disrupted (see Section 6.9.4.2). The disruption then generates the morphology that was observed during the microscopy analysis of the cellulose-derived HTC material at relatively high processing temperatures (Figure 6.8c and d). This finding is additionally well supported by XRD measurements of raw and hydrothermally treated cellulose at different temperatures (data not shown). It can be observed that between 180 and 200 °C the peaks, corresponding to crystalline cellulose [35], disappear. The structure of the material corresponds now to amorphous carbon with almost no noticeable long-range ordering.

Similarly to cellulose, abrupt changes in both the HTC yield and the carbon content are observed during HTC of rye straw at different temperatures. However, there are two main differences that need to be emphasized:

- Rye straw is more sensitive to hydrothermal treatment in the low-temperature range (120–180 °C). A probable explanation for this observation is the presence of hemicellulose- and xylose-based polysaccharides within the rye straw. These biopolymers are known to be less stable than cellulose when hydrothermally treated; as a consequence, they can undergo HTC in a lower temperature range with a consequential yield loss due to formation of liquid and volatile side products [36].
- The changes in HTC yield and carbon content are less sharp than in the case of cellulose in the region immediately above 180 °C and they appear to be slightly shifted to higher temperature values. The lignin, present within the lignocellulosic biomass, could be the cause of such differences. As can be observed in Figure 6.11, this compound is affected by hydrothermal treatment to a very limited extent. As a consequence its presence might cause the HTC yield of rye straw to decrease less abruptly as temperature increases. Furthermore, since lignin acts as support within the plant wall [37], it might stabilize the cellulose and prevent its crystalline structure disruption at lower temperatures.

The data regarding yields and elemental analysis highlight an interesting feature, which is common to all the investigated precursors: they all follow similar trends for both the HTC yield and carbon content for temperature values above 220 °C. This might be due to the fact that above this temperature value all substrates follow similar chemical reaction pathways. This hypothesis, as it will be seen in Section 6.9.4.2, is also supported ^{13}C solid-state NMR studies.

More details about the elemental composition of HTC materials obtained from different carbohydrate precursors (although very similar to glucose) together with the corresponding changed when post-treated at higher temperatures can be found elsewhere [3, 29].

6.4 FTIR

The FTIR spectra of the HTC carbons as well as those obtained after calcination at various temperatures are shown in Figure 6.12 for four precursors. Abundant functional groups can be observed on all HTC carbons. A broad absorption band between 3700 and 3100 cm^{-1} corresponds to O—H (bonded) stretching vibration. The band at around 2925 cm^{-1} indicates the presence of methylene-type groups (e.g., ν(C—H)$_{stretch}$). The bands at around 1700 and 1600 cm^{-1} are attributed to C=O and C=C stretching vibrations, respectively, supporting the concept of aromatization of biomass during hydrothermal treatment [38]. While these characters are the same for all the four HTC carbons, independent on the precursors used, there are also some differences especially in the case of xylose (HTC-X). The two weak bands in the 1520–1450 cm^{-1} region for HTC-X, which can be attributed to aromatic ring stretching vibration [39], differ from the other three HTC carbons, which

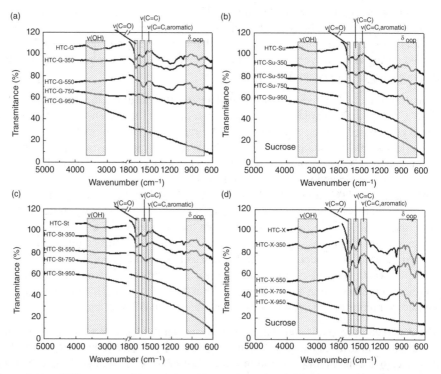

Figure 6.12   *FTIR spectra of the HTC and postcalcined carbons derived from (a) glucose, (b) sucrose, (c) starch, and (d) xylose. (Reproduced with permission from [32]. © 2011 Royal Society of Chemistry.)*

only have one band each. A stronger aromatic character for HTC-X, as shown by the more-resolved bands at 880 and 752 cm^{-1} (δ(C—H)$_{oop}$, oop = out of plane), is also suggested in good agreement with the solid-state NMR results presented in Section 6.9.4.1 [40]. These differences are again related to the fact that xylose is a 5C sugar and the formation of hydrothermal carbon takes place via the formation of furfural and not HMF. Therefore, a more condensed structure with a slightly more "aromatic" character is obtained upon HTC.

The carbons calcined at 350 °C have a similar spectra to the HTC materials. This is in good agreement with the TGA data (Figure 6.16), which shows that the main decomposition event (for glucose) starts only above 300 °C.

Increasing the postcalcination temperature to 550 °C, the intensity of most oxygen-containing related bands decrease, while the C=C band becomes of stronger intensity. Some new bands corresponding to aromatic features, such as bands at 876, 814, and 748 cm^{-1} (δ(C—H)$_{oop}$), appear at 550 °C [40]. These are due to further dehydration and condensation reactions towards arene-like aromatic domains [41]. As the temperature further increases to 750 and 950 °C, the degree of carbonization increases, and therefore no adsorption bands can be detected and the spectra become smooth. All these results are also nicely confirmed by a gradual increase in the elemental composition of the HTC carbons upon further thermal treatment [3], XPS results (Section 6.5), and ^{13}C solid-state NMR on postpyrolyzed HTCs (Section 6.9.4.2).

6.5 XPS: Surface Groups

XPS offers information about the surface functional groups and for all the investigated HTC and postcarbonized HTC materials. The XPS spectra show two peaks corresponding to C1(s) and O1 (Figure 6.13 and Table 6.1). The high resolution C1(s) envelope of glucose- and xylose-derived HTC carbons are characterized by three main contributions, at 285.0 (C1, C—C and C—H$_x$), 286.3/286.6 (C2, C—O—H (hydroxyl), C—O—C (ether)), and 287.9/288 eV (C3, C=O (carbonyl)) with a minor shoulder at 289.3/289.44 eV (C4, O=C–O (acid or ester)). The relatively high intensity of the peaks at C2 and C3 indicates the presence of a considerable amount of oxygenated functionalities, mostly related to the furan and carbonyl moieties present in the HTC carbon structure [41]. In good agreement with the previously presented FTIR results, pyrolysis at 550 °C causes a major reduction of the oxygen-related contributions, indicating the loss of such functionalities and a resultant increase of surface hydrophobicity. The residual C3 peaks can be attributed to the presence of phenolic groups forming during the pyrolysis process as confirmed by ^{13}C solid-state NMR [41]. The peak at 290.1 eV (C4(b)) for the glucose-derived carbon is attributed to adsorption of CO$_2$. Increasing the temperature to 950 °C leads to a further loss of oxygenated functional groups present at this stage only in minor quantities. In addition, the simultaneous increase of the peak at 291.1 eV (C5) for xylose-derived carbon, corresponding to $\pi \rightarrow \pi^*$ shake-up satellites, suggests the presence of extended pregraphinic polyaromatic domains as the major building unit of the carbon scaffold. The peak at 290.4 eV (C4(b)) for glucose-derived carbon is also attributed to adsorption of CO$_2$.

In good agreement with elemental composition, the HTC from glucose contains less sp^2 graphitic carbon and more carbon bound to oxygen then the xylose material.

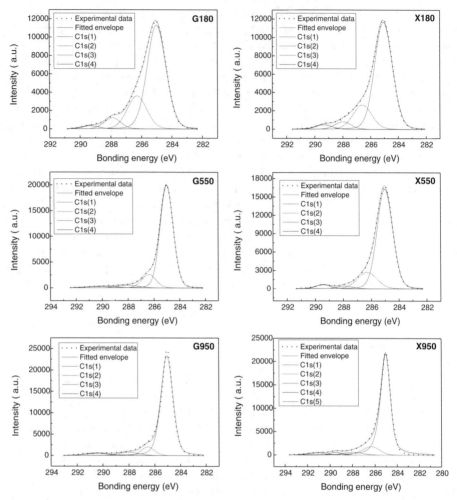

Figure 6.13 *High-resolution XPS scans of C1(s) photoelectron envelope for the HTC and postcalcined carbons derived from glucose and xylose. (Reproduced with permission from [32]. © 2011 Royal Society of Chemistry.)*

Once the materials are calcined at higher temperatures the differences between them become negligible.

6.6   Zeta Potential: Surface Charge

The development of a net charge at the particle surface affects the distribution of ions in the surrounding interfacial region, resulting in an increased concentration of counter-ions (ions of opposite charge to that of the particle) close to the surface. Thus, an electrical double layer exists around each particle.

The liquid layer surrounding the particle exists as two parts: an inner region, called the "Stern layer," where the ions are strongly bound and an outer, diffuse, region where they

Table 6.1 *Experimental C1(s) binding energy (eV)/chemical state assignments for glucose- and xylose-derived carbons. (Reproduced with permission from [32]. © 2011 Royal Society of Chemistry.)*

Samples	C1 (sp^2 graphitic or C—C/C—H$_x$)	C2 (C—O)	C3 (C=O)	C4 (O=C—O)	C4(b) (carbonate)	C5 ($\pi \rightarrow \pi^*$ shake-up satellite)
HTC-G-180	285/69.4%	286.3/22.5%	287.9/6.4%	289.3/1.7%	X	X
HTC-G-550	285/82.5	286.4/11.6	288/3.3%	X	290.1/2.6%	X
HTC-G-950	285/85.1%	286.54/7.3%	288/4.2%	X	290.4/3.4%	X
HTC-X-180	285/74.4%	286.6/18%	288/4.7%	289.44/2.8%	X	X
HTC-X-550	285/78.9	286.3/16.3	287.9/2.1%	289.4/2.8%	X	X
HTC-X-950	285/81.4%	286.33/10%	288/3.8%	289.5/2.1%	X	291.1/2.7%

are less firmly attached. Within the diffuse layer there is a notional boundary inside which the ions and particles form a stable entity. When a particle moves (e.g., due to gravity), ions within the boundary move with it, but any ions beyond the boundary do not travel with the particle. This boundary is called the surface of hydrodynamic shear or slipping plane. The potential that exists at this boundary is known as the "zeta potential." If all the particles in suspension have a large negative or positive zeta potential then they will tend to repel each other and there is no tendency to flocculate. However, if the particles have low zeta potential values then there is no force to prevent the particles coming together and flocculating.

The most important factor that affects zeta potential is pH. A zeta potential value on its own without a quoted pH is a virtually meaningless number.

Imagine a particle in suspension with a negative zeta potential as is the case with HTC particles. If more alkali is added to this suspension then the particles will tend to acquire a more negative charge. If acid is then added to this suspension a point will be reached where the negative charge is neutralized. Any further addition of acid can cause a build-up of positive charge. Therefore, a zeta potential versus pH curve will be positive at low pH and lower or negative at high pH. The point where the plot passes through zero zeta potential is called the isoelectric point and is the point where the colloidal system is least stable.

For the HTC particles, zeta potential measurements Figure 6.14 support very well the conclusions drawn from the FTIR and XPS experiments. The surface charge of the materials

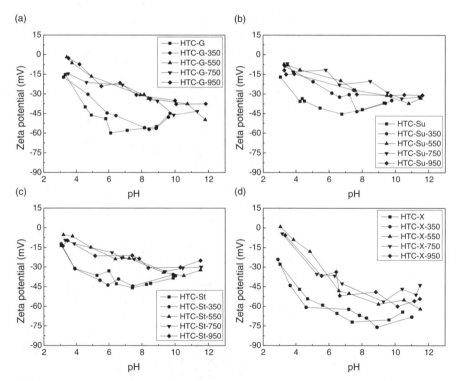

Figure 6.14 *Zeta potential as a function of pH for HTC and postcalcined carbons derived from (a) glucose, (b) sucrose, (c) starch, and (d) xylose. (Reproduced with permission from [32]. © 2011 Royal Society of Chemistry.)*

is in all cases negative over all the chosen pH range. This is expected for HTC materials due to the acidic oxygenated groups on the surface, behaving like a weak acid [42]. It can be clearly observed that the materials obtained after HTC have highly negative zeta potential values similar to those further calcined at 350 °C. Upon further heat treatment at 550 °C there is a significant increase in the zeta potential value in good agreement with the TGA profile as well as FTIR experiments. This is due to the loss of functional groups as described previously. This effect is even more pronounced upon further thermal treatment at even higher temperatures, although not such great differences in the zeta potential values are noticed between 550 and 950 °C. This is because at 550 °C almost all the functional groups have been lost. The values are still negative due to the acidic character of the carbon materials with still some oxygen atoms incorporated in the final structure [42].

6.7   XRD: Degree of Structural Order

XRD is a powerful tool for the characterization of crystalline carbon materials such as graphite, graphene, carbon nanotubes (CNTs), and so on [43, 44]. HTC materials are amorphous, independent of the temperature at which they are synthesized or the post-treatment applied (Figure 6.15). The observed X-ray reflections for all HTC materials are indeed very broad, suggesting that the materials have very disordered and amorphous structures.

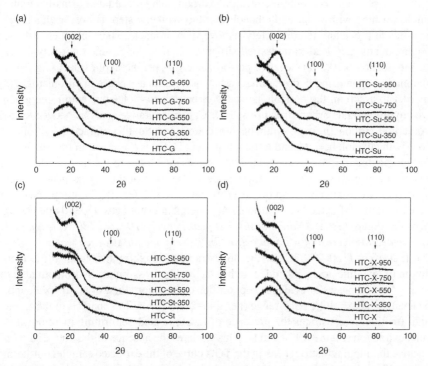

Figure 6.15   *XRD patterns of the HTC and postcalcined carbons derived from (a) glucose, (b) sucrose, (c) starch, and (d) xylose. (Reproduced with permission from [3]. © 2012 American Chemical Society.)*

The reflection at higher scattering angles is typically found in amorphous materials (e.g., polymers) and might reflect the average separation distance (4.0–4.5 Å) between segments. The reflection could also be related to a highly disturbed (002) interlayer carbon packing. Interestingly, these carbonaceous materials also present a reflection at lower scattering angles, potentially originating from either intramolecular repeat distances or large inter-segmental distances. The presence of these two reflections is analogous to observations on some intrinsically microporous polymers [45] and is in accordance with the observation of microporosity in the carbonaceous materials.

Raman spectroscopy is also not suitable for the characterization of such carbon materials as they are polymer-like and will tend to further aromatize/graphitize under the laser beam.

Upon increasing the temperature, the level of structural order increases, which suggests that indeed aromatization towards a more turbostratic-like structure occurs. Starting at 750 °C, the 002 reflection corresponding to the interlayer scattering as well as the intralayer 100 reflection become obvious, as a proof of aromatization/pregraphitization. There are no visible differences in XRD patterns between different precursors, all indicating the same general trend.

6.8   Thermal Analysis

During HTC formation, side-products such as levulinic acid or formic acid coexist and are generally physiosorbed onto the resulting carbons. The procedure normally employed is simple washing with water and ethanol after the synthetic step. However, this may not be an efficient procedure to completely remove such products. The subsequent release of such compounds may lead to undesired structural changes (e.g., microporosity development; see Section 6.10.2) or negative performance for specific applications (e.g., analyte contamination upon using hydrothermal carbon as a stationary phase in chromatography).

In order to ascertain the possible presence of entrapped molecules within the glucose-derived hydrothermal carbon structure at 180 °C, a comparative analysis between washed hydrothermal carbon (normal synthetic procedure) and Soxhlet-extracted hydrothermal carbon (with ethanol, hexane, and tetrahydrofuran (THF)) is presented in this section.

Thermogravimetric-infrared (IR) analysis was used to analyze the thermal decomposition behavior of both samples. The thermal decomposition of the nonextracted sample is composed of two relatively broad events (160–270 and 350–600 °C; Figure 6.16a), whilst in the case of the extracted HTC carbon, only one main event between 350 and 600 °C is observable (Figure 6.16b). (*Note that the first peak between 100 and 180 °C is ignored, as it is attributed to removal of residual extraction solvent and physisorbed water.*)

The first event (160–270 °C) of the nonextracted sample is attributed to the thermal evolution of levulinic acid, embedded within the highly cross-linked hydrothermal carbon structure. Several pieces of evidence support this observation. First of all the peak appears at the levulinic acid boiling point range (i.e., 245–246 °C). GC-MS analysis of the extracted ethanol fraction demonstrates the presence of levulinic acid confirming its removal upon Soxhlet extraction (Figure 6.17a and b). This explains the absence of the 160–270 °C dTG (derivative thermogravimetric) peak in the TGA curve of the extracted sample. Furthermore gas-phase IR analysis of the evolved species during the thermal degradation of HTC carbon indicates that over the temperature range 160–270 °C, the detected peaks are all ascribable to

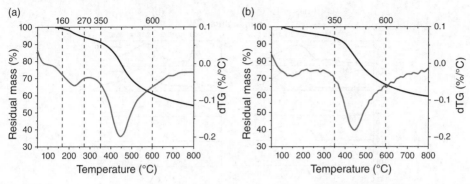

Figure 6.16 *TGA of glucose-derived HTC carbon at 180 °C: (a) before extraction and (b) after three consecutive extractions with ethanol, hexane, and THF. (Reproduced with permission from [3]. © 2012 American Chemical Society.)*

levulinic acid. In particular, extract gas-phase IR traces at 1770 and 1740 cm^{-1} correspond to the carbonyl stretching of ketones and carboxylic acids. Such traces are not detected during the thermal decomposition of the extracted sample.

The decomposition event over the range 350–600 °C is observed for both samples. This indicates a restructuring of the carbon motifs with simultaneous loss of volatile species, as supported by the corresponding IR analysis of the evolved gases (Figure 6.18). At 400 °C the peaks with the strongest intensities correspond to traces of CO_2 and CO. Subsequently, at around 450 °C, methane evolution is detected and the intensity of its corresponding peak increases until 550 °C. Above this temperature threshold all the detected signals fade away at different rates.

The formation of microporosity during the thermal degradation of HTC carbon is due to the combined effects of volatile species elimination (e.g., CO, CO_2, and CH_4) and formation

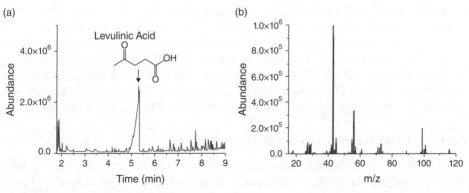

Figure 6.17 *(a) Gas chromatograph of the extract obtained after Soxhlet extraction with ethanol of glucose-derived HTC carbon at 180 °C. (b) Mass spectrum of the peak indicated in (a) (grey, literature data [46]; black, experimental data). (Reproduced with permission from [3]. © 2012 American Chemical Society.)*

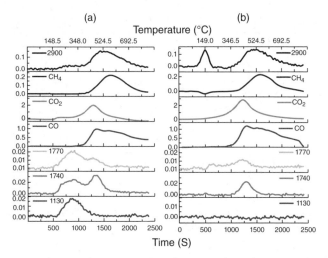

Figure 6.18 *Extracted traces from TGA-IR analysis at selected wavenumbers (cm⁻¹) for (a) hydrothermal carbon synthesized at 180 °C, and (b) hydrothermal carbon synthesized at 180 °C after extraction with ethanol, hexane, and THF. The panels indicate either the name of the gas, the trace can be attributed to, or the wavenumber the gas is detected at. (Reproduced with permission from [3]. © 2012 American Chemical Society.)*

of condensed aromatic intermediates. Both these processes lead to the development of voids (i.e., pores) between the isotropic pregraphinic structures present at this stage. We will return to the issue of porosity in hydrothermal carbons in Section 6.10.

6.9   Structure Elucidation of Carbon Materials Using Solid-State NMR Spectroscopy

Here, we want to show how the proper manipulation of advanced solid-state NMR tools can provide access to a more detailed understanding of carbon-based materials chemical structure, with a particular focus on hydrothermal carbons. As a first step, we provide a short summary of the principal NMR spectroscopic tools and their usefulness. Then, we introduce how solid-state NMR was used in the structural characterization of various families of carbon-based materials Finally, a description of how solid-state NMR was particularly useful in determining the structure of various HTC materials is provided.

6.9.1   Brief Introduction to Solid-State NMR

NMR is commonly associated to the fields of organic chemistry and structural biology, where all analysis are generally performed in liquid solutions. On the contrary, NMR is less known as a routine technique for solid-state matter studies. The absence of Brownian motion, which generally averages out specific magnetic interactions in solution (e.g., dipolar coupling), allows their observation in the solid state, leading to a significant loss of signal intensity and resolution if well-adapted technical countermeasures are not employed. Two of the most relevant interactions to deal with in the solid state are chemical shift anisotropy (CSA), which is directly related to the chemical environment of the nuclei, and dipolar coupling, which is a through-space interaction and directly related to the internuclear distance. These interactions may be extremely strong (up to several hundreds of megahertz) and have a dramatic effect on the spectral resolution. For an overview on NMR, see Levitt [47], while for more insights on solid-state NMR, see Schmidt-Rohr and Spiess [48]. A fair compromise between a broader explanation and shorter reading of basic NMR principles and their application in chemistry can be found in Andrew and Szczesniak [49], Blanc *et al.* [50] and, in particular, Laws *et al.* [51]. We now provide a brief description of the main interactions in 1/2-spin systems and of the main tools used in order to improve spectral resolution necessary for structural characterization.

6.9.1.1   *Relevant Interactions in Solid-State NMR*

CSA is responsible for the large broadening of NMR resonances in the solid state. It can be efficiently removed by spinning the sample holder (commonly a zirconia rotor) around its axis at a 54.74° angle to the external magnetic field. This is generally addressed as MAS and it only refers to a mechanical treatment of the sample. Special probes, whose technology is much different from solution NMR probes, need to be employed.

Dipolar coupling depends on the internuclear distance (around $1/r^3$) and it can be a source of extreme line broadening for rigid solids. Two main ways exist to average the dipolar interaction and recover acceptable resolution: via MAS and/or via use of specific radiofrequency pulse sequences. In both cases, the characteristic frequency associated to MAS or radiofrequency pulses must be larger than the characteristic frequency of the inter-action, where homonuclear ^{1}H–^{1}H dipolar coupling is by far the strongest interaction in 1/2-spin solid systems (up to 100 kHz and depending on internuclear distance). Despite its many drawbacks, motivating the development of several tools to achieve its elimination, dipolar interaction can actually be controlled and wisely exploited for a number of infor-mative experiments, which produce valuable pieces of structural information. For instance, CPI is a dipolar coupling-related technique that consists of transferring the magnetization between abundant (I) and dilute (S) nuclear spins close in space. Together with MAS, CP is routinely used to enhance the sensitivity of rare, low-γ, (S) nuclei (e.g., ^{13}C, ^{29}Si) using, most commonly, the magnetization transfer from abundant nuclei, like ^{1}H (Figure 6.19). This has several benefits: (i) long spin-lattice relaxation times (T_1) are lowered, thus reducing the overall acquisition times of the experiment, (ii) sensitivity is increased, (iii)

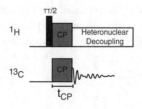

Figure 6.19   *Classical pulse sequence for the basic CP-NMR experiment where S = ^{13}C and I = ^{1}H. t_{CP} is a variable parameter that controls the magnetization transfer between I and S and its manipulation allows us to probe I–S through-space proximities.*

the characteristic time of the S–I interaction can be manually tuned via the adjustment of the cross-polarization contact time (t_{CP}), and (iv) valuable information on the structure and chemistry of the sample (sensitivity to a protic environment, molecular mobility) can be extracted by manipulating the t_{CP} time. Additional dipolar coupling-based experiments can be set to specifically address the general problem of structural resolution in solids, particularly amorphous solids. For instance, excitation of multiple quantum coherences is a more and more developed way to let spins interact and consequently obtain unique information on through-space coupled spin pairs – most of the times via easy-to-read two-dimensional homonuclear and heteronuclear correlation maps.

6.9.2   Solid-State NMR of Crystalline Nanocarbons: Fullerenes and Nanotubes

6.9.2.1   *Fullerenes*

Fullerenes, discovered in 1985 [52], are a family of carbon allotropes, molecules composed entirely of carbon, in the form of a hollow sphere, ellipsoid, tube, or plane. Spherical fullerenes are also called buckyballs and cylindrical fullerenes are called CNTs or buckytubes. Fullerenes are similar in structure to graphite, which is composed of stacked sheets of linked hexagonal rings, but may also contain pentagonal (or sometimes heptagonal) rings that would prevent a sheet from being planar. Applications vary from medicinal use to heat-resistant devices and superconductivity.

^{13}C solid-state NMR studies have been performed on fullerenes since the early 1990s, and confirmed the chemical homogeneity of the 60 carbon atoms for the C_{60} [53] and the expected inhomogeneity for the C_{70} material [54]. Initial structural studies (bond length calculation, molecular motion) [55–57] including spin relaxation dynamics of C_{60} under different external conditions (e.g., pressure, temperature) [58, 59] were followed by more detailed investigations on the interactions between fullerene and intercalation compounds, focusing on molecular mobility and van der Waals interactions [60, 61]. Finally, recent works directed more efforts to the understanding of molecular entrapping within fullerene cages [54]. Recent review papers [62, 63] have already shown some of these aspects, and for this reason we limited ourselves here to a short and broad description.

Most of the structural studies have been performed using ^{13}C NMR under both static and MAS conditions. Due to the high molecular mobility of the fullerene C_{60} cage in solid state at ambient conditions, static NMR is largely sufficient to evidence the characteristic isotropic peak at 143 ppm. At low temperature, on the contrary, part of the CSA is reintroduced, as expected, but a small fraction of a mobile phase is kept at temperatures as low as 100 K [64]. Spin lattice T_1 relaxation times have been largely investigated under different conditions. The first study proposed by Tycko [58] revealed discontinuous values of T_1 as a function of temperature. This is due to a phase transition from the face centered cubic to simple cubic phase, which was already seen from differential calorimetry and X-ray powder diffraction experiments at about 250 K.

6.9.2.2   CNTs

CNTs are allotropes of carbon and members of the fullerene structural family having a diameter of few nanometers, while they can be up to several millimeters in length. Nanotubes, categorized as either single-walled and multiwalled, are entirely composed of C-sp^2 bonds, similar to those of graphite, providing the molecules with their unique strength. Under high pressure, nanotubes can merge together, trading some sp^2 bonds for sp^3 bonds, giving the possibility of producing strong, "unlimited-length" wires through high-pressure nanotube linking. These cylindrical carbon molecules exhibit extraordinary strength and unique electrical properties, and are efficient heat conductors, which makes them potentially useful in many applications in nanotechnology, electronics, optics, and other fields of materials science, as well as having potential uses in structural materials.

Until the work of Tang *et al.* [65], solid-state NMR of nanotubes was very challenging due to some intrinsic problems in the production process, allowing relatively small and polluted (with paramagnetic species from metal catalysts) sample amounts of CNTs [62]. Initial data reported static and MAS-NMR spectra. The former highlighted the nonisotropic and nonplanar behaviors of the chemical shift tensor, while the latter showed a single, multicomposite, peak centered at 124 ppm, suggesting a metallic and semiconducting character of the material. Confirmation for the existence of the electron-conducting behavior is also provided by the linear relationship between the spin lattice T_1 relaxation time and temperature, as described by the Korringa relationship [66]. After this pioneering study, several others started to appear and focused their interest towards a better characterization of the magnetic properties of the CNTs as a result of their metallic behavior. ^{13}C NMR both under static and MAS conditions and T_1 analysis constitute the main tools for investigating the precise nature of the metallic and semiconducting properties of CNTs. More details on this topic have been already reviewed [66]

Functionalization of single-walled nanotubes also constitutes also a domain in which solid-state NMR was successfully employed. A number of studies report on oxidation [67], fluorination [68], protonation [69], and grafting of large polymeric moieties [70] on the surface of CNTs, but the complete potential of NMR has not been fully exploited yet, due to the lack of protons, which generally prevents the use of CP-based techniques. Engtrakul *et al.* [69] used ^{1}H–^{13}C CP to show protonation of CNTs after a liquid (sulfuric acid) and solid-state (sulfonated polymers, Nafion, and AQ-55 were used) acidic treatment. Reversibility was proved after a second treatment under basic conditions. Cahill *et al.* [70] performed a nice study on poly(methyl methacrylate) (PMMA)-functionalized CNTs in which both ^{1}H and ^{13}C nuclei were studied using homonuclear and heteronuclear correlation experiments. Figure 6.20 shows the ^{1}H–^{13}C two-dimensional correlation map of PMMA-CNT and recorded using the dipolar coupling-based TEDOR (transferred echo double resonance) pulse sequence [70]. The low intensity cross peak centered at $\delta(^{13}$C) = 121 ppm and $\delta(^1$H) = 0.5 ppm shows the existence of a spatial proximity between the aliphatic protons of PMMA and the nanotube ^{13}C signal; even if this does not prove the direct functionalization between PMMA and CNTs, it strongly suggests that part of the PMMA is very close to the surface of the CNTs.

6.9.3   Solid-State NMR Study of Biomass Derivatives and their Pyrolyzed Carbons

The structural resolution of complex, carbon-rich, biopolymers and biopolymer-derived chars is an ongoing task, which has always represented a major challenge for solid-state NMR analysis. First of all, these materials are most of the time noncrystalline, they are characterized by many similar chemical sites, inducing a dispersion in chemical shift values

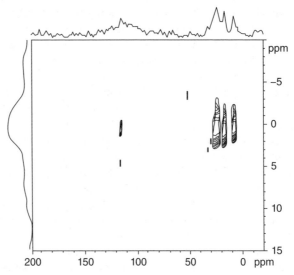

Figure 6.20 *^{1}H–^{13}C two-dimensional correlation spectra of PMMA-NT. (Reproduced with permission from [70]. © 2004 American Chemical Society.)*

upon NMR analysis, and, very importantly, some may suffer from lack of protons limiting the use of CP. Last, but not least, paramagnetic centers can occasionally exist and, in some cases, they contribute to broadening of the NMR signal. HTC shares most of these features with classical pyrolyzed chars and a comparison between them is herein presented. Below, we briefly focus on lignin (a widespread phenylpropanoid-based biopolymer), cellulose (a crystalline D-glucose-based polysaccharide), and their derived chars.

6.9.3.1   Lignin and its Derivatives

Lignification [71] is the polymerization process in plant cell walls transforming phenolic monomers into radicals and coupling them with other monomer radicals (only during initiation reactions) or, more typically, cross-coupling them with the growing lignin polymer/oligomer to build up a phenylpropanoid polymer [72]. Even though extensive research efforts have been made to elucidate the finer structural details of the highly complex polyaromatic lignin, a definitive model does not yet exist. This task is further complicated by the wide natural variation in lignin structure, with the main difficulties arising during characterization due to the high level of chemical and structural heterogeneity of its bonding patterns. Nonetheless, the lignin polyphenolic nature has been ascertained, and the most abundant constituent monomers characterized as *p*-coumaryl, coniferyl, and synapyl alcohols [73].

Solution NMR spectroscopy is a proven tool in the analysis of lignin, but most of the time the biopolymer and its derivatives must be fractioned into model compounds, which are eventually isolated and accurately characterized by several liquid-state NMR techniques. In particular, two-dimensional ^{13}C–^{1}H correlated (heteronuclear multiple quantum coherence and heteronuclear single quantum coherence) spectroscopy continues to be the method of choice to identify unambiguously the different lignin units and the subunit bonding patterns [74, 75]. The entire lignin fraction can also be analyzed in the so-called "cellulolytic enzyme lignin," in which large fractions of the polysaccharides are removed by enzymatic (cellulases) digestion of crude wood, for instance [64, 76]. Of course, isolation or fractionation may cause significant modification of the original structure, yielding unrepresentative final results. This is not the case if solid-state NMR spectroscopy is used instead; however, the analysis becomes more challenging. Interesting ^{13}C-enrichment techniques have been developed to allow a direct study of protolignin in the cell walls [77, 78]. Selective ^{13}C enrichment can be obtained by using ^{13}C-enriched compounds (e.g., monolignol glucosides, ferulic acid, phenylalanine) in seedlings cultures, and tissue-cultured cells, for instance. The achievement of selective ^{13}C enrichment at a specific carbon has been confirmed by solid-state NMR [79] and structural studies were proposed, for instance, by Terashima *et al.* [80]. They evaluated specific alkyl–alkyl and alkyl–aryl ether linkages on enriched wheat straw via ^{13}C CP/MAS experiments. Evaluation of alkyl–aryl ethers has also been a matter of debate in both lignin and lignin-derived polymers. In lignins, for instance, their amount has evolved from 63% to 80% and eventually to 74% [81, 82]. On the contrary, type III kerogens, which are lignin geo-derived coals, were shown not to contain significant amounts of alkyl–aryl ethers by means of chemical shift analysis of the corresponding ^{13}C solid-state NMR spectra and density functional theory (DFT) calculations [83]. These pieces of information are extremely helpful in the study of structural evolution during the lignin coalification process and further works using CP/MAS have contributed to the study of the structure of natural coals, particularly the problem of signal attribution to aromatic and aliphatic species [84].

6.9.3.2 Cellulose and Cellulose-Derived Char

Cellulose is a polysaccharide consisting of a linear chain of $\beta(1 \rightarrow 4)$ linked D-glucose units (which differ from the $\alpha(1 \rightarrow 4)$-glycosidic bonds in starch), and is the structural component of the primary cell wall of green plants, many forms of algae, and the oomycetes as well as a secretion product of some species of bacteria (e.g., *Gluconacetobacter xylinus*). Cellulose is a straight-chain polymer where no coiling or branching occurs, with an extended and stiff rod-like conformation. The multiple hydroxyl groups on the glucose monomers form hydrogen bonds with oxygen atoms on the same or on a neighboring chain, holding them firmly together side-by-side and forming microfibrils with high tensile strength. From a structural point of view, native cellulose is a crystalline solid with two allomorphs, I_α and I_β, where the former is the metastable, low-density form while the latter is the thermodynamically most stable, high-density form. ^{13}C solid-state CP/MAS NMR has been crucial in the discovery and identifications of I_α and I_β [85, 86]. In terms of relative abundance, I_α and I_β are generally found in differing mixtures and the proportions depend on the origin of the cellulose biopolymer. For instance, *Valonia* and bacterial cellulose are rich in I_α, while animal cellulose contains more I_β structure [87, 88].

Upon pyrolysis, cellulose undergoes thermal decomposition leading to the elimination of small volatile species (e.g., CO_2, CO, CH_4, H_2O) and condensation reactions that produce a complex polyaromatic network, commonly referred to as char. Early FTIR studies can be traced back to the 1960s [89], while solid-state ^{13}C NMR started to be employed much later [90, 91]. However, its use has become rapidly widespread and actually necessary to complement FTIR data, so that up-to-date models of the early stages of the carbonization process could be proposed [92]. In 1994, Pastorova *et al.* [90] presented one of the first studies, where ^{13}C NMR was combined with GC-MS and FTIR to elucidate the structure of char obtained from pyrolyzing cellulose between 250 and 400 °C for 150 min. The authors showed that cellulose keeps its initial structure up to 250 °C, while major chemical modifications occur at 270 °C; both phenolic and furanyl groups were detected as volatile compounds. The idea that the structure of char from cellulose or other biopolymers (pectins, wood) is mainly constituted of an aromatic motif including furanoic compounds connected via aliphatic bridges is generally accepted [93, 94]. For instance, Zhang *et al.* [93] suggested the presence of furfuryl motifs in char obtained from pyrolyzed starch. Nevertheless, in most studies, probably due to the lack of a clear-cut proof, the structures were rather interpreted as being composed of polyaromatic hydrocarbons [84, 92, 95–97], as in lignins or coal [84, 98]. The mechanism of formation of char and the fate of the polysaccharide network at medium/high pyrolysis temperatures (below 500 °C) has also attracted much attention and the elucidation was mainly possible via a fine NMR study. Wooten *et al.* [92] have shown that, after 30 min at 300 °C, cellulose undergoes depolymerization to form an "intermediate cellulose" product, which then transforms into a "final carbohydrate" before aromatization and which was associated with large amount of oligo- and polysaccharides.

6.9.4   Solid-State NMR Study of Hydrothermal Carbons

The degree of structural complexity in hydrothermal carbons is very similar to the that found in coals, lignins and related derivatives. Elemental analysis indicates their predominantly carbonaceous nature (C% > 60 wt%). In the context of HTC carbon characterization, FTIR (Section 6.4), and XPS (Section 6.5) are of poor resolution, whilst the absence of diffraction peaks in XRD (Figure 6.15) indicates that the material is amorphous [3]. Even if a basic study, using solid-state NMR in a similar way to the studies on coal mentioned above (i.e., employing only MAS and CP), can probably be sufficient to obtain a preliminary analysis of the HTC carbon chemical structure, it does not provide a definitive model as in the case of lignins, kerogens, or any other biopolymer-derived char. To go beyond, a different analytical strategy has been employed. Here, we show how adopting advanced ^{13}C solid-state NMR techniques combined with isotopic enrichment contributes to a detailed understanding of the HTC carbons' chemical structure and help draw some conclusions on their formation mechanism as well.

Two structural models have been proposed in literature to interpret FTIR, Raman, XPS, and even preliminary standard ^{13}C solid-state NMR spectroscopic results [25, 99–102]. As shown in Figure 6.21, the carbonaceous scaffold has been considered as being composed of "small clusters of condensed benzene rings forming stable groups with oxygen in the core" [99] or condensed polyaromatic structures [25]. As we can notice, both structural models show great similarities.

Even if these models could satisfactorily interpret a combination of experimental data (i.e., FTIR, XPS, and XRD), they cannot explain the findings highlighted by ^{13}C solid-state NMR experiments and presented in the following sections. In particular, NMR has the possibility and advantage of using selective pulse sequences to identify and separate unresolved chemical species without ambiguity, which can also lead towards better quantitative analysis, as shown hereafter.

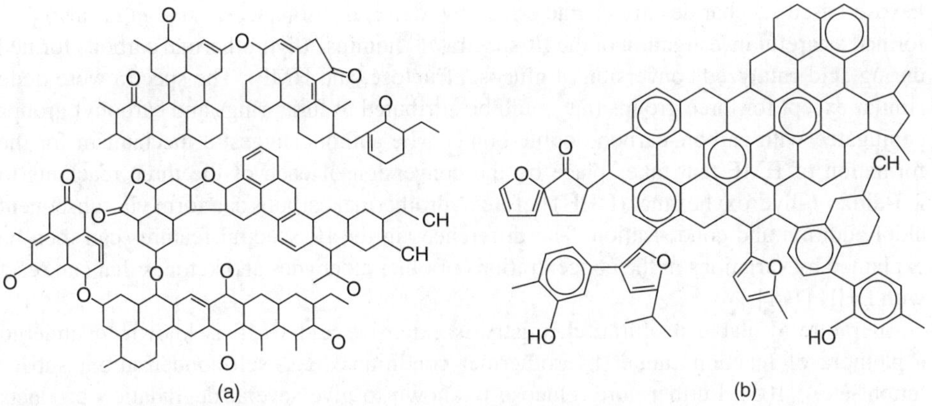

(a)   (b)

Figure 6.21 *Existing structural models proposed for hydrothermal carbons. ((a) Adapted with permission from [99]. © 2009 WILEY-VCH Verlag GmbH & Co. KGaA, Weinheim; (b) adapted with permission from [25]. © 2009 American Chemical Society.)*

6.9.4.1 Carbohydrate-Derived Hydrothermal Carbons

As previously mentioned (see Section 6.2), the formation of HTC carbons was proposed to take place via dehydration of the carbohydrate precursors to furanic species (i.e., HMF, furfural), the parent intermediates of HTC carbon [12]. The first, very important, question to which we must provide an answer is to establish a relationship between the type of carbohydrates, their complexity, and the final HTC carbon structure. A preliminary ^{13}C solid-state NMR study on HTC carbons derived from different mono- and polysaccharides (i.e., fructose, glucose, starch, xylose) highlighted that the main factor affecting the chemical nature of the HTC product is the structure of the parent sugar [29, 101, 103]. Pentose (e.g., xylose)-derived HTC carbons possess a more marked aromatic character than hexoses (e.g., glucose). Such a difference is demonstrated by a more intense peak at $\delta = 125$–129 ppm in the ^{13}C CP/MAS solid-state NMR spectrum in the former case, which is characteristic of aromatic carbons belonging to graphitic or long-range conjugated double bonds structures. Different HTC reaction intermediates may be the explanation for such a finding. Pentose sugars are expected to be dehydrated to furfural when hydrothermally treated. On the other hand, HMF is the main dehydration product derived from hexoses. As reported in literature, the reactivity of these two intermediates is indeed different; as a consequence, it also reflects in the chemical structure of the respective HTC carbons [104]. Further GC-MS and solution ^{13}C NMR experiments on the glucose system confirmed that the major intermediate in the reaction mixture is HMF (Figure 6.4). This finding, coupled to the evidence that the ^{13}C solid-state NMR spectrum of HMF-derived HTC carbon is very similar to all HTC spectra obtained from different types of mono- and hexose polysaccharides, led to the conclusion that polymerization/condensation reactions involving HMF monomers are the route of formation of HTC carbon. Furthermore, by a simple comparison of ^{13}C CP/MAS NMR spectra of various HTCs, it was also observed that the degree of initial polymerization of the hexose-based saccharides (mono-, di-, or polysaccharides) does not influence the final structure, since all the ^{13}C spectra of HTC carbons derived from hexose-based saccharides are characterized by identical resonances. Lund *et al.* also performed a careful investigation of the IR spectra of "humins" (hydrothermal carbons formed during acid-catalyzed conversion of glucose, fructose, and HMF). The spectra were quite similar except for three groups that could be attributed to furan rings and carbonyl groups conjugated with carbon–carbon double bonds. The authors suggest a mechanism for the formation of HMF that takes place by the conversion of each of the three reactants to 2,5-dioxo-6-hydroxyhexanal (DHH) before hydrothermal carbons can form via subsequent aldol addition and condensation. The differences in the IR spectral features can then be explained by variations in the concentrations of other aldehydes and ketones that can react with DHH [101].

Literature available on furan chemistry is extensive and HMF is known to undergo a plethora of reactions under hydrothermal conditions (e.g., self-condensation, substitution, etc.) [105]. Furthermore, glucose is known to give several degradation products (e.g., formic acid, levulinic acid) that are present within the reaction mixture and may at the same time react with HMF units (e.g., aldol condensation) as suggested by Patil *et al.* [19]

For these reasons, despite the evidence that HMF is involved in the formation of hydrothermal carbons, a complete proof regarding the incorporation of this compound within the HTC carbon structure still needs to be provided, fostering the development of further studies. In this sense, NMR is an ideal technique. The use of fully labeled [^{13}C]glucose in the initial reaction batch allows obtaining an isotopically enriched HTC carbon, which is the ideal starting point for a thorough NMR-based analysis.

Figure 6.22 shows several ^{13}C NMR spectra of the same isotopically enriched glucose-derived hydrothermal carbons sample; differences in relative intensities depend on the type of pulse sequence employed (mentioned next to each spectrum), which act as spectroscopic filters:

- *SP*: this spectrum is quantitative. By comparing the relative intensities of each (deconvoluted) peak, it allows a precise estimate of the relative amount of each chemical group corresponding to a known chemical shift value.
- *CP*: this spectrum was acquired using a ^{13}C–^{1}H-CP-filtered pulse sequence at a contact time of $t_{CP} = 3$ ms. Interestingly, its envelope is very close to the SP spectrum, which indicates that the magnetization transfer from protons towards carbons involves all chemical groups at this specific t_{CP}. As a partial conclusion, one can say that the material is macroscopically homogeneous from a chemical point of view.
- *INEPT*: this typical liquid-state NMR technique explores through-bond *J*-coupling interactions between heteroatoms (e.g., ^{1}H and ^{13}C). In this case, this pulse sequence allowed filtering out from the spectrum all nonprotonated carbons, thus leaving carbon atoms with direct C—H bonds.
- *DQ-SQ*: the DQ-SQ one-dimensional spectrum presented here is extracted from the two-dimensional experiment discussed later. In this case, the very good resolution of the two-dimensional experiment allows us to identify, in the one-dimensional spectral projection here, a larger number of peaks (A–M) corresponding to specific chemical groups, whose abundance might be quite low and for this reason not detected in none of the previously commented spectra.

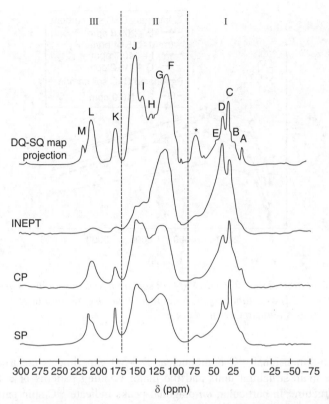

Figure 6.22 ^{13}C *MAS NMR spectra of HTC obtained from glucose. SP = single-pulse exper-*
iment; CP = cross-polarization experiment; INEPT = insensitive nuclei enhanced by polar-
ization transfer experiment; DQ-SQ = double quantum/single quantum, projection of the
two-dimensional experiment presented in Figure 6.24. (Reproduced with permission from
[103]. © *2009 American Chemical Society.)*

Analyzing these results, it is evident that each NMR spectrum can be divided in three separate regions: region I, identifying the aliphatic region (0–80 ppm), region II, the aromatic
region (100–160 ppm) characteristic of sp^2 hybridized aromatic carbons, where oxygen-
bound sp^2 carbons are detected, and region III (175–225 ppm) typical of carbonyls, such
as carboxylic acids, ketones, aldehydes, and esters.

Using the experimental evidence extracted from the spectra in Figure 6.22, it is possible
to assign each signal to a specific functional group. Variations of t_{CP} values in CP and
CP-derived experiments can be very helpful to accomplish this task. For instance, in Figure
6.23 we show the variation of the intensity of selected peaks (F, G, I, J, and L in regions I
and III of Figure 6.22) as a function of t_{CP}. All of them, except peak F at 110 ppm, have
a typical magnetization building profile in CP, typical of nonprotonated carbon groups.
Similarly, inversion recovery CP experiments (refer to [103] for more details) are able to
discriminate between CH, CH_2, and CH_3 carbons in region I, constituted by the broad
aliphatic skeleton and that would otherwise be impossible to assign.

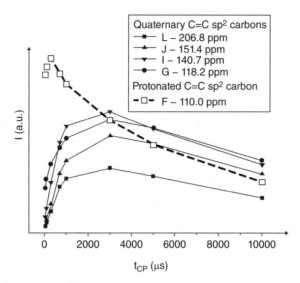

Figure 6.23 *Evolution of the ^{13}C CP/MAS NMR spectra intensities for the peaks F, G, I, J, and L (refer to Figure 6.22) as a function of the contact time, t_{CP}. (Reproduced with permission from [103]. © 2009 American Chemical Society.)*

Finally, two-dimensional ^{13}C homonuclear DQ-SQ experiments are a powerful tool to deduce the main structural units and the major bonding patterns of a carbonaceous framework structure. In particular, *on*-diagonal peaks indicate ^{13}C spin pairs, belonging to equal chemical environments, while *off*-diagonal cross-peaks show the linkage between carbons that are present in different functional groups. The two-dimensional DQ-SQ map obtained from glucose-derived HTC carbon is reported in Figure 6.24 and its detailed explanation can be found elsewhere [103]. Here, a short summary version is provided. The analysis is based on the underlining principle that lines connecting cross-peaks indicate that the corresponding carbons are covalently bonded. This allows tracing of the raw skeleton of close carbon neighbors. Two important results are extracted from this experiment and clearly show how models given in Figure 6.21 are inconsistent with the NMR results:

- J–F cross-peak connection. These two peaks, whose intensities are equivalent, are strongly connected one with the other, indicating that the HTC carbon predominantly contains a O—C=C— type of bond. O—C= and C=C— moieties, respectively, correspond to J and F. Considering the discussion before, this is the main motif of furanic species, which are the main intermediates in the HTC formation.
- H on-diagonal peak. At this chemical shift one expects intracarbon interactions in conjugated aromatic rings, as one would observe in models given in Figure 6.21. The lack of an intense peak and, in particular, the lack of a clear on-diagonal cross-peak strongly suggest that aromatic arene-like rings are a relatively rare species in glucose-derived HTC carbon at 180 °C processing temperature.

Finally, the combination of these two points suggests that furan moieties are the major constituent of HTC carbon materials obtained from glucose and, considering the

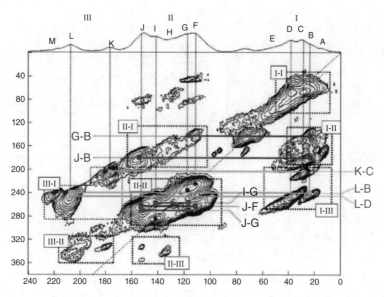

Figure 6.24   *Two-dimensional $^{13}C–^{13}C$ DQ-SQ MAS NMR correlation spectrum for HTC obtained from glucose. (Reproduced with permission from [103]. © 2009 American Chemical Society.)*

work detailed in elsewhere [29], all hexoses excluding cellulose, as presented in the following section.

6.9.4.2   Cellulose/Biomass-Derived Hydrothermal Carbons

All hexose-derived HTC materials show strong similarities in their ^{13}C NMR signature, as reported elsewhere [29]. Here, we show that this is not the case for cellulose- and lignocellulosic biomass-derived HTC carbons. Temperature-dependent ^{13}C CP/MAS NMR experiments performed on cellulose-derived HTC carbons are shown in Figure 6.25. At $T = 180$ °C, cellulose is still unaffected by hydrothermal treatment, since its characteristic resonances (i.e., $\delta = 65, 72, 75, 84, 89,$ and 105 ppm) are still present and well-resolved, while no resonances are observed in the aromatic region, indicating no relevant HTC carbon formation. At higher temperatures, dramatic differences are observed: all characteristic cellulose resonances disappear, while new resonances emerge. For instance, at $T = 200$ °C, the spectrum can easily be divided into three regions, as shown in Figure 6.22. Nevertheless, a major difference occurs. A strong resonance in the 120–130 ppm region is now observed for the samples synthesized in the temperature interval $T = 200–280$ °C, while the same peak is not observed in a pure glucose-based HTC below 200 °C [32]. Contrarily, at higher temperatures, evolution of the relative intensities is similar for both systems. These findings highlight the fact that:

- Cellulose-derived HTC carbons contain a higher amount of aromatic arene-like groups than other hexose-derived carbons. This is probably due to the higher temperature needed to degrade cellulose.

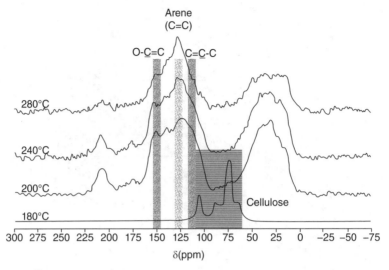

Figure 6.25 *^{13}C CP/MAS NMR spectra recorded on hydrothermal carbon obtained from cellulose and prepared as a function of temperature. (Reproduced with permission from [32]. © 2011 Royal Society of Chemistry.)*

- The identical peak evolution patterns for temperature values higher than 200 °C indicates similar chemical transformations in both systems, thus indicating that the HTC processing of both precursors is characterized by similar reaction pathways beyond this temperature threshold.

^{13}C NMR analysis of cellulose-derived HTC carbons synthesized at different reaction times confirmed the findings previously highlighted [32]. All HTC samples obtained from cellulose are characterized by the presence of the central resonance at $\delta = 125$–129 ppm. This feature is present from the early stages of the reaction contrarily to what was observed for the treatment of glucose as a function of time at the same temperature. This finding suggests that the HTC of cellulose does not proceed solely through a furane-composed intermediate (i.e., HMF), as observed in the case of the model monosaccharide (i.e., glucose) and other hexoses. The major conversion mechanism is instead thought to be the direct transformation of the cellulosic substrate into a final carbonaceous material composed of polyaromatic arene-like networks, presumably involving reactions that are normally characteristic of the pyrolysis process. During cellulose pyrolysis, char formation is attributed to numerous reactions leading to cellulose intramolecular rearrangement and formation of a cellulose-derived polymeric compound, referred to as *intermediate cellulose*. This reaction intermediate then converts to aromatic network structures at extended reaction times [90, 92, 106, 107]. This mechanistic speculation is well supported by the similar ^{13}C NMR profiles of cellulose-derived HTC carbon and char obtained from lignocellulosic biomass pyrolysis [97, 108].

According to ^{13}C NMR data, the chemical structure of the HTC carbon obtained from rye straw is practically identical to that obtained from cellulose. The only difference is the presence of two additional peaks (56 and 145–148 ppm) in the case of the lignocellulosic biomass. ^{13}C CP NMR investigations of an Alcell$^©$ lignin sample have shown that these peaks can be, respectively, assigned to the methoxy functional groups (—OCH$_3$) and to aromatic carbons bound to such moieties [97] that are present in the lignin fraction of the biomass (see also Section 6.9.3.1). This means that if not removed before (i.e., by steam explosion [109, 110] or alkaline treatment [111]) the lignin fraction will be preserved during the hydrothermal treatment [32].

We also need to mention here that cellulose can be converted into hydrothermal carbon at lower temperatures (i.e., 180 °C) in the presence of strong acid catalysts. Thus, in order to enhance the extent of cellulose hydrolysis to glucose at milder temperature values (i.e., 180 °C), HTC experiments in the presence of strong mineral acids (i.e., H$_2$SO$_4$) were performed (Figure 6.26). The ^{13}C NMR spectra obtained show that at relatively low pH (i.e., 1), cellulose fully reacts at 180 °C. Nonetheless resonances corresponding to furanic moieties are not the dominant features of the NMR spectra, as would be expected at this processing temperature value. The NMR spectrum for cellulose at pH = 1 shows a well-developed polyaromatic arene-like carbon resonance with a major shoulder at δ = 151 ppm, indicative of oxygenated sp^2 hybridized carbon environments. These features also characterize the NMR spectra of glucose-derived HTC carbon prepared at different pH. As a consequence, these results demonstrate the dependence of the chemical structure of HTC carbon upon the initial reaction mixture pH. Strong acidic solutions lead preferentially to the development of polyaromatic carbon species probably due to a higher extent of formation of furan degradation products (e.g., levulinic acid [112]) reacting via aldol condensation and increasing the degree of conjugation of the carbon framework [19, 104].

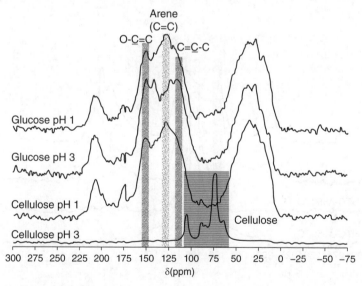

Figure 6.26 *^{13}C CP/MAS NMR spectra recorded on hydrothermal carbon obtained from glucose and cellulose at different initial pHs.*

6.9.4.3   Comparison between Hydrothermal Carbons and Pyrolyzed Char

The comparison between biomass-derived chars obtained from direct pyrolysis and the HTC process needs to be addressed in order to place HTC within the context of thermochemical processing techniques. The following section shows some elements concerning pyrolysis versus hydrothermal treatment; in particular, we show that the main difference between their carbonaceous products can be quantified using the furan-to-arene ratio calculated for each sample, as presented in detail elsewhere [113]

As described previously, pyrolysis has several stages (i.e., thermal decomposition, charring, graphitization). The following section predominantly focuses on low-temperature thermal treatment ($T < 800\,°C$) of both raw glucose and glucose-derived HTC. First of all, upon pyrolysis, the carbon content of HTC products increases from 60% to about 90%, as expected. Figure 6.27 shows the ^{13}C CP/MAS NMR of glucose-derived HTC carbon before and after pyrolysis. Even if major differences between HTC and HTC-Δ 350 °C (hydrothermal carbon further calcined at 350 °C) do not seem to occur, the aromatic regions ($\delta = 105–155$ ppm) of these carbons do differ significantly from each other. HTC-Δ 350 °C presents an enhanced level of arene groups, demonstrated by an increased intensity of the central resonance at $\delta = 125–129$ ppm and partial loss of the furanic shoulder at $\delta = 110–118$ ppm, as compared to the parent HTC material. Likewise, increasing the pyrolysis temperature from 350 to 550 °C results in the disappearance of furanic-associated resonances, as polyaromatic arene-like species become the most dominant structural motif. Increasing the temperature further to 750 °C leads to enhanced aromatization of the HTC carbon, as indicated by the drifting of the central aromatic resonance to a lower chemical shift ($\delta = 125$ ppm) and by the broadening of its profile. Both features are indicative of an extended delocalized π-system and of a reduced mobility of the carbon species [114]. To

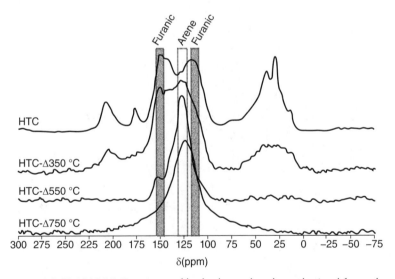

Figure 6.27   *13C SP MAS NMR spectra of hydrothermal carbon obtained from glucose and further calcined at 350, 550, and 750 °C. (Reproduced with permission from [113]. © 2011 American Chemical Society.)*

provide more details on such structures, two-dimensional DQ-SQ ^{13}C NMR experiments were performed to investigate the structural changes during pyrolysis of hydrothermal carbon and thus precisely attribute the signals in spectra given in Figure 6.27 [113]. These experiments allow a more precise interpretation of the ^{13}C signal in the aromatic regions between 100 and 150 ppm. In particular, whilst as-synthesized HTC carbon (Figure 6.24) shows large amounts of furanic groups (J–F cross-peaks in region II–II), after treatment at 350 °C, the on-diagonal H–H cross-peak at $\delta = 127$ becomes remarkable. This is a clear sign of the formation of more extended arene-like polyaromatic domains throughout the HTC carbon scaffold. Such structures (for which models given in Figure 6.21 now become more realistic) may arise due to condensation reactions, leading to the formation of a more cross-linked and thermally stable aromatic network. Similar trends have been observed during the pyrolysis of polyfurfuryl alcohol (PFA) derived furan-rich resins [93, 115]. At higher temperatures, the peak at $\delta = 127$ ppm becomes more and more intense, thus indicating that the whole material is composed of condensed aromatic rings, as expected at these temperatures.

This ^{13}C CP/MAS NMR analysis of pyrolyzed HTC samples emphasizes how examination of the aromatic resonances provides a very useful analytical handle on the development of the carbon structure as a function of pyrolysis temperature. A further comparison between low-temperature (350 °C) pyrolyzed hexose-based carbohydrates and pyrolyzed HTC carbon shows that the aromatic region (i.e., $\delta = 100$–150 ppm) of the former samples has a slightly different profile than that of pyrolyzed HTC carbon. A detailed analysis suggests that this difference arises mainly from the lack of the $\delta = 118$–110 ppm shoulder for the directly pyrolyzed samples, indicating the possible absence of furan groups. It can, however, be objected that if this was really the case, then both spectra should not show the resonance at $\delta = 148$–151 ppm, which is instead still present. Nevertheless, such a peak can also be attributed to aromatic carbons of oxygen-substituted arene-type moieties, which form during the pyrolysis process [113]. The results of the comparison between carbonaceous materials obtained from pyrolysis and glucose hydrothermal carbon strongly highlight that the hydrothermal treatment of carbohydrates offers an additional intermediate stage between the parent sugar and the polyaromatic char structure, obtained from pyrolysis. In Figure 6.28, we show a summary of the effect of hydrothermal and pyrolysis processes pathways on hexose-based carbohydrates, including cellulose. In particular, the typical ^{13}C CP/MAS NMR spectra are provided, since they can be considered as the fingerprint of the carbonaceous materials. Three main stages were identified: (A) furans are the major structural unit, (B) furans and arene groups are present within the structure, and (C) the whole material is composed of an extended aromatic network. The studies shown here underline how the furan-rich HTC carbon stage (A) can only be obtained via hydrothermal treatment of hexose-based carbohydrates, excluding cellulose.

6.9.4.4 Nitrogen-Containing Hydrothermal Carbons

The final properties of carbon materials depend to a large extent on the raw material, surface structure, and porosity, but also heteroatoms introduced into their structure may exert large effects on their physicochemical properties. Recently, nitrogen-containing carbons have attracted particular interest due to their improved performance in applications such as

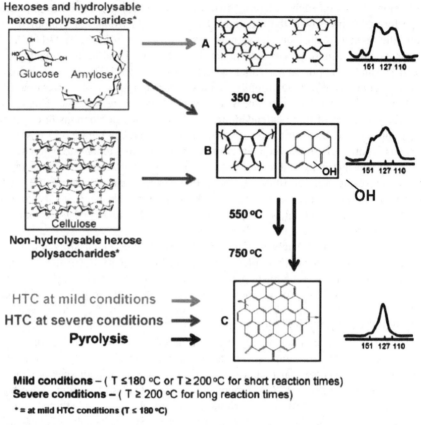

Figure 6.28 *Scheme summarizing the combined effect of hydrothermal treatment (A, B and C pathways) and pyrolysis (C pathway) on hexoses and cellulose. Typical ^{13}C NMR spectra are given on the right side. (Adapted with permission from [113]. © 2011 American Chemical Society.)*

CO_2 sequestration [116], removals of contaminants from gas and liquid phases [117], environmental protection [118], catalysts and catalysts supports [119], or in electrochemistry as supercapacitors [120], cells, and batteries [121]. The production of heteroatom-doped carbons using HTC was described in detail in Chapter 5. The methods for their production of such materials rely normally on very harsh and multistep processes, which involve high-temperature production of carbon materials followed by introduction of nitrogen into the structure using ammonia, amines, or urea [122]. Nitrogen-containing carbons have been also more readily prepared using precursors like acetonitrile, pyrrole, or polyacrylonitrile [123, 124]. However, even these precursors are less sustainable and available than carbohydrates or other biomass-derived raw materials. As described in Chapter 5, the HTC process can be used to produce, in a direct synthesis step, nitrogen-containing carbonaceous materials either from mixtures of carbohydrates and proteins [125, 126] or directly from nitrogen-containing compounds like chitosan or glucosamine [127]. The problem of structural characterization is still very important in modified carbons and in some cases it can be quite difficult to solve. For instance, nitrogen has two magnetically active isotopes, ^{14}N (spin $= 1$, abundance $= 99.6\%$) and ^{15}N (spin $= 1/2$, abundance $= 0.4\%$). The problem is that the most interesting isotope, ^{15}N (spin $= 1/2$), is also the least abundant (0.4%). For this reason, focusing on nitrogen NMR in studying nitrogen-containing compounds can be a problem if isotopic enrichment is not performed. In a recent study, we have shown that use of chicken ovalbumin in the presence of glucose induces the formation nitrogen-containing nanomaterials [125, 126]. In these works, ^{13}C CP/MAS was used as specific probe to study the effect of ovalbumin on the carbonaceous scaffold structure. Unfortunately, the low nitrogen amount (less than 10 wt%) did not allow an interesting characterization by ^{15}N NMR. For this reason, we made a similar material from glucose and a single ^{15}N-enriched amino acid, glycine, showing a ^{13}C NMR spectrum similar to the albumin-derived HTC [128].

Figure 6.29 shows several ^{15}N solid-state NMR of the glycine-derived hydrothermal carbons. Comparison of the SP (a) and CP/MAS (b) spectra shows several important features:

- Typical chemical shift analysis of the SP spectrum reveals the possible presence of various families of nitrogen-containing groups, as indicated (pyridine, pyrazine, pyrroles, amides).
- The amount of amines, initially introduced with glycine, is very low, indicating that this compound has reacted further.
- The poor efficiency of CP in the −75 ppm region suggests that protons are far from nonprotonated pyrazinic and pyridinic groups.

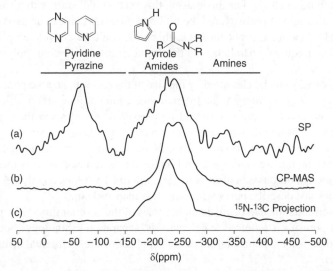

Figure 6.29 ^{15}N *SP and CP (a and b) MAS NMR spectra of hydrothermal carbons obtained from a mixture of ^{15}N-enriched glycine and glucose. (c) Projection of the two-dimensional ^{15}N–^{13}C heteronuclear correlation experiment reported elsewhere [128]. (Reproduced with permission from [128]. © 2011 American Chemical Society.)*

Even if these data are very informative, they are not sufficient to fully attribute the ^{15}N spectrum. For this reason, a heteronuclear ($^{15}N-^{13}C$) correlation spectrum has been obtained to derive a more detailed picture of the carbonaceous scaffold. The full experiment can be found elsewhere [106] while the one-dimensional projection is presented in Figure 6.29c, showing its consistency with the simple CP results. Additional $^{13}C-^{13}C$ DQ-SQ experiments on the same material were also performed in order to determine different chemical species constituting the nitrogen-doped hydrothermal carbons. The main derived information hinted to a mechanism involving a glucose reaction with the amino acid(s) via the well-known Maillard chemistry, which generally leads to melanoidin resins formations in which, on the contrary, a fraction of the amino acid is kept intact. However, under the employed hydrothermal conditions, it is not possible to precisely predict the cascade of chemical reactions taking place, which presumably encompasses a broad range of nitrogen-containing intermediates leading to the formation of nitrogen-containing aromatic species and to the disappearance of the original amino acid original structure. More insights into the Maillard chemistry and the possibilities in which nitrogen-doped HTC carbons can be formed using such chemistry were provided in detail in Chapter 5.

6.10   Porosity Analysis of Hydrothermal Carbons

6.10.1   Introduction and Definition of Porosity

Porous materials are available in a huge variety [129], such as carbons and soil, porous minerals, polymer foams, or wood, and scientists of different communities do not always understand each other easily. For this reason, this section will start with a short overview about the classification of porosity and the different methods that can be used to characterize porosity. HTC carbons, if not further carbonized, could be considered polymeric-like materials, and consequently parallels with the characterization of porous polymer networks will be drawn.

Generally, porosity can be classified by the size of the pores and by the pore accessibility (open versus closed). According to the International Union of Pure and Applied Chemistry recommendation, porous materials should be classified according to their pore sizes in the dry state [130]. Materials with pore sizes smaller than 2 nm are termed *microporous*, materials possessing pore sizes between 2 and 50 nm are *mesoporous*, and materials having pores larger than 50 nm are *macroporous*. Finally, the term nanoporous is frequently used for the description of porous thin films or materials derived by block copolymer approaches; however, it does just describe pores with sizes less than 100 nm.

Another discrimination can be made on the basis of the accessibility of the pores. While often an open porosity is required (e.g., catalysis, separation technology), there are cases where a closed porosity is beneficial for certain applications (e.g., thermal insulation, low-*k* materials, etc.).

Here, we wish to introduce the reader to the main techniques of porosity analysis. Two of the most common techniques will be discussed in more detail, namely gas adsorption and mercury intrusion. Whenever possible, the main points will be emphasized using examples based on research on hydrothermal carbons or carbons, in general.

6.10.2 Gas Physisorption

The measurement of gas adsorption/desorption isotherms is probably one of the most frequently used methods for the analysis of porous materials [129, 130]. It provides access to a number of important parameters, such as the specific surface area S, the pore volume V_p, pore size distribution (PSD), and porosity ϕ.

In a typical gas physisorption experiment, the amount of gas adsorbed on the surface is measured in dependence of the relative pressure p/p_0, where p_0 is the saturation pressure of the gas at a given temperature. Depending on the gas used, temperature, and pressure range, either monolayer formation, multilayer formation, and/or pore filling by condensation can be observed. The most widespread method is surely the analysis of nitrogen sorption isotherms measured at 77.36 K (i.e., the boiling point of nitrogen). This method allows the assessment of micro- and mesoporosity as well as the determination of the specific surface area. There are, however, some drawbacks in the analysis of very small micropores. CO_2 adsorption can be used to overcome such problems and is a commonly known technique in the analysis of microporous carbons [131–134].

A variety of methods have been developed to analyze and interpret gas adsorption isotherms. We will first focus on the "classic" methods. The use of newly developed methods, such as nonlinear density functional theory (NLDFT), will be introduced later. The section finishes with a discussion of the CO_2 adsorption behavior of (HTC) carbons and their potential in CO_2 separation.

6.10.2.1 Assessment of Specific Surface Area

The most common method for the calculation of specific surface areas is surely the use of the Brunauer, Emmett, and Teller (BET) model [135]. The model is mostly used for nitrogen adsorption, but is also applicable for other gases such as argon, krypton, or CO_2. The BET model assumes the formation of multilayers by the adsorbate, which is more realistic than the assumption of monolayers. At this stage, we will skip the derivation of the BET equation (6.2), but just discuss the equation itself:

$$\frac{p/p_0}{n_a(1 - p/p_0)} = \frac{1}{n_m C} + \frac{C - 1}{n_m C} \frac{p}{p_0} \tag{6.2}$$

where n_a denotes the adsorbed amount, n_m is the capacity of a monolayer, and C is a constant that takes the heats of adsorption (first layer) and condensation (multilayers) into account. Commonly, the range $0.05 < p/p_0 < 0.3$ is used for application of the BET equation. From a linear fit of the experimental data, it is possible to extract the monolayer capacity (which allows calculation of the surface area, given the molecular cross-section σ of the adsorbate is known, $\sigma_{N2} = 16.2$ Å^2) and the C constant.

At this stage, it is worth reconsidering the basis of the BET model, which was developed for the analysis of planar surfaces. It works reasonably well for powders and meso- and macroporous materials, as the curvature of the surface is far larger than the molecular scale of the adsorbate, but care has to be taken when analyzing microporous materials [136]. Here, an overlap of mono- or multilayer formation with pore filling is plausible. Certain criteria have been proposed to judge the quality of the BET analysis [136–138]. This includes cross-checking, whether the calculated C constant has a positive value. A negative value has no physical meaning. Another point that requires attention is the choice of the correct

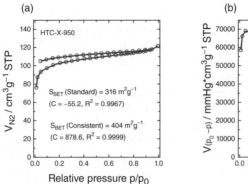
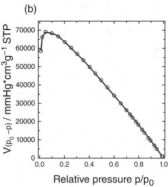

Figure 6.30 *(a) Nitrogen adsorption/desorption isotherm of a xylose-derived HTC carbon, which was postprocessed at 950 °C, including BET values derived using either the standard conditions or the consistency criteria. (b) Plot of V(p₀ − p) versus p/p₀ using the respective adsorption branch; the grey line indicates the upper limit for points suitable for determination of the BET plot.*

pressure range for application of the BET theory. A plot of $V(p_0 - p)$ versus p/p_0 indicates the correct pressure range for application of the BET equation. Often this plot exhibits a maximum at a certain p/p_0 that describes the upper limit for the applicability of the BET equation. Using the "standard" pressure range often results in an underestimation of the specific surface area of microporous materials. This is exemplarily shown in Figure 6.30, where the BET surface areas of a postcarbonized HTC carbon are calculated following the standard or the consistency criteria. It becomes obvious that the error is roughly 25%, which is a tremendous underestimation of the surface area if the standard range is used.

Hence, it is recommended to talk about *apparent* BET surface areas when discussing microporous materials. There is nevertheless a good reason to use the BET model, as it still gives a common basis for the comparison of different materials. Indeed, it was shown by modeling of the adsorption process that there is a good agreement between calculated and experimental BET surface areas, even in ultramicroporous materials [137, 138].

In the case of micro- and mesoporous materials, the BET criteria can also be applied, but it is advisable to use DFT-based data evaluation in addition, as it allows the attribution of the fractional surface area to the pore sizes (see below).

Another way to determine specific surface areas in microporous materials is the use of the Langmuir method. The drawbacks using this approach are somewhat comparable to the usage of the BET model, but as only monolayer formation on homogenous places is assumed, it is even further away from reality. Hence, surface areas derived from the Langmuir approach are reported frequently, but should only be used for comparative reasons [136].

There are a number of other (mostly semiempirical) methods for the determination of specific surface areas [139], but these can be considered as outdated and will consequently not be discussed here. Finally, specific surface areas can also be derived from DFT approaches, as will be discussed below. DFT-type approaches have become more and more important over recent years and can be considered as a highly valuable addition to the classic

BET analysis. Due to their high importance, they will be discussed in more detail later in this section.

6.10.2.2 Assessment of Pore Size Distribution and Pore Volume

From the total uptake of nitrogen at $p/p_0 = 0.995$, one can estimate the total pore volume and it is possible to calculate the porosity ϕ, which is given as the volume fraction of pores within the material. This is, however, only possible if no macropores are present. The presence of macropores or large void spaces between nanoparticles is indicated by the absence of a clear plateau of the volume uptake as the pressure reaches the saturation pressure. The pore volume could also be derived from integration of the PSD, which itself can be calculated in a number of ways. Analysis of mesopore size distribution can be done on the basis of the Kelvin equation (6.3). The equation relates the relative vapor pressure to the mean radius of curvature of the meniscus of a liquid confined within a capillary:

$$\ln \frac{p}{p_0} = -\frac{\gamma v^l}{r\,RT} \qquad (6.3)$$

where γ is the surface tension of the liquid–gas interface and v^l is the molar volume of the liquid. Both quantities are typically assumed to be independent on the relative pressure. Let us assume a set of cylindrical pores of different radius r_i. As the relative pressure will be increased within the system, a thin layer of adsorbate will first cover all of the pore walls. As p/p_0 is further increased the system will reach a point where the liquid would be in equilibrium with its vapor, given a full thermodynamic reversibility. According to Eq. (6.3) this point will be reached first for the smallest pores, while the largest pores will be filled last. The adsorbate condensation within the mesopore is accompanied by a steep increase of the uptake volume, which gives rise to the common isotherm shape of mesoporous materials. Figure 6.31 illustrates the process schematically. The adsorption process in an idealized cylindrical pore is illustrated, showing the different stages (multilayer formation, pore filling, plateau). It also becomes clear that the meniscus radius r_m is not equal to the pore radius r as the preformed adsorbate multilayer of thickness t must be considered, hence $r = r_m + t$. In reality, mesopore sizes always have a certain distribution, which is reflected in the steepness of the volume uptake. Thus, already the shape of the adsorption isotherm gives reasonable information about the width of the PSD. One of the most common (and commercialized) methods for the evaluation of mesopore size distribution is the Barrett, Joyner, and Halenda (BJH) approach, which also relies on the well-known Kelvin equation (6.3) [139]. The method also takes the aforementioned pre-existing multilayer as well as its thinning during the desorption process into account. A detailed derivation of the calculation method can be found in the relevant literature [139].

A common feature that is frequently observed for mesoporous materials is the presence of a hysteresis loop between the adsorption branch (mesopore filling) and desorption branch (pore emptying). The origin of the hysteresis, which can (by nitrogen adsorption at 77.4 K) only be observed in mesopores larger than approximately 5 nm, is of thermodynamic reason. Typically, condensation of the adsorbate is delayed due to the presence of metastable states, while the desorption branch is believed to reflect the thermodynamic equilibrium. Hence, the desorption branch is commonly used for the calculation of the PSD using the BJH or other approaches. There are, however, cases (disordered pores, pore blocking,

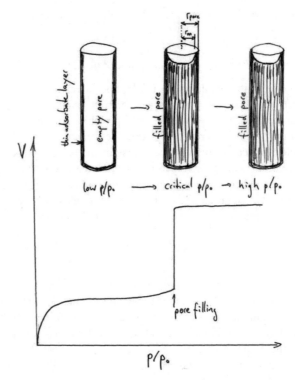

Figure 6.31   *Schematic drawing of the pore-filling process of a model pore and the corresponding schematic adsorption isotherm.*

cavitation, or percolation effects), where the desorption branch should not be used for the assessment of the mesopore size [140]. Generally, these effects are manifested in nonparallel adsorption/desorption branches (classified as type H2 hysteresis) due to a very abrupt desorption at a certain relative pressure (Figure 6.32). This abrupt pore emptying can have two origins, either pore blocking or cavitation, both of which are related to pore entrances, which are much smaller than the pore itself.

Pore blocking is present when the whole pore is just emptied after the pore opening could be emptied (i.e., pores and pore openings are emptied simultaneously). In this case, the pore size analysis of the desorption branch can give information about the neck size, as the desorption from the pore neck follows the formation of a meniscus, as predicted by the Kelvin equation. However, if the pore opening is very small (typically smaller than 4 nm), the pore will be emptied while the pore opening is still filled with adsorbate. This effect is known as "cavitation" – the spontaneous nucleation of a gas bubble due to the thermodynamic instability of the condensed adsorbate. The pore is emptied by diffusion. In this case, no physically meaningful values can be derived from analysis of the desorption branch.

Taking these considerations into account, one should carefully consider whether the adsorption or desorption branch is used for pore size calculation. As a rule of thumb, it

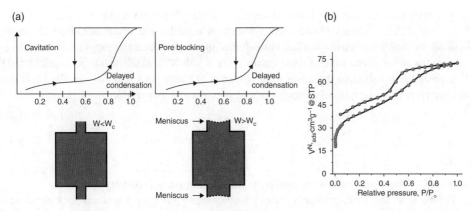

Figure 6.32 *(a) Schematic illustration of pore-blocking and cavitation phenomena. (Reproduced with permission from [140].) (b) Example of a mesoporous HTC carbon showing a type H2 hysteresis. (Reproduced with permission from [141]. © 2011 American Chemical Society.)*

is advisable to use the adsorption branch for the calculation of the PSD combined with a DFT methodology developed for adsorption data if a type H2 hysteresis is present. A representative example of a meso-microporous HTC carbon showing a type H2 hysteresis is shown in Figure 6.32b. In this example the adsorption branch was used to derive the PSD by means of DFT, see below and Chapter 2 for details on the synthesis. Finally, it should be mentioned that there is still no closed discussion on the pore blocking/cavitation phenomena and research on this topic is ongoing [142, 143], given its relevance not only for porosity analysis, but also for applications such as in gas storage.

The classic macroscopic models, such as methods based on the Kelvin approach, have some drawbacks when it comes to the analysis of small or spherical mesopores. For instance, it was pointed recently by Neimark and Ravikovitch that the BJH method can underestimate pore sizes by up to 100% [144].

A recent development for an advanced analysis of PSDs are the so-called DFTs [140, 145], which can be classified as microscopic models taking fluid–wall interactions explicitly into account. DFTs also represent the state-of-the art methodology for the analysis of microporosity and are about to replace semiempirical methods (e.g., Dubinin–Raduskevich and Horvath–Kawazoe). The method will be discussed within the following sections, which is also a supplement to the classic analysis of specific surface area, PSD, and pore volume.

The basis of DFT methods is to take the local ordering of adsorbed fluids near the pore wall and its impact on the density profile into account. Pioneering work by Tarazona *et al.* showed that nonlocal density functional (also known as smoothed density approximation (SDA)) [146, 147] could take short-range correlation more realistically into account and give better descriptions on the equilibrium states of confined adsorbates. The local short-range ordering near the wall leads to characteristic oscillations of the fluid density near the wall. This effect is especially important for small mesopores and micropores, where the pore size is only of a few molecular diameters σ. The NLDFT idea has since then been developed by many researchers and is now a common tool for porosity analysis. Early studies could show that the density profile $\rho(r)$ for a given adsorbate/adsorbent system (e.g., nitrogen/silica) can be calculated based on validated intermolecular interaction parameters [148, 149]. Typically, the adsorbate/adsorbent as well as the adsorbate/adsorbate interactions are described through a Lennard–Jones approach. Once the pressure dependence of $\rho(r)$ for an idealized pore of fixed radius R (or width W) and shape has been calculated, it is possible to calculate the adsorption isotherm per unit pore volume, see Eq. (6.4). or alternatively the adsorption isotherm per unit pore area by integration:

$$N_v \left(\frac{p}{p_0} \right) = \frac{2}{R^2} \int_0^R r \, dr \, (\rho(r) - \rho_g) \tag{6.4}$$

where N_v is the adsorbed amount and ρ_g is the equilibrium gas density

Figure 6.33 shows exemplarily model isotherms of nitrogen on silica for cylindrical (meso)pores with different diameters, which reproduce the capillary condensation phenomenon quite well. Also shown are model isotherms of hydrogen in carbon micropores. It could be shown by simulation methods, such as grand-canonical Monte-Carlo (GCMC) methods, that the NLDFT approach could correctly reproduce the local density profiles and also describe the phase transition of the fluid within the pores [151].

The NLDFT methodology could now be used to describe the experimental isotherms by a set of model isotherms (i.e., a fit of the experimental data with a given kernel (fixed pore shape and interaction parameters)), according to:

$$N \left(\frac{p}{p_0} \right) = \int_{W_{\min}}^{W_{\max}} N \left(\frac{p}{p_0}, W \right) f(W) \, dW \tag{6.5}$$

where $N(p/p_0)$ is the experimental data, $f(W)$ is the PSD function, and $N(p/p_0, W)$ is the isotherm on a single pore of width W.

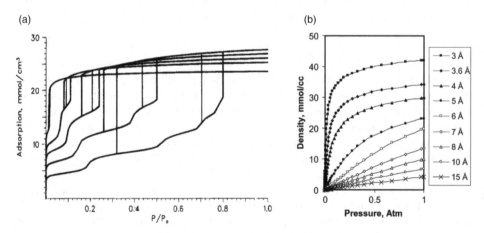

Figure 6.33   *(a) NLDFT model isotherms in cylindrical silica pores at 77.35 K. From left to right: 18.2, 25.4, 32.6, 45, and 79 Å. (Reproduced with permission from [148].) (b) Selected NLDFT isotherms of H_2 at 77 K in carbon micropores of different sizes. (Reproduced with permission from [150]. © 2004 Elsevier.)*

This procedure allows finally the calculation of the PSD, but also of the surface area distribution and the specific surface area independently from the BET methodology.

One of the main advantages of NLDFT methodologies over classic methods is their use for the analysis of micro- and mesopores at the same time. For example, the PSD of a micro-mesoporous silica obtained by NLDFT could be verified by scattering experiments [152]. Different models are available for different pore morphologies (e.g., cylindrical, slit-like, spherical) as well as for different pore wall chemistries (carbon, zeolites, or silica). Additionally, models are available that use the adsorption branch for calculation of the PSD [140], which is useful if the desorption branch cannot be used due to pore-blocking or cavitation effects (see discussion above). Nowadays, NLDFT or GCMC analysis of adsorption isotherms is supported by most manufacturers of adsorption machines.

Recent developments of the DFT and GCMC methodology are related to the understanding of defects the investigation of adsorption-induced deformations (which have impact on, for example, carbon capture and storage technologies) [153–156]. Another recent development is the so-called quenched-solid DFT (QSDFT), which was developed for adsorption of nitrogen or argon on carbon surfaces [157, 158]. This new methodology takes the surface roughness and heterogeneity of the adsorbent into account. This is closer to the reality of carbon surfaces as they are found in HTC carbons or activated carbons. Another advantage of the QSDFT model is the absence of a prominent minimum in the micropore PSD (for nitrogen typically around 1 and 2 nm), which is nearly always present in standard NLDFT PSDs. The research on DFT models is highly active currently and that is why there cannot be a single suggestion on which model to use for porosity analysis. Currently, QSDFT methodology seems to be the state-of-the-art method for the analysis of carbons by nitrogen or argon adsorption. An example of the porosity analysis of a HTC-derived meso- and microporous carbon presented in detail in the soft templating section in Chapter 2 (Section 2.2.2) is shown in Figure 6.34, which shows clearly the advantages of QSDFT analysis over a common BJH analysis, which would not be able to resolve micro- and small mesopores.

It should be noted that there are other useful analyte gases beside nitrogen, and DFT models have been developed for argon, hydrogen, and CO_2 as well [133, 134, 150]. It was shown that hydrogen could be used as a probe molecule for the detection of ultramicropores that were not accessible by nitrogen [159–161]. Although the hydrogen adsorption capacity is measured for many samples nowadays as a consequence of the search for hydrogen storage systems, the use of hydrogen adsorption as an analytical tool has not found widespread application. The use of CO_2 adsorption at ambient temperatures also has advantages for the analysis of ultramicroporous materials [132, 160]. The advantages of CO_2 adsorption are well known within the carbon community, and will be discussed in the next section along with some considerations on the adsorption strength and selectivity.

6.10.2.3   *CO₂ Adsorption*

CO_2 adsorption in carbonaceous materials is interesting from both pure analytical and industrial viewpoints. Adsorption-induced deformations need consideration when discussing CO_2 storage or use in subterrestrial reservoirs. It was shown that the acting solvation

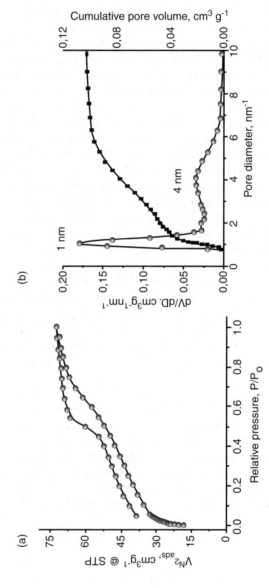

Figure 6.34 (a) Nitrogen sorption isotherm and (b) QSDFT PSD of as-synthesized HTC carbon made out of fructose in the presence of an amphiphilic block copolymer (Pluronic® F127). (Reproduced with permission from [141]. © 2011 American Chemical Society.)

pressures and deformations can be quite enormous, especially at high loadings [156, 162, 163]. The effect is, however, not yet fully understood and is still under investigation.

CO_2 can also be used as a simple analytical probe to investigate the porosity of materials. Typical adsorption experiments are undertaken at 273 or 298 K (i.e., at much higher temperatures compared to classic nitrogen adsorption experiments). Accordingly, the probe gas has a higher thermal energy and diffusional problems in very narrow micropores are less problematic. Due to the high saturation pressure at ambient temperatures (around 26 140 mmHg at 273 K), CO_2 adsorption up to 1 atm can only give information about very narrow micropores (less than 1.5 nm). However, if high-pressure adsorption is employed, mesopore filling due to capillary condensation can also be observed. CO_2 adsorption as a versatile tool for the analysis of microporous carbons has been known for quite some time [132, 131, 164], but has recently gained increased attention. CO_2 adsorption experiments on ultramicropores are typically faster compared to nitrogen adsorption experiments. Due to the high saturation pressure at 273 K the instrumental demands necessary to reach very low relative pressures are also minimized, which makes CO_2 adsorption attractive. As stated in the previous section, NLDFT or GCMC models are also available to analyze the adsorption data with regard to specific surface area and PSD.

An example of the use of CO_2 adsorption in the analysis of HTC carbons is shown in Figure 6.35 and discussed in the following part.

Figure 6.35a shows the nitrogen adsorption/desorption isotherms (77.3 K) of a typical HTC carbon (here derived from xylose) and carbons derived from this sample by further pyrolysis at higher temperatures [3]. From the nitrogen data, one would judge that the carbons as obtained after HTC and even after treatment at 350 °C are essentially nonporous. This is the general case with all the HTC carbons unless templates, additives, or activation agents are used (see Chapters 2 and 3). Higher temperature treatment results in a measurable porosity; however, a nonclosing isotherm is observed. This is a common feature due to either so-called swelling/rearrangement events and/or diffusional problems. The high-temperature

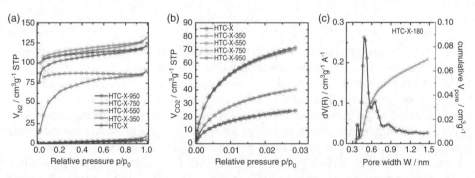

Figure 6.35 *(a) Nitrogen adsorption/desorption isotherms of a xylose-derived HTC carbon (HTC-X-180) and carbons derived from it by pyrolysis (HTC-X-350, . . . , HTC-X-950) obtained at 77.3 K. (b) CO_2 adsorption/desorption isotherms of the same samples (measured at 273 K). (c) PSD and cumulative pore volume (calculated from the CO_2 adsorption isotherm using the GCMC analysis method) of the carbon as obtained after HTC. (Reproduced with permission from [3]. © 2012 American Chemical Society.)*

treated materials seem to be microporous only, as evidenced by the type I-like isotherm shape.

An analysis of the same samples by CO_2 adsorption results in a somewhat different picture: (i) no hysteresis is observed in any sample pointing to optimal analysis conditions and (ii) the low-temperature-treated samples show a significant CO_2 uptake, indicating the presence of porosity even in those samples. An analysis of the PSD of the carbon as obtained by HTC shows indeed the presence of ultramicropores (Figure 6.35c). The major fraction of the pores has a width of approximately 0.4–0.7 nm. That is exactly the size range that can be difficult to access by nitrogen adsorption at 77.3 K [132, 165]. Additionally, HTC carbons are much less rigid materials compared to high-temperature-treated carbons, which have typically higher degrees of condensation. Such increased rigidity might facilitate the transport of CO_2 trough the micropore systems, as less deformations/diffusional problems are expected. This is in analogy to the difference between rigid microporous polymer networks and more "soft" non-cross-linked microporous polymers. Stiff networks show typically significant CO_2 (at 273 K) and nitrogen (at 77.3 K) uptake, while the more soft linear microporous polymers can often be analyzed by CO_2 adsorption at 273 K only [166]. Indeed, the carbon materials which were postcarbonized at higher temperatures show a significant nitrogen uptake, but also an increased CO_2 uptake (see Figure 6.35). It could, however, be shown that CO_2 adsorption gives still higher specific surface areas. This is in line with the fact that the PSD does not change significantly upon thermal treatment (see Figure 6.36c) – very narrow pores are opened up, but apparently not enlarged in size, which results in the same drawbacks for nitrogen adsorption at 77.3 K as discussed before.

The combined use of CO_2 adsorption at 273 and 283 K and nitrogen adsorption at 273 K can give further information about the pore characteristics. It is possible to determine the isosteric heat of adsorption q_{st} from CO_2 adsorption measurements at 273 and 283 K. Various calculation protocols have been suggested to access q_{st} [167, 168]. The classic methodology is to apply the Clausius–Clapeyron equation (6.6), which gives a relation between the pressure and the temperature at constant loading. The slope of the plot of $\ln p$ versus $1/T$ gives access to the heat of adsorption:

$$(\ln p)_V = -\frac{q_{st}}{R}\frac{1}{T} + D \tag{6.6}$$

where R is the universal gas constant and D is a constant. Modern gas adsorption software environments often have an implementation for the determination of q_{st} as a function of the gas loading. It is, however, advisable to cross-check the results doing the calculation by "hand," following accepted guidelines [168]. The isosteric heat of CO_2 adsorption is not only influenced by the presence of functional groups, such as open metal centers, amine-, hydroxyl-, carboxylic acid groups, or counter-ions [169–172], but to some extent also by the pore width and heterogeneity. An example is shown in Figure 6.36d, where the heats of CO_2 adsorption for HTC-X, HTC-X-950 (X = xylose, 950 °C = postcalcination temperature), and an activated carbon derived from a HTC carbon (AC-HTC-G4) (G = glucose; 4 is the ratio of KOH to HTC) by chemical activation using KOH are shown. A comparison of the CO_2 adsorption/desorption isotherm shape of the activated carbon and the HTC-X-950 shows already that the activated carbon contains larger pores. This is evident from the much lower slope at low pressures compared to HTC-X-950 and is confirmed by the PSDs calculated using the GCMC method (Figure 6.36c). We can now compare the heats of

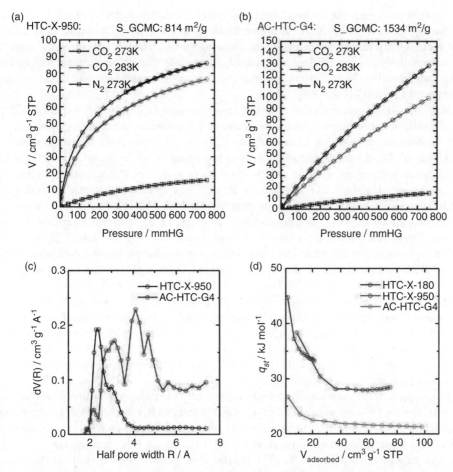

Figure 6.36 *(a) CO$_2$ and nitrogen adsorption/desorption isotherms of HTC-X-950 measured at 273 and 283 K. (b) CO$_2$ and nitrogen adsorption/desorption isotherms of an activated carbon (chemical activation by KOH) derived from a HTC carbon measured at 273 K and 283 K. (c) PSDs of the two samples, derived from the CO$_2$ adsorption isotherms (273 K) shown in (a) and (b) by application of the GCMC method. (d) Isosteric heats of adsorption for HTC-X, HTC-X-950, and the activated carbon described in (b).*

adsorption of a highly functionalized carbon with small pores (HTC-X-180), a carbon with almost no surface functionalities (HTC-X-950), and a carbon with large pores and many functionalities (AC-HTC-G4) in order to identify the effects of surface functionalities and pore size.

The effect of pore size becomes evident when comparing HTC-X-180 (after HTC of xylose) and AC-HTC-G4 (after HTC of glucose and chemical activation), both of which have surface functionalities, but different pore sizes. The heat of adsorption of HTC-X-180 is significantly higher than that of AC-HTC-G4, thus illustrating that small pores can indeed be beneficial for a high q_{st}. The comparison between HTC-X-180 and HTC-X-950 (i.e., same pore sizes but different degree of surface functionalization) shows that q_{st} is somewhat lower for HTC-X-950 at higher loadings (at 1 bar: $q_{st} \sim 28$ versus 33 kJ mol^{-1} for HTC-X-180) as can be seen from the crossing curves. This indicates that the surface functionalities (e.g., hydroxyl groups) indeed enhance the interaction strength, which is in line with reports on hydroxyl-group-containing microporous polymers and metal organic frameworks [173, 174].

Additional information can be obtained from the nitrogen adsorption measured at 273 K. No diffusional limitations should be active at this temperature and nitrogen should be able to enter the micropores. Comparing the CO_2 and nitrogen uptake of a material (e.g., at 1 bar and 273 K) allows the calculation of the apparent single gas selectivity. Next to its stronger interactions with the pore wall (as a consequence of its quadrupole moment and easier polarizability), CO_2 has also a somewhat smaller kinetic diameter than nitrogen (0.33 versus 0.364 nm). In the limiting case this leads to molecular sieving (presence of pores that allow only CO_2 but not nitrogen to enter) [175], which is characterized by tremendously high selectivities (above 100) while maintaining high capacities. Hence, both factors (pore size and adsorbate/adsorbent interactions) influence the observed selectivity. In the HTC example above, the measured apparent CO_2/nitrogen selectivities are circa 20 for HTC-X-180, circa 10 for AC-HTC-G4, and 5.5 for HTC-X-950. These values illustrate that specific interactions obviously can strongly increase the gas selectivity. Although the CO_2 heat of adsorption of HTC-X-950 was rather high, its selectivity for CO_2 was the lowest. This illustrates nicely that a high heat of CO_2 adsorption alone does not guarantee a high selectivity. On the contrary, HTC-X-180 shows a rather good selectivity having almost the same PSD shape as HTC-X-950, which highlights again the importance of functional groups.

In summary, this example demonstrates quite nicely the additional information that can be gained using CO_2 and nitrogen adsorption at 273 and 283 K. Finally, it must be noted that such estimations as done here need additional verification when it comes to the discussion about real applications in gas separation tasks [168]. Indeed, HTC carbons are discussed for CO_2 capture applications [176, 177], but this topic will not be discussed here in detail, but in the application chapter (Chapter 7).

6.10.3   Mercury Intrusion Porosity

Pore size analysis by mercury intrusion porosimetry is a rather old and well-established technique for the analysis of macro- and partly also for mesoporous materials. The technique is based on the nonwetting behavior of mercury. An external pressure (i.e., a Δp to the ambient pressure) needs to be applied if the pores of a material should be filled with mercury. This pressure is related to the inverse pore radius by the Young–Laplace equation:

$$\Delta p = \gamma \left(\frac{1}{r_1} + \frac{1}{r_2} \right) \tag{6.7}$$

where γ denotes the interfacial tension, and r_1 and r_2 are the curvature radii. In the case of cylindrical pores, mercury would form a spherical meniscus ($r_1 = r_2 = r$). Taking the contact angle θ between mercury and the pore wall into account, one arrives at the Washburn equation (6.8), which was named after Edward Washburn, who suggested to use mercury intrusion as a method for pore size determination [57]:

$$\Delta p = \frac{2\gamma}{r} \cos \theta \qquad (6.8)$$

The contact angle could either be determined experimentally by different methods, but it is also tabulated for a large number of substrates, including carbon [57].

From Eq. (6.8) it is obvious that the pressure that needs to be applied relates to the inverse of the pore radius – for the characterization of very small pores, huge pressures have to be applied, while large pores can be characterized by application of moderate pressures. The technique is commercialized and machines are available from a variety of manufacturers.

Mercury intrusion allows the determination of the total pore volume of a sample, the specific surface area, and the PSD. Pores between 500 μm and 4 nm can be characterized, spanning a huge size range from large macropores down to small mesopores. The PSD can be calculated directly from the Washburn equation. Attention should be paid to the fact that the pore sizes do represent the size of the smallest entrance (throat opening) to a pore, as this dictates the pressure needed to press the mercury in. This could falsify the PSD especially in the case of so-called ink-bottle-type pores and a cross-check by microscopy could be useful. The total pore volume is, however, not affected by this effect and could, for instance, be determined directly from the volume change of the mercury column. Specific surface areas can be determined under the assumption of fixed pore geometry (e.g., a cylindrical pore). A good overview about the technique is available, for instance, in [178]

Drawbacks of the method are the use of toxic mercury. Often, it is also not possible to recover the mercury completely from the sample upon pressure release (hysteresis effects due to connectivity issues). The mercury trapped in the material makes them toxic waste themselves, classifying the method as destructive. Other drawbacks, especially in the analysis of "soft" matter, are the application of very high pressures, which ultimately could lead to material damage.

Nevertheless, mercury intrusion represents a useful technique for the analysis of macroporous materials, especially for small macropores, which are hard to access by classic gas adsorption. An example is given in Figure 6.37 applied to hollow carbon spheres prepared by HTC combined with polymer latex templating as described in detail in Chapter 2 and elsewhere [179]. The template particles had diameters of 100 and 130 nm. Nitrogen adsorption shows a high surface area of 370 and 470 m^2 g^{-1}, respectively. From the adsorption isotherms shown in Figure 6.37a and b, however, it is not possible to determine a mean pore size or a pore volume, as only a steep increase at very high relative pressures is observed. Contrarily, mercury intrusion can probe the macropores (i.e., the interior of the hollow particles) reliably and the determined pore size agrees well with the size of the used template particles (Figure 6.37c and d). Mercury intrusion can in this case also give information about the pore volume, which enables the calculation of the overall porosity.

Interestingly, the specific surface determined by mercury intrusion also corresponds quite well with the determined BET surface areas, which indicates that the materials have only a low fraction of micropores.

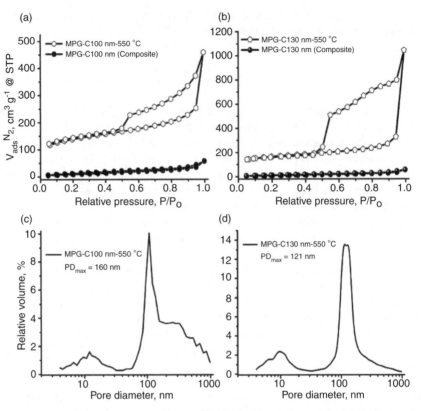

Figure 6.37   *(a and b) Nitrogen sorption profiles for hollow sphere materials before and after removal of a polystyrene latex template. (c and d) Mercury sorption PSDs for as-prepared hollow carbon sphere materials prepared at 550 °C (i.e., after template removal; PD$_{max}$ indicates pore size maxima). (Reproduced with permission from [179]. © 2010 American Chemical Society.)*

6.10.4 Scattering Methods

Scattering methods, especially small-angle X-ray but in principle also neutron scattering (SAXS and SANS) are versatile methods for the analysis of porous materials [180]. They allow the analysis of either solvent-filled or dry networks, can probe closed porosity, and allow also the monitoring of processes (such as adsorption events) *in situ* [181, 182].

References

(1) Radovic, L.R. (2012) *Chemistry & Physics of Carbon*, vol. **31**, CRC Press, Boca Raton, FL.
(2) Dela Rosa, L., Pruski, M. *et al.* (1992) *Energy & Fuels*, **6**, 460–468.
(3) Yu, L., Falco, C., Weber, J. *et al.* (2012) *Langmuir*, **28**, 12373–12383.

(4) Bergius, F. (1928) *Naturwissenschaften*, **16**, 1–10.
(5) Bergius, F. (1937) *Industrial and Engineering Chemistry*, **29**, 247–253.
(6) Harris, E.E. and Beglinger, E. (1946) The Madison Wood-Sugar Process. Report R1617. US Department of Agriculture, Forest Service, Forest Products Laboratory, Madison, WI.
(7) Kusama, J. (1960) Technical Panel on Wood Chemistry. Report FAD/WC/60/WH-1. UN FAO, Tokyo.
(8) Thompson, D.R. and Grethlein, H.E. (1979) *Industrial & Engineering Chemistry. Product Research and Development*, **18**, 166–169.
(9) Erckel, R., Franz, R., Woernle, R., and Riehm, T. (1985) US Patent 4,556,431.
(10) Bröll, D., Kaul, C., Krämer, A. *et al.* (1999) *Angewandte Chemie International Edition*, **38**, 2998–3014.
(11) Jin, F. and Enomoto, H. (2009) *Bioresources*, **4**, 704–713.
(12) Chheda, J.N., Roman-Leshkov, Y., and Dumesic, J.A. (2007) *Green Chemistry*, **9**, 342–350.
(13) Scallet, B.L. and Gardner, J.H. (1945) *Journal of the American Chemical Society*, **67**, 1934–1935.
(14) Asghari, F.S. and Yoshida, H. (2007) *Industrial & Engineering Chemistry Research*, **46**, 7703.
(15) Sun, X. and Li, Y. (2004) *Angewandte Chemie International Edition*, **43**, 597–601.
(16) Titirici, M.-M., Thomas, A., and Antonietti, M. (2007) *Advanced Functional Materials*, **17**, 1010–1018.
(17) Antonietti, M., Thomas, A., and Titirici, M. (2007) *Physical Chemistry Chemical Physics*, **9**, T45–T45.
(18) Titirici, M.M., Thomas, A., and Antonietti, M. (2007) *New Journal of Chemistry*, **31**, 787–789.
(19) Patil, S.K.R. and Lund, C.R.F. (2011) *Energy & Fuels*, **25**, 4745–4755.
(20) Cakan, R.D., Titirici, M.-M., Antonietti, M. *et al.* (2008) *Chemical Communications*, 3759–3761.
(21) Fuertes, A.B., Arbestain, M.C., Sevilla, M. *et al.* (2010) *Australian Journal of Soil Research*, **48**, 618–626.
(22) Angyal, S.J. (2001) in *Glycoscience: Epimerization, Isomerization and Rearrangement Reactions of Carbohydrates* (ed. A.E. Stulz), Springer, Berlin, pp. 1–14.
(23) Qi, X., Watanabe, M., Aida, T.M., and Smith, R.L., Jr. (2008) *Catalysis Communications*, **9**, 2244–2249.
(24) Yu, S.H., Cui, X.J., Li, L.L. *et al.* (2004) *Advanced Materials*, **16**, 1636–1640.
(25) Chuntanapum, A. and Matsumura, Y. (2009) *Industrial & Engineering Chemistry Research*, **48**, 9837–9846.
(26) LaMer, V.K. (1952) *Industrial & Engineering Chemistry*, **44**, 1270–1277.
(27) Sakaki, T., Shibata, M., Miki, T. *et al.* (1996) *Bioresource Technology*, **58**, 197–202.
(28) Bremholm, M., Becker-Christensen, J., and Iversen, B.B. (2009) *Advanced Materials*, **21**, 3572–3575.
(29) Titirici, M.-M., Antonietti, M., and Baccile, N. (2008) *Green Chemistry*, **10**, 1204–1212.
(30) Wohlgemuth, S.-A., White, R.J., Willinger, M.-G. *et al.* (2012) *Green Chemistry*, **14**, 1515–1523.

(31) Kumar, P., Barrett, D.M., Delwiche, M.J., and Stroeve, P. (2009) *Industrial & Engineering Chemistry Research*, **48**, 3713–3729.

(32) Falco, C., Baccile, N., and Titirici, M.-M. (2011) *Green Chemistry*, **13**, 3273–3281.

(33) O'Sullivan, A.C. (1997) *Cellulose*, **4**, 173–207.

(34) Kruse, A.G.A. (2003) *Industrial & Engineering Chemistry Research*, **42**, 267–279.

(35) Mansikkamaki, M.L.P. and Rissanen, K. (2007) *Carbohydrate Polymers*, **68**, 235–241.

(36) Mok, W.S.L. and Antal, M.J. (1992) *Industrial & Engineering Chemistry Research*, **31**, 1157–1161.

(37) Kumar, P., Barrett, D.M., Delwiche, M.J., and Stroeve, P. (2009) *Industrial & Engineering Chemistry Research*, **48**, 3713–3729.

(38) Sun, X.M. and Li, Y.D. (2004) *Angewandte Chemie International Edition*, **43**, 597–601.

(39) Titirici, M.M., Thomas, A., and Antonietti, M. (2007) *Journal of Materials Chemistry*, **17**, 3412–3418.

(40) White, R.J., Antonio, C., Budarin, V.L. *et al.* (2010) *Advanced Functional Materials*, **20**, 1834–1841.

(41) Falco, C., Perez Caballero, F., Babonneau, F. *et al.* (2011) *Langmuir*, **27**, 14460–14471.

(42) Baccile, N., Antonietti, M., and Titirici, M.M. (2010) *Chemsuschem*, **3**, 246–253.

(43) Fitzer, E., Fritz, W., Christu, N., and Overhoff, D. (1968) *Carbon*, **6**, 236–237.

(44) Endo, M., Kim, Y.A., Fukai, Y. *et al.* (2001) *Applied Physics Letters*, **79**, 1531–1533.

(45) Ritter, M.A.N., Thomas, A., Senkovska, I. *et al.* (2009) *Macromolecules*, **42**, 8017–8020.

(46) NIST (2011) Pentanoic acid, 4-oxo-. Material Measurement Laboratory, Gaithersburg, MD. http://webbook.nist.gov/cgi/cbook.cgi?ID=C123762&Mask=200#Mass-Spec.

(47) Levitt, M.H. (2001) *Spin Dynamics: Basics of Nuclear Magnetic Resonance*, John Wiley & Sons, Ltd, Chichester.

(48) Schmidt-Rohr, K. and Spiess, H.W. (1994) *Multidimensional Solid State NMR and Polymers*, Academic Press, London.

(49) Andrew, E.R. and Szczesniak, E. (1995) *Progress in Nuclear Magnetic Resonance Spectroscopy*, **28**, 11–36.

(50) Blanc, F., Coperet, C., Lesage, A., and Emsley, L. (2008) *Chemical Society Reviews*, **37**, 518–526.

(51) Laws, D.D., Bitter, H.-M.L., and Jerschow, A. (2002) *Angewandte Chemie International Edition*, **41**, 3096–3129.

(52) Kroto, H.W., Heath, J.R., O'Brien, S.C. *et al.* (1985) *Nature*, **318**, 162–163.

(53) Taylor, R., Hare, J.P., Abdul-Sada, A.K., and Kroto, H.W. (1990) *Journal of the Chemical Society, Chemical Communications*, 1423–1425.

(54) Carravetta, M., Murata, Y., Murata, M. *et al.* (2004) *Journal of the American Chemical Society*, **126**, 4092–4093.

(55) Yannoni, C.S., Bernier, P.P., Bethune, D.S. *et al.* (1991) *Journal of the American Chemical Society*, **113**, 3190–3192.

(56) Fowler, P.W., Lazzeretti, P., Malagoli, M., and Zanasi, R. (1991) *Journal of Physical Chemistry*, **95**, 6404–6405.

(57) Yannoni, C.S., Johnson, R.D., Meijer, G. *et al.* (1991) *Journal of Physical Chemistry*, **95**, 9–10.

(58) Tycko, R., Dabbagh, G., Fleming, R.M. *et al.* (1991) *Physical Review Letters*, **67**, 1886–1889.

(59) Walton, J.H., Kamasa-Quashie, A.K., Joers, J.M., and Gullion, T. (1993) *Chemical Physics Letters*, **203**, 237–242.

(60) Williams, R.M., Zwier, J.M., Verhoeven, J.W. *et al.* (1994) *Journal of the American Chemical Society*, **116**, 6965–6966.

(61) Kolodziejski, W., Corma, A., Barras, J., and Klinowski, J. (1995) *Journal of Physical Chemistry*, **99**, 3365–3370.

(62) Orendt, A.M. (2007) *Encyclopedia of Magnetic Resonance*, John Wiley & Sons, Ltd, Chichester.

(63) Carravetta, M., Danquigny, A., Mamone, S. *et al.* (2007) *Physical Chemistry Chemical Physics*, **9**, 4879–4894.

(64) Kim, H. and Ralph, J. (2010) *Organic & Biomolecular Chemistry*, **8**, 576–591.

(65) Tang, X.-P., Kleinhammes, A., Shimoda, H. *et al.* (2000) *Science*, **288**, 492–494.

(66) Slichter, C.P. (1996) *Principles of Magnetic Resonance*, Springer, Berlin.

(67) Goze Bac, C., Bernier, P., Latil, S. *et al.* (2001) *Current Applied Physics*, **1**, 149–155.

(68) Alemany, L.B., Zhang, L., Zeng, L. *et al.* (2007) *Chemistry of Materials*, **19**, 735–744.

(69) Engtrakul, C., Davis, M.F., Gennett, T. *et al.* (2005) *Journal of the American Chemical Society*, **127**, 17548–17555.

(70) Cahill, L.S., Yao, Z., Adronov, A. *et al.* (2004) *Journal of Physical Chemistry B*, **108**, 11412–11418.

(71) Vanholme, R., Demedts, B., Morreel, K. *et al.* (2010) *Plant Physiology*, **153**, 895–905.

(72) Harkin, J.M. (1967) in *Oxidative Coupling of Phenols* (eds W.I. Taylor and A.R. Battersby), Dekker, New York, pp. 243–321.

(73) Ralph, J., Lapierre, C., Marita, J.M. *et al.* (2001) *Phytochemistry*, **57**, 993–1003.

(74) Ralph, J., Marita, J.M., Ralph, S.A. *et al.* (1999) in *Advances in Lignocellulosic Characterization* (ed. D. Argyropoulos), TAPPI Press, Atlanta, GA, pp. 55–108.

(75) Ralph, J. and Landucci, L.L. (2010) in *Lignins* (ed. C. Heitner and D.R. Dimmel), Dekker, New York, pp. 137–234.

(76) Yelle, D.J., Ralph, J., and Frihart, C.R. (2008) *Magnetic Resonance in Chemistry*, **46**, 508–517.

(77) Lewis, N.G., Yamamoto, E., Wooten, J.B. *et al.* (1987) *Science*, **237**, 1344–1346.

(78) Lewis, N.G., Razal, R.A., Dhara, K.P. *et al.* (1988) *Journal of the Chemical Society, Chemical Communications*, 1626–1628.

(79) Eberhardt, T.L., Bernards, M.A., He, L. *et al.* (1993) *Journal of Biological Chemistry*, **268**, 21088–21096.

(80) Terashima, N., Atalla, R.H., and Vanderhart, D.L. (1997) *Phytochemistry*, **46**, 863–870.

(81) Adler, E. (1977) *Wood Science and Technology*, **11**, 169–218.

(82) Fukagawa, N., Meshitsuka, G., and Ishizu, A. (1992) *Journal of Wood Chemistry and Technology*, **12**, 91–109.

(83) Hatcher, P.G. (1990) *Organic Geochemistry*, **16**, 959–968.

(84) Supaluknari, S., Burgar, I., and Larkins, F.P. (1990) *Organic Geochemistry*, **15**, 509–519.

(85) Atalla, R.H. and Vanderhart, D.L. (1984) *Science*, **223**, 283–285.

(86) VanderHart, D.L. and Atalla, R.H. (1984) *Macromolecules*, **17**, 1465–1472.

(87) Belton, P.S., Tanner, S.F., Cartier, N., and Chanzy, H. (1989) *Macromolecules*, **22**, 1615–1617.

(88) Larsson, P.T., Westermark, U., and Iversen, T. (1995) *Carbohydrate Research*, **278**, 339–343.

(89) Tang, M.M. and Bacon, R. (1964) *Carbon*, **2**, 211–220.

(90) Pastorova, I., Botto, R.E., Arisz, P.W., and Boon, J.J. (1994) *Carbohydrate Research*, **262**, 27–47.

(91) Kolodziejski, W., Frye, J.S., and Maciel, G.E. (1982) *Analytical Chemistry*, **54**, 1419–1424.

(92) Wooten, J.B., Seeman, J.I., and Hajaligol, M.R. (2003) *Energy & Fuels*, **18**, 1–15.

(93) Zhang, X., Golding, J., and Burgar, I. (2002) *Polymer*, **43**, 5791–5796.

(94) Soares, S., Ricardo, N.M.P.S., Jones, S., and Heatley, F. (2001) *European Polymer Journal*, **37**, 737–745.

(95) Link, S., Arvelakis, S., Spliethoff, H. *et al.* (2008) *Energy & Fuels*, **22**, 3523–3530.

(96) Sharma, R.K., Wooten, J.B., Baliga, V.L., and Hajaligol, M.R. (2001) *Fuel*, **80**, 1825–1836.

(97) Bardet, M., Hediger, S., Gerbaud, G. *et al.* (2007) *Fuel*, **86**, 1966–1976.

(98) Wikberg, H. and Liisa Maunu, S. (2004) *Carbohydrate Polymers*, **58**, 461–466.

(99) Sevilla, M. and Fuertes, A.B. (2009) *Chemistry – A European Journal*, **15**, 4195–4203.

(100) Yao, C., Shin, Y., Wang, L.-Q. *et al.* (2007) *Journal of Physical Chemistry C*, **111**, 15141–15145.

(101) Patil, S.K.R., Heltzel, J., and Lund, C.R.F. (2012) *Energy & Fuels*, **26**, 5281–5293.

(102) Zhang, M., Yang, H., Liu, Y. *et al.* (2012) *Carbon*, **50**, 2155–2161.

(103) Baccile, N., Laurent, G., Babonneau, F. *et al.* (2009) *Journal of Physical Chemistry C*, **113**, 9644–9654.

(104) Gandini, A. and Belgacem, M.N. (1997) *Progress in Polymer Science*, **22**, 1203–1379.

(105) Rosatella, A.A., Simeonov, S.P., Frade, R.F.M., and Afonso, C.A.M. (2011) *Green Chemistry*, **13**, 754–793.

(106) Knežević, D., van Swaaij, W., and Kersten, S. (2009) *Industrial & Engineering Chemistry Research*, **49**, 104–112.

(107) Sanders, E.B., Goldsmith, A.I., and Seeman, J.I. (2003) *Journal of Analytical and Applied Pyrolysis*, **66**, 29–50.

(108) David, K., Pu, Y., Foston, M., Muzzy, J., and Ragauskas, A. (2008) *Energy & Fuels*, **23**, 498–501.

(109) Schell, D., Nguyen, Q., Tucker, M., and Boynton, B. (1998) *Applied Biochemistry and Biotechnology*, **70–2**, 17–24.

(110) Glasser, W.G. and Wright, R.S. (1998) *Biomass & Bioenergy*, **14**, 219–235.

(111) Martinez, J.M., Granado, J.M., Montane, D. *et al.* (1995) *Bioresource Technology*, **52**, 59–67.

(112) Mosier, N.S., Ladisch, C.M., and Ladisch, M.R. (2002) *Biotechnology and Bioengineering*, **79**, 610–618.

(113) Falco, C., Perez Caballero, F., Babonneau, F. *et al.* (2011) *Langmuir*, **27**, 14460–14471.

(114) Freitas, J.C.C., Emmerich, F.G., Cernicchiaro, G.R.C. *et al.* (2001) *Solid-State Nuclear Magnetic Resonance*, **20**, 61–73.

(115) Shindo, A. and Izumino, K. (1994) *Carbon*, **32**, 1233–1243.

(116) Plaza, M.G., Pevida, C., Arenillas, A. *et al.* (2007) *Fuel*, **86**, 2204–2212.

(117) Stavropoulos, G.G., Samaras, P., and Sakellaropoulos, G.P. (2008) *Journal of Hazardous Materials*, **151**, 414–421.

(118) Jia, Y.F., Xiao, B., and Thomas, K.M. (2001) *Langmuir*, **18**, 470–478.

(119) Shao, Y., Sui, J., Yin, G., and Gao, Y. (2008) *Applied Catalysis B: Environmental*, **79**, 89–99.

(120) Kawaguchi, M., Itoh, A., Yagi, S., and Oda, H. (2007) *Journal of Power Sources*, **172**, 481–486.

(121) Weydanz, W.J., Way, B.M., van Buuren, T., and Dahn, J.R. (1994) *Journal of The Electrochemical Society*, **141**, 900–907.

(122) Nxumalo, E.N., Nyamori, V.O., and Coville, N.J. (2008) *Journal of Organometallic Chemistry*, **693**, 2942–2948.

(123) Yang, Z., Xia, Y., Sun, X., and Mokaya, R. (2006) *Journal of Physical Chemistry B*, **110**, 18424–18431.

(124) Hou, P.-X., Orikasa, H., Yamazaki, T. *et al.* (2005) *Chemistry of Materials*, **17**, 5187–5193.

(125) Baccile, N., Antonietti, M., and Titirici, M.-M. (2010) *ChemSusChem*, **3**, 246–253.

(126) White, R.J., Yoshizawa, N., Antonietti, M., and Titirici, M.-M. (2011) *Green Chemistry*, **13**, 2428–2434.

(127) Zhao, L., Baccile, N., Gross, S. *et al.* (2010) *Carbon*, **48**, 3778–3787.

(128) Baccile, N., Laurent, G., Coelho, C. *et al.* (2011) *Journal of Physical Chemistry C*, **115**, 8976–8982.

(129) Rouquerel, F., Rouquerel, J., and Sing, K.S.W. (2002) in *Handbook of Porous Solids* (eds F. Schüth, K.S.W. Sing, and J. Weitkamp), Wiley-VCH Verlag GmbH, Weinheim, pp. 236–275.

(130) Sing, K.S.W., Everett, D.H., Haul, R.A.W. *et al.* (1985) *Pure and Applied Chemistry*, **57**, 603–619.

(131) Cazorla-Amorós, D., Alcañiz-Monge, J., de la Casa-Lillo, M.A., and Linares-Solano, A. (1998) *Langmuir*, **14**, 4589–4596.

(132) Lozano-Castelló, D., Cazorla-Amorós, D., and Linares-Solano, A. (2004) *Carbon*, **42**, 1233–1242.

(133) Vishnyakov, A., Ravikovitch, P.I., and Neimark, A.V. (1999) *Langmuir*, **15**, 8736–8742.

(134) Ravikovitch, P.I., Vishnyakov, A., Russo, R., and Neimark, A.V. (2000) *Langmuir*, **16**, 2311–2320.

(135) Brunauer, S., Emmett, P.H., and Teller, E. (1938) *Journal of the American Chemical Society*, **60**, 309–319.

(136) Rouquerol, J., Llewellyn, P., and Rouquerol, F. (2005) *Studies in Surface Science and Catalysis*, **160**, 49–56.

(137) Walton, K.S. and Snurr, R.Q. (2007) *Journal of the American Chemical Society*, **129**, 8552–8556.

(138) Bae, Y.-S., Yazaydın, A.O., and Snurr, R.Q. (2010) *Langmuir*, **26**, 5475–5483.

(139) Rouquerol, J., Rouquerol, F., and Sing, K. (1999) *Adsorption by Powders and Porous Solids*, Academic Press, London.

(140) Thommes, M., Smarsly, B., Groenewolt, M. *et al.* (2005) *Langmuir*, **22**, 756–764.

(141) Kubo, S., White, R.J., Yoshizawa, N. *et al.* (2011) *Chemistry of Materials*, **23**, 4882–4885.

(142) Rasmussen, C.J., Vishnyakov, A., Thommes, M. *et al.* (2010) *Langmuir*, **26**, 10147–10157.

(143) Rasmussen, C.J., Gor, G.Y., and Neimark, A.V. (2012) *Langmuir*, **28**, 4702–4711.

(144) Ravikovitch, P.I. and Neimark, A.V. (2002) *Langmuir*, **18**, 1550–1560.

(145) Ravikovitch, P.I. and Neimark, A.V. (2001) *Journal of Physical Chemistry B*, **105**, 6817–6823.

(146) Tarazona, P., Marconi, U.M.B., and Evans, R. (1987) *Molecular Physics*, **60**, 573–595.

(147) Tarazona, P. (1985) *Physical Review A*, **31**, 2672–2679.

(148) Ravikovitch, P.I., Domhnaill, S.C.O., Neimark, A.V. *et al.* (1995) *Langmuir*, **11**, 4765–4772.

(149) Olivier, J.P. (1995) *Journal of Porous Materials*, **2**, 9–17.

(150) Jagiello, J. and Thommes, M. (2004) *Carbon*, **42**, 1227–1232.

(151) Neimark, A.V., Ravikovitch, P.I., and Vishnyakov, A. (2003) *Journal of Physics: Condensed Matter*, **15**, 347.

(152) Smarsly, B., Thommes, M., Ravikovitch, P.I., and Neimark, A.V. (2005) *Adsorption*, **11**, 653–655.

(153) Ravikovitch, P.I. and Neimark, A.V. (2006) *Langmuir*, **22**, 10864–10868.

(154) Gor, G.Y. and Neimark, A.V. (2011) *Langmuir*, **27**, 6926–6931.

(155) Berim, G.O. and Ruckenstein, E. (2011) *Journal of Physical Chemistry B*, **115**, 13271–13274.

(156) Kowalczyk, P., Furmaniak, S., Gauden, P.A., and Terzyk, A.P. (2010) *Journal of Physical Chemistry C*, **114**, 5126–5133.

(157) Neimark, A.V., Lin, Y., Ravikovitch, P.I., and Thommes, M. (2009) *Carbon*, **47**, 1617–1628.

(158) Gor, G.Y., Thommes, M., Cychosz, K.A., and Neimark, A.V. (2012) *Carbon*, **50**, 1583–1590.

(159) Weber, J., Antonietti, M., and Thomas, A. (2008) *Macromolecules*, **41**, 2880–2885.

(160) Ritter, N., Antonietti, M., Thomas, A. *et al.* (2009) *Macromolecules*, **42**, 8017–8020.

(161) Germain, J., Svec, F., and Fréchet, J.M.J. (2008) *Chemistry of Materials*, **20**, 7069–7076.

(162) Balzer, C., Wildhage, T., Braxmeier, S. *et al.* (2011) *Langmuir*, **27**, 2553–2560.

(163) Brochard, L., Vandamme, M., Pellenq, R.J.M., and Fen-Chong, T. (2011) *Langmuir*, **28**, 2659–2670.

(164) Garrido, J., Linares-Solano, A., Martin-Martinez, J.M. *et al.* (1987) *Langmuir*, **3**, 76–81.

(165) Rodriguez-Reinoso, F., Lopez-Gonzalez, J. de D., and Berenguer, C. (1982) *Carbon*, **20**, 513–518.

(166) Ritter, N., Senkovska, I., Kaskel, S., and Weber, J. (2011) *Macromolecules*, **44**, 2025–2033.

(167) Mason, J.A., Sumida, K., Herm, Z.R. *et al.* (2011) *Energy & Environmental Science*, **4**, 3030–3040.

(168) Sumida, K., Rogow, D.L., Mason, J.A. *et al.* (2011) *Chemical Reviews*, **112**, 724–781.

(169) Bae, Y.-S. and Snurr, R.Q. (2011) *Angewandte Chemie International Edition*, **123**, 11790–11801.

(170) Dawson, R., Adams, D.J., and Cooper, A.I. (2011) *Chemical Science*, **2**, 1173–1177.

(171) Kiskan, B., Antonietti, M., and Weber, J. (2012) *Macromolecules*, **45**, 1356–1361.

(172) Bacsik, Z., Ahlsten, N., Ziadi, A. *et al.* (2011) *Langmuir*, **27**, 11118–11128.

(173) Katsoulidis, A.P. and Kanatzidis, M.G. (2011) *Chemistry of Materials*, **23**, 1818–1824.

(174) Gassensmith, J.J., Furukawa, H., Smaldone, R.A. *et al.* (2011) *Journal of the American Chemical Society*, **133**, 15312–15315.

(175) Liu, Q., Mace, A., Bacsik, Z. *et al.* (2010) *Chemical Communications*, **46**, 4502–4504.

(176) Zhao, L., Bacsik, Z., Hedin, N. *et al.* (2010) *ChemSusChem*, **3**, 840–845.

(177) Sevilla, M. and Fuertes, A.B. (2011) *Energy & Environmental Science*, **4**, 1765–1771.

(178) Giesche, H. (2008) in *Handbook of Porous Solids* (eds F. Schüth, K.S.W. Sing, and J. Weitkamp), Wiley-VCH Verlag GmbH, Weinheim, pp. 309–353.

(179) White, R.J., Tauer, K., Antonietti, M., and Titirici, M.-M. (2010) *Journal of the American Chemical Society*, **132**, 17360–17363.

(180) Smarsly, B., Groenewolt, M., and Antonietti, M. (2005) *Progress in Colloid and Polymer Science*, **130**, 127–140.

(181) Smarsly, B., Göltner, C., Antonietti, M. *et al.* (2001) *Journal of Physical Chemistry B*, **105**, 831–840.

(182) Mascotto, S., Wallacher, D., Kuschel, A. *et al.* (2010) *Langmuir*, **26**, 6583–6592.

7

Applications of Hydrothermal Carbon in Modern Nanotechnology

Marta Sevilla[1], Antonio B. Fuertes[1], Rezan Demir-Cakan[2], and Maria-Magdalena Titirici[3]

[1] *National Council for Scientific Research (CSIC), Instituto Nacional del Carbon (INCAR), Spain*
[2] *Department of Chemical Engineering, Gebze Institute of Technology, Turkey*
[3] *School of Engineering and Materials Science, Queen Mary, University of London, UK*

7.1 Introduction

Starting with the discovery of fullerenes [1] and carbon nanotubes (CNTs) [2], materials science related to valuable carbon materials has become a hot area, motivated by its potential applications [3]. Many synthetic methods, such as carbonization, high-voltage-arc electricity, laser ablation, or hydrothermal carbonization (HTC), have been reported for the preparation of amorphous, porous, or crystalline carbons with different sizes, shapes, and chemical compositions.

During recent years, carbon materials have undergone great development because of their numerous applications in energy storage and conversion, adsorption, catalysis, and many other fields of current interest. There is no wonder that a number of novel nanostructured carbon materials have been processed, many of them from the molecular level using bottom-up strategies. To date, it has been possible to create materials with well-defined nanostructures, morphologies, and tunable surfaces [4].

We already know from the previous chapters that the HTC technique is a sustainable and flexible alternative to synthesize carbon materials. It uses low-cost precursors, water

Sustainable Carbon Materials from Hydrothermal Processes, First Edition. Edited by Maria-Magdalena Titirici.
© 2013 John Wiley & Sons, Ltd. Published 2013 by John Wiley & Sons, Ltd.

as a carbonization media, and lower temperatures in comparison with other carbonization methods, thus enabling the presence of polar surface groups that can be modulated by additional heat treatment.

Chapter 2 showed how it is possible to introduce tailored porosities into these types of materials and change their morphologies, Chapter 3 showed how chemical activation is a suitable method to introduce microporosity, Chapter 4 showed how HTC-derived materials can be combined with inorganics resulting in hybrid materials with novel properties, and Chapter 5 described how these materials (and their properties) can be modified with various functionalities and heteroatoms.

It can thus be foreseen that HTC-derived materials are suitable for a wide range of important applications from adsorption and catalysis to energy storage and generation. Indeed, the most appealing feature of HTC is the fact that it represents an easy, green, and kilogram-scalable process, allowing the production of various carbon and hybrid nanostructures with practical applications on a price basis that is comparatively lower then corresponding petrochemical processes. Even though relatively in their infancy, HTC-derived materials have already found numerous applications, including soil enrichment (see Chapter 8), catalysis, water purification, energy and gas storage, sensors, and bioapplications. In this chapter, examples where HTC-derived materials have proved to be not only sustainable, but to possess extraordinary properties, which in some cases surpass those of current "gold standards," are introduced and discussed. We give some examples describing the application of carbonaceous materials made using the HTC process in the previously mentioned fields.

7.2   Energy Storage

Climate change and the decreasing availability of fossil fuels require society to move towards sustainable and renewable resources. As a result, we are observing an increase in renewable energy production from sun and wind, as well as the development of electric vehicles with low CO_2 emissions. As the sun does not shine during the night, wind does not blow on demand, and we all expect to drive our car with a few hours of autonomy, energy storage systems are starting to play a larger role in our lives.

Therefore, in response to the needs of modern society and emerging ecological concerns, it is now essential that new, low-cost, and environmentally friendly energy conversion and storage systems are found. This is why over the past few years the number of publications and research groups working in the field of "sustainable materials for energy storage" applications has exploded. The performance of these energy storage devices depends initially on the properties of the materials they are made off. Innovative materials chemistry lies at the heart of the advances that have already been made in energy conversion and storage, such as the introduction of the rechargeable Li-ion battery, development of supercapacitors, and the emerging new and high-energy-density technologies such as Li–S and Li–air batteries.

Further breakthroughs in materials, not incremental changes, hold the key to new generations of energy storage and conversion devices. Additionally, it is strongly desired that these materials should be easy to make, cheap, and if possible based on renewable resources.

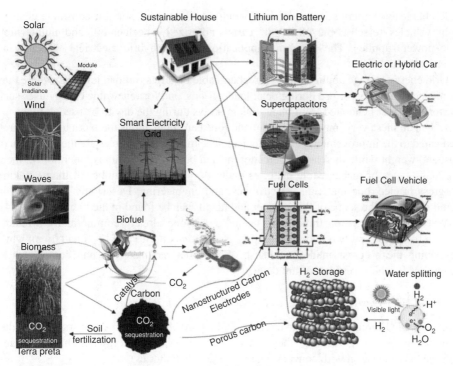

Figure 7.1 *Schematic representation of renewable energy generation, storage, and utilization using modern electrochemical devices based on "green" carbon materials.*

Figure 7.1 shows a schematic representation on how such high-performance green materials such as hydrothermal carbons should have specific characteristics and be designed for performing specific tasks, either as electrode materials for Li-ion batteries and supercapacitors, catalysts for the oxygen reduction reaction (ORR) in fuel cells, or host materials for hydrogen storage.

Within this section we will provide a brief overview on the use of HTC-derived materials for such energy storage-related applications when used as electrodes in rechargeable batteries or supercapacitors.

7.2.1 Electrodes in Rechargeable Batteries

Rechargeable batteries are made up of one or more electrochemical cells and are also known as secondary cells because the electrochemical reactions are electrically reversible. Several different combinations are currently commercially used, including lead acid, nickel cadmium (NiCd) [5], nickel metal hydride (NiMH) [6], Li-ion [7], and Li-ion polymer [8]. Many other are under development, such as thin-film lithium [7f, 9], Li–S [10], Na–S [11], Mg-ion [12], Li-air [13], and others. Although such rechargeable batteries have a higher initial cost, their main advantage is that they can be recharged cheaply and used many times.

Rechargeable batteries are used today for automobile starters, portable consumer devices, light vehicles (electric bicycles, golf carts, motorized wheelchairs), and uninterruptable power supplies. The emerging applications are in hybrid electrical vehicles and electrical vehicles.

Grid energy storage applications use rechargeable batteries for load leveling, where they store electric energy for use during peak load periods, and for renewable energy uses such as storing power generated from photovoltaic arrays during the day to be used at night.

Due to all these very important current and foreseen applications there is a lot of research dedicated to the improvement of these devices. The main driving forces are the reduction of cost and weight while increasing their lifetime and power. As previously mentioned, there are many types of rechargeable batteries under development, a number of them, making progress in electrode materials, electrolytes, and engineering. Describing all of them in detail is beyond the scope of this chapter and details can be found in the literature [7e, 14].

We will focus here on three main technologies, one already commercially available (Li-ion batteries) and two under development (Na-ion batteries and Li–S batteries), while describing the progress made in the use of HTC-derived materials and composites as electrodes for these very important devices.

7.2.1.1   HTC-Derived Materials as Electrodes in Li-Ion Batteries

Rechargeable lithium batteries have reformed the portable electronics industry. This is due to their superior energy density (they can store 2–3 times more energy per unit weight and volume in comparison with conventional rechargeable batteries).

Li-ion batteries are Li-ion devices comprising a graphite negative electrode (anode), a nonaqueous liquid electrolyte, and a positive electrode (cathode) of a spinel-type oxide $LiMO_2$ (where M = Co, and sometimes Mn and Ni). On charging, Li-ions are desintercalated from the layered $LiMO_2$ intercalation host, pass across the electrolyte, and intercalated between the graphite layers in the anode. Discharge reverses the process (Figure 7.2). The electrons of course pass around the external circuit.

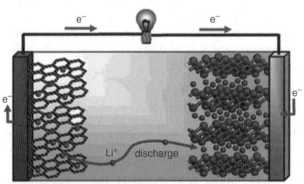

Figure 7.2 *Schematic representation of a Li-ion battery. Negative electrode (graphite) and positive electrode (LiCoO₂), separated by a nonaqueous liquid electrolyte. (Reproduced with permission from [7b]. © 2008 WILEY-VCH Verlag GmbH & Co. KGaA, Weinheim.)*

7.2.1.1.1 Anodes Based on Intercalation/Desintercalation The storage of one lithium atom between every six carbon atoms is only valid for graphite. This LiC_6 stoichiometry permits graphite a storage capacity of 372 mAh g^{-1} – a value lower than can be obtained with other materials, as we will see later. However, this low storage capacity results in only a small volumetric change of about 10% and allows for a life of at least 500 cycles depending on the current rate used. The anode reaction based on intercalation/desintercalation is:

$$yC + xLi + e = Li_xC_y \qquad (7.1)$$

Disordered carbons (the so-called hard carbons) can easily exceed this theoretical value reported for graphite. This phenomenon is still hard to explain by classical graphite intercalation science and new mechanisms are under development. A more detailed discussion about the possible lithium storage in carbon nanostructures can be found elsewhere [15]. The main problem associated with carbon materials is that the values for higher capacities are obtained when the potential is close to 0 V versus Li/Li^+, which is not safe, especially for high-power applications such as electrical vehicles.

There are a multitude of reports on the use of pure carbon materials as negative electrodes in Li-ion batteries as well as some very good reviews summarizing all these findings [3a, 7e, 16]. We will limit ourselves here to the use of hydrothermal carbons for such applications.

The first study using hydrothermal carbon as a negative electrode in Li-ion batteries was performed by Huang *et al.* [17]. The authors demonstrated that the reversible lithium insertion/extraction capacity of this kind of material is much higher than the theoretical capacity of graphitized carbonaceous materials. However, these materials had relatively high surface area and a high number of micropores, which is known not to be beneficial for lithium storage [18]. Therefore Titirici *et al.* limited the thickness of the carbon shell to only few nanometers, producing hydrothermal carbon hollow spheres [19]. The electrochemical behavior of the hollow carbon nanospheres was characterized by cyclic voltammetry (CV) at a scan rate of 0.1 mV s^{-1} (Figure 7.3a). The voltammogram shows two cathodic peaks at 1.3 and 0.8 V that appear only in the first cycle. It is reasonable to assume that the peak at higher potential is due to the reaction of lithium with surface functional groups located at the carbon surface, whereas the peak at 0.8 V is related to the formation of a solid electrolyte interface (SEI). This is a typical problem associated with carbon materials related to the decomposition of the electrolyte onto the electrode surface and trapping some lithium inside. This contributes to the large irreversible capacity of the first discharge process. The phenomenon was further confirmed by the galvanostatic discharge charge results (Figure 7.3b).

Two plateaus at 1.4 and 0.8 V are observed in the first discharge curve corresponding to the additional peaks in the CV curves. A large irreversible capacity of 700 mAh g^{-1} is found over the first cycle as expected. From the presented results, it is obvious that both electrolyte decomposition on the electrode/electrolyte surface and irreversible lithium insertion into potentially unique positions, such as cavities or sites in the vicinity of residual hydrogen atoms in the carbon material, cause the large irreversible capacity as reported earlier for such kind of materials [21]. The reversible capacity at 1 C rate for the hollow carbon nanospheres could reach up to 370 mAh g^{-1}, which is slightly higher than graphite (Figure 7.3c). While previous reports suggest that nongraphitic carbons show a continuous

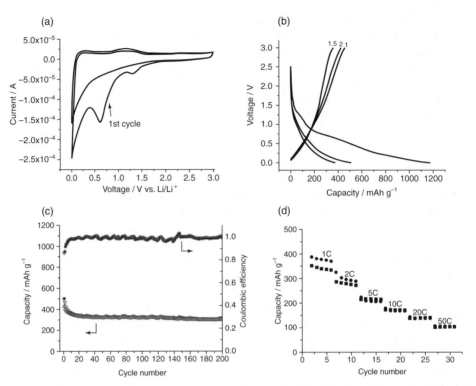

Figure 7.3 *(a) CV of hollow carbon spheres. The graph shows the first two cycles between 3 and 0 V at a scan rate of 0.1 mV s⁻¹. (b) Galvanostatic discharge/charge curves of hollow carbon nanospheres at a rate of 1 C. (c) Cycle performance of hollow carbon nanospheres cycled at a rate of 1 C. (d) Rate performance of hollow carbon nanospheres. (Reproduced with permission from [20]. © 2012 WILEY-VCH Verlag GmbH & Co. KGaA, Weinheim.)*

and progressive decay in capacity during cycling, here the investigated hollow spheres presented excellent cycling performance at 1 C rate. After 200 discharge/charge cycles, a reversible capacity as high as 310 mAh g⁻¹ was retained, while the coulombic efficiency approached almost 100%. However, the real advantage of such hollow carbon spheres besides the high cyclability is the outstanding rate capabilities (Figure 7.3d). A specific charge capacity of 200 mAh g⁻¹ is reached at 5 C and when the current density is increased from 1 to 20 C the capacity retention can approach 50%. Even at a high rate of 50 C (18.6 A g⁻¹), a capacity of 100 mAh g⁻¹ is still maintained. This value is much higher than traditional graphite electrodes (almost negligible at such a high rate). We should also mention that the rate performance is much higher than that of a similar carbon but as a full sphere, indicating that indeed the hollow nanosphere morphology plays a vital role for the excellent electrochemical performance [20].

A similar excellent performance was also confirmed by Xia *et al.* who prepared one-dimensional hierarchical porous hydrothermal carbon fibers from alginic acid fibers [22]. The carbon fibers consist of a three-dimensional network of nanosized carbon with excellent rate capability and capacity retention compared with commercial graphite.

Another material in which lithium can be inserted is TiO_2 [23, 24]. Regardless of various polymorphs of TiO_2 (rutile, anatase, brookite) the insertion reaction of Li^+ into TiO_2 can be expressed as:

$$TiO_2 + xLi^+ + xe^- = Li_x TiO_2 \qquad (7.2)$$

In this redox reaction, the insertion of positively charged Li^+ is balanced with the uptake of electrons to compensate Ti(III) cations in the Ti(V) sublatice, which usually results in a sequential phase transformation occurring in original TiO_2 as a function of Li^+ content [23a].

The theoretically calculated capacity of TiO_2 is 330 mAh g^{-1}, which is a little lower than that of graphite [25]. However, the volume change of TiO_2 is less than 4% as Li^+ inserted into TiO_2 electrodes. This affords TiO_2 an outstanding structure stability after Li^+ insertion and thus extremely long cycling life as electrodes in Li-ion batteries.

Safety is another crucial concern in Li-ion batteries. However, the working potential for most high-capacity materials including graphite is in the region of 0–0.5 V (versus Li/Li$^+$). In such a low-operation-voltage region, the electrolyte is prone to decompose and form the SEI on the anode surface, as we could observe in the example shown in Figure 7.3. Concurrent with the electrolyte decomposition, gases are released and build up pressure in the cell. This situation will endanger the safety of the battery system as gases accumulate with increasing cycling time. Another advantage of TiO_2 in this respect is that it operates at a higher voltage (1.5–1.8 V versus Li/Li$^+$) and generates less energy than other anode materials. The formation of the SEI on its surface is thus avoided at such a high potential, which greatly improves the overall safety of the battery. In addition, another great advantage is that TiO_2 is cheap. The disadvantage is that like for any inorganic material, TiO_2 is not electrically conductive (around 1×10^{-12} S m^{-1}) resulting in low energy/power densities. Its conductivity can be improved by combining it with carbon in small amounts.

Thus, carbon-coated TiO_2 nanotubes were prepared by a simple one-step hydrothermal method with the addition of glucose in the starting powder. A thin carbon coating was obtained on the nanotube surface that effectively suppressed the aggregation of TiO_2 nanotubes during postcalcination. This action resulted in better ionic and electronic kinetics when applied to Li-ion batteries. Consequently carbon-coated TiO_2 nanotubes deliver a remarkable Li-ion intercalation/desintercalation performance with reversible capacities of 286 and 150 mAh g^{-1} at 250 and 750 mAh g^{-1}, respectively [26].

In the TiO_2 family, spinel-type lithium titanates have also attracted great interest as anode materials in Li-ion batteries. This is due to their unique characteristics such as: (i) zero strain during charging/discharging, (ii) excellent cycle reversibility, (ii) fast Li$^+$ insertion/extraction, and (iv) high litigation voltage plateau at 1.55 V versus Li/Li$^+$ that avoids the formation of metallic lithium and therefore improves the safety of the battery.

Carbon coating of fine $Li_4 Ti_5 O_{12}$ was carried out by amphiphilic carbonaceous material in aqueous solution followed by carbonization at 800 °C. The materials behave well when applied as anodes in Li-ion batteries with excellent rate performance [27].

7.2.1.1.2 Anodes Based on Alloying Reactions   Some metals and semiconductors can react with lithium forming alloys according to $Li_x M_y$ through electrochemical processes [7a, 28]. The reaction is partially reversible and involves a large number of atoms per formula unit. The specific volumetric and gravimetric capacities exceed those of graphite.

For example, $Li_{4.4}Sn$ has a gravimetric capacity of 993 versus 372 mAh g^{-1} for graphite. The corresponding values for $Li_{4.4}Si$ are 4200 mAh g^{-1}. Owing to their high theoretical capacities, many efforts have been made to apply such materials as anodes in Li-ion batteries [28b]. However, such alloying reactions are normally associated with severe volume changes during lithium insertion/extraction, leading to a very fast decay in capacity with cycling. Another associated problem, as for TiO_2, is the low conductivity of such materials.

In order to overcome such problems, Titirici *et al.* successfully coated "*in situ*" commercially available silicon nanoparticles with a thin layer (25 wt%/10 nm) of HTC-derived material via the conversion of D-glucose (Figure 7.4a) [29]. During the hydrothermal treatment a thin layer of SiO_x was also formed on the particle surfaces. The resulting $Si@SiO_x/C$ composite was further carbonized to increase its conductivity. The resulting material showed a markedly improved cyclability compared with the pure silicon (Figure 7.4b). Especially when working in the presence of vinylene carbonate (VC) additive

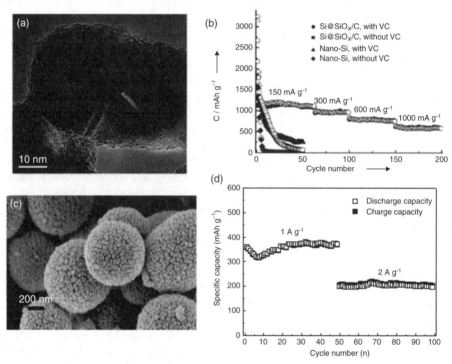

Figure 7.4 *(a) High-resolution transmission electron microscopy (TEM) of the $Si@SiO_x/C$ composite showing the crystalline silicon in the middle coated with a thin amorphous layer of SiO_x and carbon. (b) Cycling and rate performance of pure silicon nanoparticles and $Si@SiO_x/C$ nanocomposite electrodes cycled in VC-free and VC-containing electrolytes ($LiPF_6$ in ethylene carbonate/dimethyl carbonate solutions) (solid symbols: charge; empty symbols: discharge). (c) Scanning electron microscopy (SEM) micrograph of SnO_2 microspheres made out of aggregated nanoparticles. (d) Variation in discharge/charge capacity versus cycle number for the mesoporous SnO_2 sample cycled at current densities of 1 and 2 A g^{-1}. (Reproduced with permission from [29]. © 2008 WILEY-VCH Verlag GmbH & Co. KGaA, Weinheim and [30]. © 2008 American Chemical Society.)*

in the electrolyte, an excellent cycling performance was achieved. The reversible capacity was as high as 1100 mAh g^{-1} at a current density of 150 mA g^{-1}, with no further decay in capacity, even after 60 cycles.

Similarly, preformed SnO$_2$ nanoparticles were incorporated inside hydrothermal carbon spheres derived from the conversion of D-glucose. The resulting composites were carbonized and utilized as electrodes in Li-ion batteries. The carbon could be subsequently removed, resulting in mesoporous SnO$_2$ microparticles (Figure 7.4c). Here, hydrothermal carbon acts as an assembly medium of the initial nanoparticles into mesoporous SnO$_2$ nanospheres, leading to a better packing density and superior lithium insertion properties (Figure 7.4d) [30].

The reaction of SnO$_2$ with lithium takes place according to the following reaction [31]:

$$SnO_2 + 4Li \rightarrow Sn + 2Li_2O \tag{7.3}$$

$$Sn + xLi \leftrightarrow Li_xSn \tag{7.4}$$

The maximum theoretical capacity of the SnO$_2$ anode is 781 mAh g^{-1} by this mechanism. The irreversible initial capacity loss is due to the formation of amorphous Li$_2$O matrix.

Cao *et al.* have also reported on the preparation of multilayered nanocrystalline SnO$_2$ hollow spheres via a chemically induced self-assembly approach performed under hydrothermal conditions [32]. First, a multilayered, spherical SnO$_2$/hydrothermal carbon composite is produced through condensation/polymerization and carbonization of sucrose accompanied by hydrolysis of SnCl$_4$ during the hydrothermal reaction. After the removal of the carbon hollow SnO$_2$ spheres were obtained with good electrochemical performance when applied as anode materials in Li-ion batteries.

Wu *et al.* reported on the synthesis of carbon-coated SnO$_2$ nanotubes through a simple glucose hydrothermal and subsequent carbonization approach using tin nanorods as sacrificial templates. The as-synthesized SnO$_2$/C nanotubes have been applied as anodes in Li-ion batteries and exhibited improved cyclic performance compared to pure SnO$_2$ nanotubes. The hollow nanostructure, together with the carbon matrix that has a good buffering effect and high electronic conductivity, may be responsible for the improved cyclic performance [33].

In a similar fashion with coating the silicon nanoparticles, Li *et al.* synthesized SnO$_2$@C core/shell nanochains by carbonization of a SnO$_2$@carbonaceous polysaccharide (CPS) precursor at a relatively low temperature. The strategy results in the carbonization of CPS whilst avoiding the carboreduction of SnO$_2$ at 700 °C. It has been shown that moderate carbon content contributes to the one-dimensional growth of SnO$_2$@carbon core/shell nanochains. The thickness of the carbon shell can be easily manipulated by varying the hydrothermal treatment time. Such unique nanochain architecture could afford a very high lithium storage capacity as well as a desirable cycling performance. More than 760 mAh g^{-1} of reversible capacity was achieved at a current density of 300 mA g^{-1} and above 85% retention can be obtained after 100 charge/discharge cycles. TEM analysis of electrochemically cycled electrodes indicates that the structural integrity of the SnO$_2$@C core/shell nanostructure is retained during the electrochemical cycling, contributing to the good cycling performance demonstrated by the robust carbon shell [34].

Following a procedure initially described by Titirici *et al.* [35], Archer *et al.* used the glucose-mediated hydrothermal method for a gram-scale synthesis of nearly monodisperse hybrid SnO_2 nanoparticles [36]. Here, glucose was playing a dual role: (i) facilitation of rapid precipitation of polycrystalline SnO_2 nanocolloids and (ii) creation of a uniform carbon coating on the SnO_2 cores. The thickness of the coated layer could be easily manipulated by variation of the glucose concentration in the synthesis medium. The resulting SnO_2 colloids coated with carbon exhibited significantly enhanced cycling performance for lithium storage. By reduction with hydrogen the authors demonstrated a simple route to carbon-coated tin nanospheres. Lithium storage properties of the later materials are also reported in the paper, suggesting that the large irreversible losses in these materials are caused not only by the initial irreversible reduction of SnO_2 as generally perceived in the field, but also by the formation of the SEI [36].

7.2.1.1.3 Anodes Based on Conversion Reactions   Another possibility to increase the capacity of an anode in Li-ion batteries is by the so-called "conversion reactions" as first reported by Poizot in 2000 [37]. Lithium can thus be stored reversibly in a transition metal oxide through a heterogeneous conversion reaction according to:

$$M_a X_b + (b \cdot n)\, Li \leftrightarrow aM + bLi_n X \qquad (7.5)$$

where M = transition metal, X = anion, and n = formal oxidation state of X.

Later, reversible lithium storage was also observed in transition metal fluorides, sulfides, phosphides, and nitrides [38, 39]. It has been also shown that a small nanoparticle size is beneficial for such conversion reactions [7b].

Taking in consideration these arguments and with additional benefits from the conductive carbon phase, many carbon–metal oxides and nitrides have been prepared as successful electrodes in Li-ion batteries. We will again only mention here those prepared using the HTC of inorganic precursors and carbohydrates.

Tu *et al.* prepared a spherical NiO/C composite by dispersing NiO in glucose solution and subsequent carbonization under hydrothermal conditions at 180 °C. Electrochemical tests showed that the NiO/C composite exhibited higher initial coulombic efficiency (66.6%) than the pure NiO (56.4%) and better cycling performance. The improvement of these properties is attributed to the carbon as it can reduce the specific surface area and enhance the conductivity [40].

CoFe layered double hydroxide (LDH) nanowall arrays have been grown directly from a flexible alloy substrate by a facile hydrothermal method in the presence of glucose. After annealing treatment in argon atmosphere, carbon-coated CoFe mixed-oxide nanowalls arrays were further investigated as anode materials for Li-ion batteries. They exhibited improved electrical conductivity and superior electrochemical performance in terms of specific capacity and cyclability as compared to a carbon-free sample and a sample made by a previous carbon-coating method [41].

Co_2SnO_4@C core/shell nanostructures were prepared through a simple glucose hydrothermal and subsequent carbonization approach. The as-synthesized Co_2SnO_4@C core/shell nanostructures have been applied as anodes for Li-ion batteries with improved cycling performance compared to the pure materials [42].

7.2.1.1.4 Positive Electrodes   In addition to negative electrodes in Li-ion batteries, the development of improved cathode materials is recognized as even more challenging [43]. Although transition metal oxides (with the golden standard being $LiCoO_2$; see Figure 7.2) are commonly used as energy storage materials in today's modern portable devices, concerns over safety and cost have prompted research into other new positive electrode materials for Li-ion batteries. Several new compounds have been explored as possible alternatives, including those obtained by introducing large polyanions of the form $(XO_4)^{y-}$ ($X = S$, P, Si, As, Mo, W) into the lattice. An inductive effect from $(PO_4)^{3-}$ and $(SO_4)^{2-}$ ions raises redox energies compared to those in oxides and stabilizes the structure [44]. The presence of polyanion $(XO_4)^{y-}$ with strong X—O covalent bonds increases the potential as a result of the strong polarization of oxygen ions toward the X cation, which lowers the covalency of the M—O bond. Research shows that most of the lithium metal phosphate and sulfate compounds containing FeO_6 octahedra as the redox center have potentials in the range of 2.8–3.5 V versus Li/Li^+ [45]. Another advantage of using iron-based compounds is that, in addition to being naturally abundant and inexpensive, they are less toxic then other transitional metal compounds.

The focus of the Li-ion batteries community further intensified on this class of compounds with Pahadi's report on the electrochemical properties of $LiFePO_4$ [46]. $LiFePO_4$ satisfies many of the criteria for an electrode material in a Li-ion battery: it can reversibly intercalate lithium at a high voltage (3.5 V) and has a gravimetric capacity of 170 mAh g^{-1}, which gives a cell a high energy density. $LiFePO_4$ is stable against overcharge or discharge and is compatible with most electrolyte systems [47]. $LiFePO_4$ is also environmentally friendly, as it is found in nature as the mineral triphylite and is made from abundant elements, reducing its production costs.

Some drawbacks associated with $LiFePO_4$ are its low electronic conductivity (1×10^{-9} S cm^{-1} at room temperature), which limits its electrochemical performance as the electrons cannot easily transport through the material. $LiFePO_4$ becomes conductive in the presence of small amounts of carbon [48] as well as when doped with various cations, forming compounds of the type $Li_{1-x}M^{z+}{}_xFePO_4$ ($z \geq 2$) with exceptional conductivities when sintered at 800 °C, due to charge compensation (10 times higher than pure $LiFePO_4$) [49].

Titirici and Popovic in collaboration with Demir-Cakan *et al.* synthesized in one step a hierarchical mesocrystal of $LiFePO_4$ coated with a thin layer of nitrogen-doped carbon (Figure 7.5a, b). Due to an increased conductivity of the nitrogen-doped carbon, the coated material exhibited a superior performance compared with the pure $LiFePO_4$ [50].

Paranthaman *et al.* modified the surface of rod-like $LiFePO_4$ with a conductive nitrogen-doped carbon layer using hydrothermal processing followed by postannealing in the presence of an ionic liquid. The conductive surface-modified rod-like $LiFePO_4$ exhibits good capacity retention and high rate capability as the nitrogen-doped carbon layer improves conductivity and prevents aggregation of the rods during cycling [51].

Similar to phospholivines, the orthosilicates (Li_2MSiO_4, where M = Fe, but also Mn and Co) have been employed recently as alternative cathodes in Li-ion batteries. Their main advantages are cell safety, the possibility of extraction of more than one Li^+ ion per unit formula, and high theoretical capacity (333 mAh g^{-1}) [52]. Moreover, the orthosilicate group material renders excellent thermal stability offered through strong Si—O bonding. However, they have the same disadvantage suffered similarly by their related olivine phosphates – very poor electronic conductivity. Using the advantages of hydrothermal synthesis, Aravindan *et al.* prepared in one step carbon-coated Li_2MnSiO_4 with a flower-like morphology similar to that of $LiFePO_4$ in Figure 7.5a and b, and good electrochemical performance (100 mAh g^{-1}) [53]. The improvement of the electronic conductivity due to the carbon coating was validated through electrochemical impedance spectroscopy.

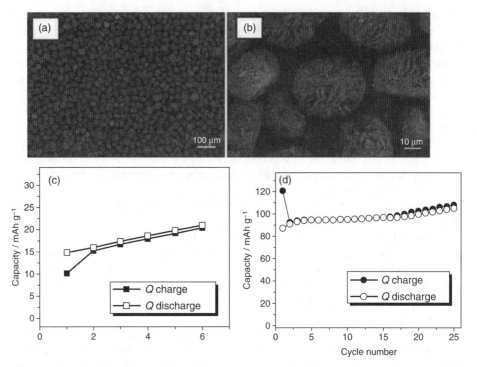

Figure 7.5   *(a and b) SEM micrographs of the LiFePO$_4$@C mesocrystals. (c) Charge/discharge capacity versus cycle number plot for the pristine LiFePO$_4$. (d) Charge/discharge capacity versus cycle number plot for the LiFePO$_4$@C showing superior performance. (Adapted with permission from [50]. © 2011 WILEY-VCH Verlag GmbH & Co. KGaA, Weinheim.)*

Since Sony announced the first version of commercialized Li-ion batteries in 1991, Li-ion batteries have rapidly penetrated into everyday life. Here, we have tried to show the main research directions taken for improving the performance of the electrode materials in Li-ion batteries with a focus on HTC technology. We have learned that Li-ion technology relies on a rich and versatile chemistry, leading to a wide range of attractive electrode materials for both positive (LiCoO$_2$, LiMn$_2$O$_4$, LiFePO$_4$) and negative electrodes (C, Sn, Si, etc.). Much progress has been achieved in this field and the HTC-derived materials proved to be successful candidates for this task.

7.2.1.2   HTC-Derived Materials as Electrodes in Na-Ion Batteries

There has been recent concern that the amount of the lithium resources that are buried in the Earth would not be sufficient to satisfy the increased demands for Li-ion batteries [54]. While there is ample evidence that this is not a cause for immediate concern, a very large market share of electric vehicles can put a strain on lithium production capabilities [55].

Sodium is located below lithium in the periodic table and they share similar chemical properties in many aspects. The fundamental principles of Na-ion and Li-ion batteries are identical – the chemical potential difference of the alkali-ion (lithium or sodium) between two electrodes (anode and cathode) creates a voltage in the cell. During charge and discharge the alkali ions shuttle back and forth between the two electrodes. There are several reasons to investigate Na-ion batteries. Recent computational studies by Ceder *et al.* [56] on voltage, stability, and diffusion barriers of Na-ion and Li-ion materials indicate that Na-ion systems can be competitive with Li-ion systems. In any case, Na-ion batteries would be interesting for very-low-cost systems for grid storage, which could make renewable energy a primary source of energy rather than just a supplemental one.

As battery applications extend to large-scale storage, such as electric buses or stationary storage connected to renewable energy production, high energy density becomes less critical. Moreover, the abundance and low cost of sodium in the Earth can become an advantage when a large amount of alkali is demanded for large-scale applications, although at this point the cost of lithium is not a large contribution to the cost of Li-ion batteries. However, most importantly, there may be significant unexplored opportunity in sodium-based systems. Sodium intercalation chemistry has been explored considerably less than lithium intercalation and early evidence seems to indicate that structures that do not function as well as lithium intercalation compounds may work well with sodium. Hence, there may be opportunities to find novel electrode materials for Na-ion batteries [57].

Many materials have been proposed in the literature as possible cathodes for Na-ion batteries, whereas only some carbon-based anodes have been pointed out for this storage technology. The best candidates for cathodic materials in a Na-ion battery are phosphate-based materials, because of their thermal stability and higher voltage due to the inductive effect [58]. We can mention olivine $NaFePO_4$ (with the highest theoretical specific capacity), $NaVPO_4F$, $Na_3V_2(PO_4)_2F_3$, Na_2FePO_4F, and $Na_3V_2(PO4)_3$. These phosphates require conductive coating and nanostructured morphology in order to improve their electrochemical performance [59].

The development of negative electrodes is, however, more challenging. Graphite, the standard anode material in current Li-ion batteries, seems not to be suited for a sodium-based system, as sodium hardly forms staged intercalation compounds with graphite [60]. The situation is different when amorphous carbons are used as intercalation media, and it is suggested that the storage mechanisms for sodium and lithium are similar, although the capacity is smaller in the case of sodium [61]. Thus, several types of nongraphitic carbon materials have been tested, and capacities between 100 and 300 mAh g^{-1} under differing conditions were found [62, 63]. Although these capacities are promising, the cells were only cycled for a few times and, more importantly, could only be obtained at extremely low currents (typically C rates between C/70 and C/80) or at elevated temperatures (above 60 °C), which indicates very slow kinetics for the sodium storage process Thus, alternative carbon materials are needed in order to achieve satisfactory performance at room temperature and at higher currents. A recent and very important report in the field describes clear improvements with faster kinetics and higher capacity by introducing nanoporosity and a hierarchical pore system into the carbon anode. High capacities can be achieved at room temperature at high currents (e.g., C rate of C/5, 15 times faster than C/75), while also exhibiting long cycling stability. The outstanding performance of the templated carbon is illustrated by comparing with several commercially available porous carbon materials and nonporous graphite as reference [64]. There is great progress in the field, although a great deal of research still needs to be dedicated to the production and improvement of carbon materials and carbon composites for Na-ion batteries, and more and more reports are being published in the literature on this topic [65–67].

As Na$^+$ can be inserted in disordered amorphous carbons, HTC-derived materials should also be considered as a class of suitable candidates as anodes for Na-ion batteries. In this respect, Titirici and Tang have applied hydrothermal carbon hollow spheres as a negative electrode in Na-ion batteries [68].

The materials have been prepared as similarly described by coating latex nanoparticles with glucose followed by HTC, and further calcination to remove the latex and increase the conductivity of carbon (Figure 7.6a and b) [19]. The cycling stability during sodium insertion/extraction in the hollow carbon nanospheres was investigated at a current density of 50 mA g^{-1} for the first 10 cycles and then 100 mA g^{-1} for subsequent cycles (Figure 7.6c). It is thought that the observed capacity loss over the initial cycling steps stems from SEI film stabilization and irreversible Na-ion insertion. After 100 cycles, a reversible capacity of around 160 mAh g^{-1} is stably maintained. The coulombic efficiency approaches around 94% after several cycles, whilst the observed irreversible capacity during each cycle is attributed to the incomplete stabilization of the SEI for the presented sodium battery system. The electrochemical impedance spectra of the hollow-carbon-nanosphere electrode was measured (Figure 7.6d). The Nyquist plots consist of a depressed semicircle

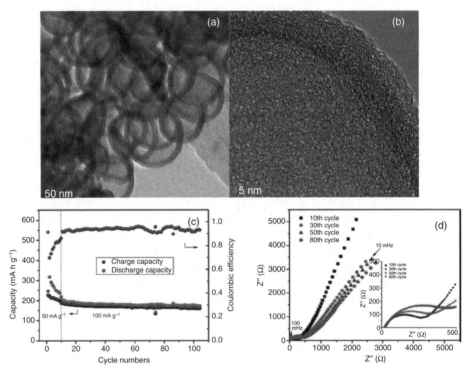

Figure 7.6   (a) TEM image and (b) high-resolution TEM image of hollow carbon nanospheres. (c) Cycle performance of hollow carbon nanospheres. (d) Impedance spectra of the hollow carbon nanospheres electrode after 10, 30, 50, and 80 cycles (inset: enlarged spectra). (Reproduced with permission from [68]. © 2012 WILEY-VCH Verlag GmbH & Co. KGaA, Weinheim.)

in the high- and middle-frequency regions, and a straight line in the low-frequency region. The semicircle can be attributed to the SEI film and contact resistance at high frequencies, and a charge transfer process in the middle frequency, while the linear increase in the low-frequency range may reflect Warburg impedance associated with Na^+ diffusion in the carbon electrode [69]. The rate performance of these materials was also excellent and clearly superior to nonporous hydrothermal carbon spheres obtained from glucose without the latex templates, proving the important role played by the hollow sphere morphology [68].

7.2.1.3 HTC-Derived Material Cathodes for Li–S Batteries

Li-ion batteries have transformed portable electronics and will play a key role in the electrification of transport. However, the highest energy storage possible for Li-ion batteries is insufficient for the long-term needs of society (e.g., extended-range electric vehicles). To go beyond the horizon of Li-ion batteries is a formidable challenge; there are few options and one of them is Li–S batteries [10c, 70].

Under intense examination for well over two decades, the cell in its simplest configuration consists of sulfur as the positive electrode and lithium as the negative electrode [71]. Li–S cells operate on a principle very different from the Li-ion battery described above. The redox couple described by the reversible reaction lies near 2.2 V with respect to Li^+/Li^0 – a potential about two-thirds of that exhibited by conventional positive electrodes [72]:

$$S_8 + 16Li = 8Li_2S \tag{7.6}$$

However, this is compensated by the very high theoretical capacity afforded by the nontopotactic "assimilation" process, of 1675 mAh g^{-1}. Thus, compared with intercalation batteries, Li–S cells have the opportunity to provide a significantly higher energy density (a product of capacity and voltage). Values can approach 2500 Wh kg^{-1} or 2800 Wh l^{-1} on a weight or volume basis, respectively, assuming complete reaction to Li_2S. Despite its considerable advantages, the Li–S cell is plagued with problems that have slowed down its widespread practical realization.

One of the problems in Li–S batteries is sulfur's low electrical conductivity. Another problem is the fact that the polysulfides that are generated at the cathode during discharging are soluble into most of the utilized electrolytes and thus they migrate to the anode where they react with the lithium electrode to form lower-order polysulfides, which are then transported back to the sulfur cathode and regenerate the higher form of polysulfides [10f]. Such a polysulfide "shuttle" process decreases the utilization of the overall active material mass during discharge, and triggers current leakage, poor cyclability, and reduced columbic efficiency of the battery [10e].

Some significant progress in overcoming these two very important challenges associated with Li–S batteries commenced after the pioneering work of Nazar *et al.* [73]. They demonstrated for the first time that those cathodes based on nanostructured/sulfur/mesoporous carbon materials can overcome these challenges to a large degree, and the Li–S cell can exhibit stable, high, reversible capacities (up to 1320 mAh g^{-1}) with good rate properties and cycling efficiency. The proof-of-concept studies are based on CMK-3 – the most well-known member of the mesoporous carbon family obtained from replication of SBA 15 silica [74].

The melt is imbibed into the channels by capillary forces, whereupon it solidifies and shrinks to form sulfur nanofibers that are in intimate contact with the conductive carbon walls. Such an intimate contact of the insulating sulfur and discharge-product sulfides with the retaining conductive carbon framework at nanoscale dimensions affords excellent accessibility of the active material. The carbon framework not only acts as an electronic conduit to the active mass encapsulated within, but also serves as a minielectrochemical reaction chamber. The entrapment ensures that a more complete redox process takes place and results in enhanced utilization of the active sulfur material. This is vital to the success of all conversion reactions to ensure full reversibility of the back-reaction. Following this report, the development of novel cathodes for Li–S cells based on nanostructured carbon/sulfur composites has flourished, and more and more research interest is dedicated to improving the performance of Li–S cells [10a, 75].

Among these carbon materials, hydrothermal carbons have just started to find their role. The first HTC-derived materials tested as cathodes in Li–S batteries following sulfur infil-tration were the carbon hollow spheres previously applied as electrodes in lithium [20] and sodium [68] storage. This time they were prepared using silica templates instead of latex (Figure 7.7a and b). Three different hydrothermal carbon/S composites were compared. The first two were hollow spheres, although prepared using two different infiltration techniques for sulfur: melt diffusion and simple physical mix. The third tested material was a non-porous hydrothermal carbon infiltrated with sulfur by simple physical mixing. As shown in Figure 7.7c, stable cycling properties can be obtained when using the nanostructured hollow spheres. This indicates that the highly porous shells of hollow spheres acts as an absorbent for soluble polysulfides, especially since a very small amount of silica remaining from the template is also present in this composite, which was previously proved to be benefic for polysulfide adsorption [76].

The higher specific capacity obtained via the melt diffusion method reveals the formation of a better conductive contact in the hydrothermal carbon hollow spheres/S composite compared to the physical mixture of hollow spheres and sulfur. The results obtained for nonhollow microspheres emphasized two benefits related to the use of hollow spheres in Li–S cells. (i) The nanosized spheres improve the specific capacity by ensuring formation of a conductive network in the cathode made by a composite consisting of carbon spheres, sulfur, and binders. If we compare the specific capacity at the first cycle of the mixtures made by hollow spheres and nonhollow microspheres, the hollow spheres show a capacity of more than $800 \, \text{mAh} \, \text{g}^{-1}$ while it remains at only $300 \, \text{mAh} \, \text{g}^{-1}$ for nonhollow hydrothermal carbon microspheres. (ii) While the capacity decreases rapidly for the nonhollow microspheres, the hydrothermal carbon hollow spheres/S composite, which displays a discharge capacity of $1000 \, \text{mAh} \, \text{g}^{-1}$ at the first cycle, maintains a discharge capacity of $600 \, \text{mAh} \, \text{g}^{-1}$ at the 50th cycle. The drop in specific capacity during the first few cycles can be explained by the formation of sulfur layers on the surface of hollow spheres due to the strong affinity of sulfur for carbon. This result reveals that the unique nanostructure of the hollow nanospheres is the essence of the high-performance electrochemical properties.

Finally, the rate capability of the hydrothermal carbon hollow spheres/S composite cath-ode was investigated and a remarkably high specific power was observed. The hydrothermal carbon hollow spheres/S composite cathode showed an initial discharge capacity of 1000, 700, and $400 \, \text{mAh} \, \text{g}^{-1}$ at 1, 2.5, and 5 C, respectively. Even at a very high current density of 10 C $(=16.75 \, \text{A} \, \text{g}^{-1})$, our cathode showed a discharge capacity of $170 \, \text{mAh} \, \text{g}^{-1}$. If we

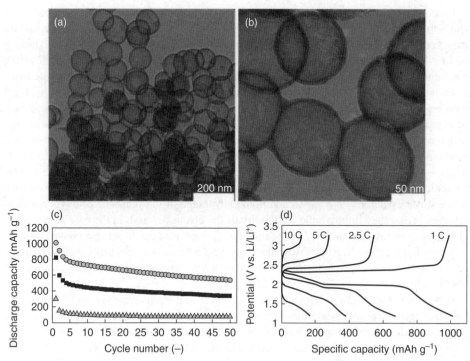

Figure 7.7 *Micrographs obtained by TEM of (a and b) carbon-based hollow spheres obtained after hydrothermal treatment at 180°C in the presence of glucose, pyrolysis at 950°C, and silica removal, 950-CarbHS-G. Arrows highlight the presence of residual silica within the carbon hollow spheres. (c) Cycling performance at 1 C for various electrodes using hydrothermal carbons. The cycling stability of the hydrothermal carbon hollow spheres/S composite made by the melt diffusion method is indicated by circles. The cycling performance of the simple mixture of hydrothermal carbon hollow spheres/S is indicated by squares, while that associated with the mixture of hydrothermal carbon nonhollow microspheres and sulfur is indicated by triangles. (d) Rate performance for hydrothermal carbon hollow spheres/S composite prepared by melt diffusion. The specific capacity was calculated based on the mass of sulfur. (Reproduced with permission from [77].*

assume that a full Li–S cell using hollow spheres contains 25 wt% Li_2S, this full cell will provide a specific energy of 460 Wh kg^{-1} and a specific power of 5000 W kg^{-1} [77].

7.2.2 Electrodes in Supercapacitors

Electrochemical capacitors, also called supercapacitors, are of great interest as high-power electrochemical energy storage devices due to the fact that they combine high power density with a long cycle life and wide operational temperature range – properties that are currently unattainable in Li-ion batteries. However, the energy density achieved by commercial products (about 5 Wh kg^{-1}) limits their use to a few seconds of charge/discharge, hampering their utilization in energy-harvesting applications. Therefore, efforts are focused now on increasing the energy density of supercapacitors.

Based on the charge storage mechanism, electrochemical capacitors can be divided as: (i) electrochemical double-layer capacitors (EDLCs), where the energy storage is based on the electrostatic adsorption of electrolyte ions on the surface area of electrically conductive porous electrodes, and (ii) pseudocapacitors, where the energy is stored through redox reactions at the electrode/electrolyte interface [78]. Porous carbon materials are the main candidate for supercapacitors in terms of cost, availability, large surface area, versatility with regard to porosity development and surface chemistry, good conductivity, and lack of negative environmental impact. They behave mainly as EDLCs, with their large surface area providing high capacitance values. Nevertheless, many carbons possess surface functional groups, like activated carbons, which give rise to an additional pseudocapacitance contribution. Thus, nitrogen and oxygen functionalities are known to give rise to Faradaic redox reactions, increasing the capacitance values and thus the energy density of the supercapacitor [79–82]. An additional advantage of the presence of oxygen and/or nitrogen is that they improve the wettability of the electrodes and, in the case of nitrogen, also the electronic conductivity of the material. Taking into account these considerations, the common methods for increasing the capacitance of carbon electrode materials focus on the preparation of high-surface-area carbons with appropriate pore size and surface functionalities.

As already described in other chapters, hydrothermal carbons are oxygen-rich by nature, with the oxygen content being tuned through the operating conditions (temperature, time, precursor, and concentration) [83–85]. In addition, by selecting the appropriate initial carbon precursor, operating conditions, or additional additives, other functionalities can be introduced in the final carbonaceous materials. Thus, as described in Chapter 3, Titirici *et al.* produced nitrogen-rich carbon materials by carrying out the HTC process in the presence of ovalbumin [86] or by using nitrogen-containing carbohydrates as carbon precursor [87]. Unfortunately, as mentioned in Chapter 2, HTC-derived materials generally possess almost no open porosity and their heat treatment at elevated temperatures can only lead to moderate increases in surface area, which hampers their application in surface-area-sensitive applications, such as supercapacitors. However, this limitation has been circumvented recently by the chemical activation of hydrothermalcarbons, as thoroughly described in Chapter 3.

Thus, Wei *et al.* analyzed the supercapacitor performance in organic electrolyte (1 M TEABF$_4$ in acetonitrile) of hydrothermally carbonized cellulose (C), starch (S), and sawdust (W) activated at 700 and 800 °C with a KOH/sample weight ratio = 4 [88]. Those carbon materials possessed BET (Brunauer, Emmett, and Teller) surface areas around 2100–3000 m^2 g^{-1} and pore volumes up to around 1.4 cm^3 g^{-1}, with a relatively narrow pore size distribution (PSD) in the supermicropore range with virtually no pores greater than 3 nm. The performance of those materials was spectacular, recording the highest capacitance ever reported for porous carbons in a symmetric two-electrode configuration using such electrolyte: 236 F g^{-1} (100 F cm^{-3}) at 1 mV s^{-1} (Figure 7.8a). It exceeded the specific capacitance of commercial activated carbons optimized for EDLC applications, such as YP-17D, by 100%. What is more, the samples were capable of retaining 64–85% of the capacitance when the current density was increased from 0.6 to 20 A g^{-1} (Figure 7.8b). Among the tested carbons, the samples activated at the highest temperature (800 °C), which were those with the largest volume of small mesopores in the range of 2–3 nm, showed better capacitance retention at high sweep rates in the CV measurements or higher current

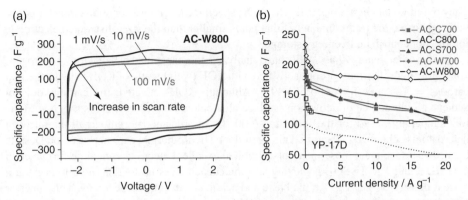

Figure 7.8 *Electrochemical characterization of activated carbons derived from hydrothermally synthesized carbon materials in 1 M TEABF$_4$ solution in acetonitrile at room temperature: (a) CV of the activated carbon obtained from sawdust at 800 °C with KOH/sample = 4 (AC-W800) and (b) capacitance retention with current density in comparison with that of commercially available YP-17D activated carbon. (Reproduced with permission from [88]. © 2011 WILEY-VCH Verlag GmbH & Co. KGaA, Weinheim.)*

densities in the charging/discharging tests (Figure 7.8b). The small reduction/oxidation (redox) peaks visible in CV at around 0 and 2 V at the slowest sweep rate (Figure 7.8a) are believed to originate from oxygen-containing functional groups remaining in the carbon samples. Such peaks completely disappear at the rate of 10 mV s^{-1}, suggesting relatively slow redox reaction kinetics. This may further suggest that the pure EDLC capacitance (without any pseudocapacitance contribution) in these materials exceeds 193 F g^{-1}. This combination of very high specific and volumetric capacitance and good rate capability of the hydrothermally synthesized porous carbons is unmatched by state-of-the-art activated carbons, CNTs, carbon onions, and graphene. Taking into account these results, the study of their capacitance behavior in aqueous and/or ionic liquids, where pseudocapacitance effects due to oxygen functionalities are larger, is highly desirable.

Wang *et al.* have also applied chemical activation to hydrothermal carbons, in this case using phosphoric acid as activating agent and rice husk as carbon precursor [89]. The authors varied the activation temperature between 300 and 700 °C and the weight ratio of phosphoric acid to hydrothermal carbons between 1 and 6. In this way, the obtained porous materials ranged from supermicroporous to mesoporous, with BET surface areas around 700–2700 m^2 g^{-1} and pore volumes up to around 2 cm^3 g^{-1}. Although no PSDs are shown, from the shape of the isotherms it can be envisaged that they are broader than those of the activated carbons obtained by Sevilla *et al.* It should be noted that several authors have pointed out that KOH allows us to obtain narrower PSDs in comparison with other activating agents [79, 90, 91]. The specific capacitance of these activated carbons reached 130 F g^{-1} (measured at 2 mV s^{-1}) in an aqueous solution of 6 M KOH – a value quite below that measured for activated carbons with similar BET surface area, but with narrower PSD in the micro/supermicropore range (235–286 F g^{-1}) [92]. This is due to the fact that, as shown by Chmiola *et al.* [93], a good matching between the electrode pore size and the dimensions of the electrolyte ions is critical for an optimal performance of supercapacitors.

They observed an anomalous increase on the specific capacitance for subnanometer pores, which defeated the traditional belief that pores smaller than the size of solvated electrolyte ions do not contribute to energy storage.

On the other hand, Zhao *et al.* have analyzed the capacitance behavior of nitrogen-containing hydrothermal carbons activated with KOH (weight ratio of KOH to hydrothermal carbons = 1–4 and T = 600 °C) [94]. Although KOH-activated, the carbons did not achieve a high degree of activation (surface area below 600 m^2 g^{-1}). This was due to the low temperature used for the activation (600 °C) in order not to remove all the nitrogen heteroatoms contained in the initial nitrogen-doped hydrothermal carbon precursor. Despite their low surface area, those materials exhibited excellent electrochemical performance in 6 M KOH and 1 M H$_2$SO$_4$, achieving specific capacitances up to 220 and 300 F g^{-1} at a current density of 0.1 A g^{-1} in the basic and acidic electrolyte, respectively. This superior capacitance is due to the combination of EDLC capacitance and pseudocapacitance arising from redox reactions of the nitrogen functionalities. Humps can be detected at around 0.5 V versus saturated calomel electrode (Figure 7.9a), being more obvious in acidic than in basic medium, probably because of the basic character of the functionality in these carbon materials [80]. When compared with the nonactivated hydrothermal carbon, the dependence of the electrosorption process on surface area is obvious. Thus, in spite of having a higher nitrogen content than the activated samples (i.e., 6.7% versus 2.3–5.2%), the hydrothermal carbon exhibits very low specific capacitance (below 5 F g^{-1}), which is a clear reflection of its low specific surface area (i.e., 18 m^2 g^{-1}). It is worth noting that the chemical activation process leads to a reduction of the nitrogen content of the materials, this reduction being larger with the hardening of the activation conditions (i.e., increase of the amount of KOH used). Thus, it decreases from 6.7% in the hydrothermal carbon to 2.3–5.2% for the activated carbons obtained with a ratio varying from 1 to 4. This is a consequence of preferential oxidation of nitrogen functionalities during the activation

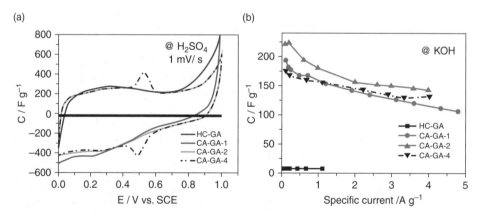

Figure 7.9   *Electrochemical performance of different carbons using a three-electrode cell: (a) CV at a scan rate of 1 mV s^{-1} in 1 M H$_2$SO$_4$ and (b) capacitance retention with current density in 6 M KOH. (Adapted with permission from [94]. © 2010 WILEY-VCH Verlag GmbH & Co. KGaA, Weinheim.)*

process, as observed for the activation of other types of nitrogen-containing precursors [95, 96]. This indicates the necessity of optimization of the activation parameters in order to obtain a compromise between surface area and nitrogen content. Nevertheless, the values of capacitance achieved with these materials are comparable to activated carbons reported in the literature with a much higher surface area (above 3000 m^2 g^{-1}) [92], and higher than that of a high-surface-area and high-nitrogen-content carbon derived from polyacrylonitrile (160 F g^{-1}) [97]. Additionally, good capacitance retention at high current density (4 A g^{-1}) is observed (Figure 7.9b), which proves good conductivity and quick charge propagation in both acid and base electrolytes.

Recently, Zhu *et al.* have prepared electrodes for supercapacitors from hydrothermally carbonized fungi (HTC at 120 °C for 6 h) [98]. The product resulting from the HTC process was characterized by small particles (50–200 nm) fused together to give rise to sheets or tubular networks, the oxygen content being 13.4% (50% decrease from raw fungi). The value of the BET surface area was typical for hydrothermal carbons, around 14 m^2 g^{-1}. With the aim of creating porosity, as well as improving the electronic conductivity, the authors pyrolyzed the hydrothermal carbon at 700 °C for 3 h. As a result, the oxygen content decreased to 5%, but the surface area increased to 80 m^2 g^{-1}. Surprisingly, even though the oxygen content is not high and the surface area is really low, the material stored up to 196 F g^{-1} at a scan rate of 1 mV s^{-1} – a value comparable to high-surface-area carbons like Maxsorb or highly doped carbons obtained from seaweed [99]. However, with the increase in the scan rate, the decrease of the capacitance was quite high (around 40% decrease at 50 mV s^{-1}). The authors explain the large specific capacitance of the material on the basis of its small particle size, which reduces the ion and electron transport path, and increases the material usage, as pseudocapacitors store charges in the first few nanometers below the surface. Furthermore, it showed a high electrochemical stability (after repeated galvanostatic cycling, the specific discharge capacitance decreased slowly and stabilized at 180 F g^{-1} after 1000 cycles) and the maximum cell voltage reached 1 V for reversible charging as a consequence of the existence of oxygen – a fact previously observed by other authors for carbons with a large oxygen content [99]. This material therefore exhibits a promising behavior (stability, energy density, power density, surface capacitance, and volumetric capacitance) as an electrode in supercapacitors.

To sum up this subsection, the results of the works described above clearly indicate that the postsynthesis activation of hydrothermal carbons is a powerful tool to increase their porosity, thus enabling their use in an application in which otherwise they could not be used.

The oxygen functionalities present in those materials represent a clear advantage providing an additional pseudocapacitance contribution to the EDLC capacitance, which can enhance their performance. Furthermore, the use of nitrogen-containing substances as precursors in the HTC process allows the introduction of nitrogen functionalities in the activated carbons, thus providing additional pseudocapacitance effects. However, optimization of the activation conditions (temperature and amount of KOH) is necessary to achieve a good compromise between surface area and pore volume, pore size, and oxygen and nitrogen contents.

In addition to doping with heteroatoms, another possibility for inducing pseudocapacitance in carbon materials when used as electrodes in supercapacitors is their modification with certain metal oxide nanoparticles such as manganese or ruthenium oxide. The pioneering work on the pseudocapacitive behavior of manganese oxide in an aqueous solution was published in 1999 by Lee and Goodenough [100]. Both charge storage mechanisms involve a redox reaction between the III and IV oxidation states of manganese ions. In general, hydrated manganese oxides exhibit specific capacitances within the range of 100–200 F g^{-1} in alkali salt solutions, which values are much lower than those for RuO_2. As MnO_2 has no electrochemical conductivity, very often MnO_2/C structures are utilized [101]. Furthermore, it is possible to combine the supercapacitive properties of MnO_2 with the double-layer storage mechanism of microporous carbons for increasing the capacity value in the so-called "asymmetric capacitors" [101].

Applied to HTC, Zhang *et al.* used carbonaceous sphere@MnO_2 rattle-type hollow spheres synthesized under mild hydrothermal conditions [102]. The as-prepared materials showed a mesoporous MnO_2 shell and a carbonaceous sphere core. The composition and shell thickness of the hollow spheres can be controlled experimentally. The capacitive performance of the hollow structures was evaluated using both CV and charge/discharge methods. The results demonstrated a specific capacitance as high as 184 F g^{-1} at a current density of 125 mA g^{-1}. The good electrocapacitive performance resulted from the mesoporous structure and high surface area of the MnO_2-based hollow spheres.

In the future, the discovery of novel methodologies for synthesizing HTC-derived materials in which chemical activation will no longer be necessary will be desired. This could be accomplished, for example, utilizing molten salts [103] to induce microporosity in such materials while maintaining a high level of functional groups for pseudocapacitance.

7.2.3   Heterogeneous Catalysis

Heterogeneous catalysis has always been an inherently nanoscopic phenomenon with important technological and societal consequences for energy conversion and the production of chemicals. Architectures with all of the appropriate electrochemical and catalytic requirements, including large surface areas readily accessible to molecules, may now be assembled on the benchtop [104]. Designing catalytic nanoarchitectures inspired by nature using renewable resources with control over porosity and functionalities offers the promise of even higher activity and low cost.

Carbon materials play a very important role in various heterogeneous catalytic reactions either as catalyst supports or exhibiting catalytic properties themselves following various functionalizations. There are many reports, reviews [105, 106], and books [3b] dedicated to this subject.

Here, we will describe some examples where the carbon materials prepared using HTC have found important applications in heterogeneous catalysis. One of the main advantages of HTC-derived materials compared to other "standard" catalytic carbons is the low-temperature synthesis enabling them with polar surface functionalities. These polar groups make them hydrophilic – a factor that can alter their selectivity when utilized as supports during catalytic reactions [107]. In addition, the presence of such surface groups allows their further functionalization with various catalytically active sites.

We will first describe the utilization of HTC-derived materials as supports for various nanoparticles in heterogeneous catalysis followed by a few examples were we show that various functional groups on the hydrothermal carbon surface can catalyze important chemical reactions.

7.2.4   Hydrothermal Carbon Materials as Catalyst Supports

The fusion between nanoparticle and nanoporous materials technology represents one of the most interesting and rapidly expanding areas in nanotechnology [108]. The harnessing of nanoscale activity and selectivity provides extremely efficient catalytic materials. Nanoparticles typically provide highly active catalytic centers, but are very small and are not in a thermodynamic stable state due to their high surface energies.

In the literature there are several methods to stabilize nanoparticles, such as capping with other ligands or the production of core/shell particles [34, 109]. However, when the main application targets catalysis, the application of such materials is limited due to undesired aggregation and poisoning of the catalyst. An elegant possibility to overcome all these problems is the uniform dispersion of such active monodisperse nanoparticles onto porous materials.

Porous carbon materials are ideal candidates as supports for various nanoparticles [110]. Furthermore, depending on the functional groups present on the carbon support and its porosity, it should be possible to control the size and shape of the resulting nanoparticles. This leads to the possibility of size-selective and reusable heterogeneous catalysts based on nanoparticle size compared to pore size.

HTC-derived materials, given their functional groups and the core/shell structure with more hydrophobic polyfuran compounds at the core and functional groups at the rim of the particle, are particularly interesting catalysts for the dispersion of various nanoparticles. Thus, Titirici *et al.* proved the importance of the fact that the carbon material is hydrophilic due to the high number of functional groups. They prepared a core/shell material containing palladium nanoparticles dispersed on the more hydrophobic carbon core surrounded by oxygenated functional groups by simply hydrothermally treating furfural in the presence of palladium acetylacetonate (Figure 7.10a) [107]. The furfural precursor played a double role. On the one hand, it formed the hydrophilic HTC-derived material and, on the other hand, it was able to reduce the palladium (acac)$_2$ to elemental metallic palladium nanoparticles during the HTC process.

Such a metal/carbon nanocomposite catalyst prepared in a one-step reaction was highly selective for the hydrogenation of phenol to cyclohexanone – a reaction otherwise very difficult to achieve when using more" classical" hydrophobic carbon supports (Figure 7.10b). We need to mention here that cyclohexanone is one of the main intermediates in the preparation of caprolactam and adipic acid, which are used in the manufacturing of nylon-6, nylon-6,6, and polyamide resins, for instance.

A complete conversion of the phenol into cyclohexanol was observed using commercial charcoal- and alumina-supported palladium. Even with shorter reaction times (down to 1 h), partial hydrogenation products could not be detected. Moreover, no differences in the reactivity were observed between charcoal and alumina. Both supports loaded with a similar amount of palladium led directly to cyclohexanol (Figure 7.10b). In contrast, in the case of palladium supported on hydrophilic hydrothermal carbon, the hydrogenation is selective. After 20 h of reaction, we obtained 95% of cyclohexanone with a conversion rate close to 99% (Figure 7.10b)

(a)

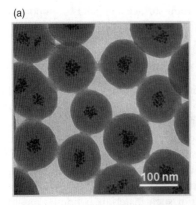

100 nm

(b)

Table 1 Catalytic activity of differently supported Pd for the hydrogenation of phenol[a]

Catalyst	Time/h	Conversion (%)	Cyclohexanol	Cyclohexanone
			Selectivity (%)	
Pd@hydrophilic-C	10	60	—	>99
Pd@hydrophilic-C	20	>99	5	95
Pd@hydrophilic-C	72	>99	50	50
Pd@hydrophilic-C[b]	20	45	30	70
10% Pd@C	20	100	100	0
10% Pd@C	1	100	100	0
10% Pd@Al$_2$O$_3$	20	100	100	0

[a] In a typical reaction, 50 mg of catalyst were added to 100 mg of phenol and the mixture was heated to 100 °C under 1 MPa of hydrogen pressure.
[b] Reference test in cyclohexane.

(c)

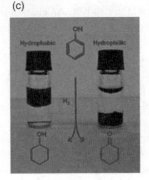

Figure 7.10 *(a) TEM micrograph of the hydrothermal carbon/Pd core/shell morphology. (b) Table describing the catalytic activity of our hydrothermal carbon/Pd material in comparison with other commercially available materials. (c) Picture illustrating the mechanism of conversion of phenol to cyclohexanone on hydrophilic carbon and to cyclohexanol on hydrophobic carbon. (Reproduced with permission from [107]. © 2008 Royal Society of Chemistry.)*

The lower hydrogenation speed of the ketone compared with the aromatic ring is not likely to rely on the intrinsic properties of the palladium nanoparticles as no such effect is observed with the commercial catalysts. Actually, the active particles are placed in a confined microenvironment with special chemical functionalities. The reactive pocket around our nanoparticles is very hydrophilic and decorated with —OH, C=O, and —COOH species. A possible mechanism accounting for the observed reactivity could then rely on the fact that phenol, being a strongly hydrogen-bridge interacting system, is enriched in the pores, while cyclohexanone, being more hydrophobic, is displaced from the reactive pocket and thus cannot easily further react (Figure 7.10c). Such mechanisms, involving specific interactions between the substrate and the reactive pocket, are known to control the reactivity of enzymes and are a major goal for biomimetic catalysis. This could explain why our hydrophilic support provided a more selective catalyst than the commercial hydrophobic charcoal-supported system.

Using a similar "*in situ*" procedure, Wu *et al.* prepared Ag/C hybrids using the redox properties of glucose and silver nitrate in the presence of imidazolium ionic liquid under hydrothermal conditions [111]. Monodisperse carbon hollow submicrospheres encapsulating silver nanoparticles and Ag/C cables were selectively prepared by varying the concentration of ionic liquid. Other reaction parameters, such as reaction temperature, reaction time, and the mole ratio of silver nitrate to glucose, played an important role in controlling the structures of the products. The catalytic property of the hybrid in the oxidation of 1-butanol by H_2O_2 was investigated.

Utilizing a similar procedure to that initially described by Titirici *et al.* for the one-step synthesis of carbon/inorganic nanoparticles [35], Yu *et al.* prepared Fe_xO_y@C spheres as an excellent catalyst for Fischer–Tropsch synthesis [112]. Glucose and iron nitrate were hydrothermally treated together for the fabrication of carbonaceous spheres embedded with iron oxide nanoparticles. This route is also applicable to a range of other naturally occurring saccharides and metal nitrates. A catalytic study revealed the remarkable stability and selectivity of the reduced Fe_xO_y@C spheres in the Fischer–Tropsch synthesis, which clearly exemplifies the promising application of such materials.

In addition to this one-step functionalization of HTC with nanoparticles *in situ* during the HTC process, nanoparticles can be incorporated in a postfunctionalization step, following the preparation of hydrothermal carbon, as described in Chapter 4.

White and Titirici in collaboration with Regina Palkovitz *et al.* described the preparation of porous nitrogen-doped carbon materials naturally inspired from lobster shells [113] followed by their decoration with Pt^{2+} using a postimpregnation method with $PtCl_2$ (Figure 7.11A). Thermal stability and robustness towards acidic solvents make such

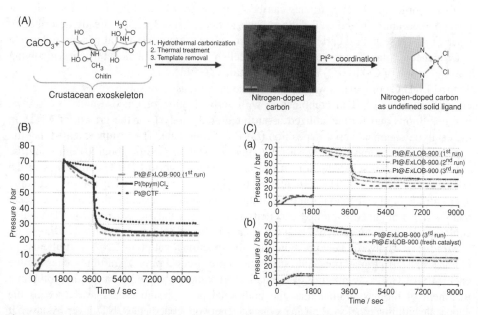

Figure 7.11 *(A) Preparation of coordinatively modified Pt@nitrogen-doped carbon materials derived from crustacean exoskeleton of lobsters (ExLOB). (B) Pressure–time plots for catalyst comparison (reaction conditions: 10 µmol Pt equivalents catalyst, 215 °C, 1000 rpm, 30.4 ml Hastelloy autoclave). (C, a) Pressure–time plots for Pt@ExLOB-900 recycling experiments. (C, b) Pressure–time plots for fresh and recycled catalyst (Pt@ExLOB-900 (third run)). (Reproduced with permission from [112].*

materials very interesting as possible solid catalysts in the Periana system for the oxidation of methane to methanol [114]. The direct utilization of methane via C—H activation remains a challenge as the development of efficient catalysts with sufficient activity and selectivity is hindered by the high binding energy of the methane C—H bond [115]. Thus, a viable methane–to–liquid process still suffers from ineffective catalytic systems. The C—H activation of hydrocarbons by transition metal complexes presents a promising pathway [116]. One of the most active (homogeneous) catalytic systems is that reported by Periana *et al.* [114, 117]. Thus, the catalytic activity of the nitrogen-doped carbons subsequently coordinated with Pt^{2+} (Pt@*Ex*Lob900) for methane oxidation was compared here to the standard Periana system (Pt@bpymCl$_2$) as well as with a nitrogen-rich covalent triazine framework (Pt@CTF), acting as solid ligand (Figure 7.11B) [118]. The initial catalytic activity of these material is superior to the molecular benchmark originally described by Periana and significantly better than that of the previously reported solid catalysts. Recycling experiments revealed that this novel solid looses activity to some extent (Figure 7.11C) [112]. This deactivation is, however, not associated with loss of platinum, since platinum leaching is negligible. The reason for the loss of activity is still under investigation. However, the remarkable catalytic performance may provide new insight for further targeted material development that in turn could bring catalytic methane utilization closer to technical feasibility.

Carbon nanofibers (CNFs) produced by HTC display remarkable reactivity, and the capability for *in situ* loading with very fine noble metal nanoparticles such as palladium, platinum, and gold. Large quantities of uniform CNFs embedded/confined with various kinds of noble metal nanoparticles can be easily prepared, resulting in the formation of the so-called uniform and well-defined "hybrid fleece" structures as already described in Chapter 4. These hybrid carbon structures embedded with noble metal nanoparticles in a heterogeneous "fleece" geometry serve as excellent catalysts for a model reaction involving the conversion of CO to CO_2 at low temperatures [119].

NiAl LDH/C composites with adjustable compositions were successfully assembled by crystallization of LDH in combination with carbonization of glucose under hydrothermal conditions and further utilized as an integrated catalyst for the growth of CNTs in the catalytic chemical vapor deposition (CVD) of acetylene. The results revealed that the supported nickel nanoparticles with the small crystallite size of about 10 nm could be obtained by *in situ* self-reduction of the as-assembled hybrid LDH/C composites in the course of catalytic CVD. The carbon in the hybrid structure played a key role as a reducing agent for the high dispersion of the resulting nickel nanoparticles. Furthermore, the nickel nanoparticles obtained exhibited excellent activity for catalytic growth of CNTs, which could be delicately tuned by varying the compositions of hybrid composites [120].

Ming *et al.* showed a new procedure for the functionalization of CNTs with the use of biomass as starting materials and introduced a novel concept of a knitting process in the chemistry of CNTs. A mixture of aromatic compounds obtained from the hydrothermal treatment of biomass, rather than the traditional polymer monomers, was used as the nanoscale building blocks to knit an oxygenated network coat on the CNTs layer-by-layer. It is an effective, mild, green, and easily controlled method for the functionalization of CNTs. The obtained functionalized CNTs were proved to be a promising catalyst support for metal catalysts, such as Ru/functionalized CNTs, and showed high activity and selectivity for the hydrogenation of citral to unsaturated alcohol [121].

7.2.5 Hydrothermal Carbon Materials with Various Functionalities and Intrinsic Catalytic Properties

From the previous subsection we learnt that the surface functional groups on hydrothermal carbon can actively participate in directing the type of nanoparticles that will be loaded on its surface (reduced metals, metal oxides), create metal complexes with metals, or control the selectivity in certain chemical reactions. In addition, the presence of the polar surface groups on the hydrothermal carbon surface offers yet another great advantage – easy chemical functionalization.

Silica is an ideal example of a material that can be easily textured at both the macroscopic and mesoscopic scales, and functionalized with various organic groups under mild conditions [122, 123]. However, easy postsynthesis surface functionalization of carbonaceous material, equivalent to that achieved on silica, is still challenging [124]. This is not the case for hydrothermal carbon as the presence of surface oxygenated groups enables its straightforward functionalization in a very similar manner to silica (see Chapter 5).

As in the case of functionalization with nanoparticles we can distinguish here "*is situ*" functionalization methods and "postmodification." Here, we will focus on a few examples of functional hydrothermal carbons with intrinsic catalytic properties.

Titirici *et al.* established a methodology to produce functional hydrothermal carbon by HTC of carbohydrates in the presence of small amounts of functional monomers [125]. For example, the HTC of glucose in the presence of small amounts of functional organic monomer (vinyl imidazole), inside a porous network of a silica template, followed by subsequent removal of the sacrificial template leads to a mesoporous hydrothermal carbon with vinylimidazole surface groups on its surface. The resulting grafted imidazole moieties were, in a second step, converted into the corresponding alkyl imidazoliums and successfully employed as catalysts for various reactions, including transesterification, Diels–Alder, or Knoevenagel condensations (see Figure 5.8) [126].

Given the depletion of fossil fuels and the alarming facts associated with global warming, there is an increasing interest in discovering new and efficient catalysts for the chemical conversion of biomass into biofuels [127]. Homogeneous acid catalysts, such as sulfuric acid, are commonly employed. However, these catalysts have several drawbacks, such as corrosion and toxicity problems, costly and inefficient procedures for separating them from the products, and the need to neutralize the waste streams. These problems could be solved by developing heterogeneous solid-acid catalysts, which could then be more easily and efficiently separated from the products, enabling their reuse. However, most solid-acid catalysts reported so far are expensive or involve complex synthetic procedures, which impede their commercialization. These include acid zeolites, mesostructured silica functionalized with sulfonic groups, tungstated zirconia, sulfated zirconia, sulfonated polymers (Amberlyst-15), and Nafion-based composites [128]. In this respect, hydrothermal carbon presents again multiple advantages. It is low cost, based on biomass precursors itself, stable, its porosity can be tuned, and it is already acidic. In addition, its surface functionality can help its further modification with stronger acidic groups, such as $-SO_3H$ groups.

Thus, Sevilla *et al.* used the sulfonation of carbonaceous microspheres obtained by the HTC of glucose (Figure 7.12A). This synthetic strategy circumvents gas-phase pyrolysis, thereby avoiding the emission of harmful gases and yields a solid acid comprising of spherical particles of uniform, micrometer-regime size (Figure 7.12B). The activity of this

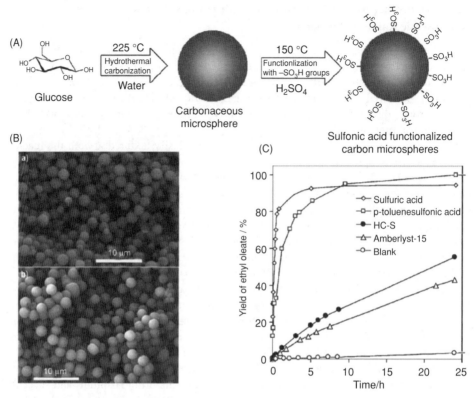

Figure 7.12   *(A) Schematic illustration of the synthetic procedure for obtaining carbon micro-spheres functionalized with —SO₃H groups. (B) SEM images of (a) the hydrothermally car-bonized glucose and (b) the sulfonated sample. (C) Time courses of the esterification of oleic acid with ethanol. The yield of ethyl oleate is based on oleic acid. Reaction temperature: 55 °C. (Reproduced with permission from [129]. © 2010 WILEY-VCH Verlag GmbH & Co. KGaA, Weinheim.)*

sulfonated carbon catalyst towards the esterification of oleic acid with ethanol, a typical reaction in the synthesis of biodiesel, was investigated. Figure 7.12C shows the formation of ethyl oleate during reaction at 55 °C. For comparison, results for equivalent amounts of sulfuric acid, *p*-toluenesulfonic acid, and Amberlyst-15 are also shown. In the absence of catalyst (blank experiment) the ethyl oleate yield was only 3.5% after 24 h. As expected, the homogeneous catalysts (sulfuric acid and *p*-toluenesulfonic acid) showed the highest activities, but they lack the advantages of solid acids pointed out before. The hydrothermal carbon with sulfonic groups sample exhibits a higher activity than Amberlyst-15 despite the fact that the latter has a higher density of sulfonic groups. This may due to the fact that the sulfonated carbon microspheres are highly hydrophilic, as pointed out before for the case of phenol hydrogenation to cyclohexanone (Section 7.2.4) [107]. This facilitates the adsorption of a large amount of hydrophilic molecules, such as ethanol, and favors the access of the reactants to the —SO₃H sites.

In a similar approach, Xiao *et al.* produced novel biacidic carbon via one-step HTC of glucose, citric acid, and hydroxyethylsulfonic acid at 180 °C for only 4 h. The novel carbon had an acidity of 1.7 mmol g^{-1} with a carbonyl/sulfonic acid group molar ratio of 1:3. The catalytic activities of the carbon were investigated through esterification and oxathioketalization. The results showed that the carbon had comparable activities to sulfuric acid, which indicated that the carbon holds great potential for the green processes related to biomass conversion [130].

It is Dr. Titirici's personal believe that, in the future, HTC-derived materials produced from biomass could play an important role in the conversion of lignocellulosic biomass into useful products and biofuels. Firstly, about 30–40% of the products resulting from HTC of biomass are left in the liquid phase [131]. Thus, if part of the HTC-derived solid materials making the other 60–70% fraction could be used to competently convert the liquid phase into useful products, the efficiency of this process will be greatly improved. Furthermore, the ratio of the liquid to solid fraction as well as their composition could be controlled throughout appropriate synthesis conditions (i.e., temperature, pressure, catalyst, additives). In this respect, it has been previously showed that various "Starbon®" materials, a process with many similarities to HTC regarding the flexibility of the resulting carbons [132], can act as efficient catalysts with tuned selectivity depending on the preparation method and the reaction conditions [133].

7.3 Electrocatalysis in Fuel Cells

Fuel cells are electrochemical devices that convert chemical energy from a fuel into electric energy continuously, as chemicals constantly flow into the cell. They consist of an anode, a cathode, and an electrolyte, as schematically shown in Figure 7.13. In the anode, the fuel is oxidized to produce electrons, which travel along an external circuit to the cathode creating an electrical current, and protons, which pass through the electrolyte to the cathode, where the oxidant combines with the protons and electrons to produce water as only byproduct.

As fuels, hydrogen, alcohols, or hydrocarbons can be used, and as oxidant, normally oxygen is used. At both the cathode and anode, catalysts are necessary for the outcome of the electrochemical reactions at low temperature. Present fuel cell prototypes often use catalysts selected more than 25 years ago. Commercialization aspects, including cost and durability, have revealed inadequacies in some of these materials. Therefore, the development of novel and alternative catalysts with a lower production cost and better efficiency then the traditional catalysts is the main driving force in the research dedicated to fuel cells.

In most cases, catalysts in fuel cells are composed of noble metal nanoparticles (platinum, palladium) dispersed over a support, which normally consists of carbon materials. Carbon is a typical electrocatalyst support for fuel cell applications due to its good electrical and mechanical properties, and versatility in pore size and distribution. However, some carbon materials, especially heteroatom-doped materials, showed intrinsic catalytic properties themselves.

For current hydrogen-gas-fed fuel cells, hydrogen production, storage, and transportation are the major challenges in addition to cost, reliability, and durability issues. Direct methanol fuel cells, using liquid and renewable methanol fuel, have been considered to be a favorable option in terms of fuel usage and feed strategies compared to hydrogen-fed fuel cells, which

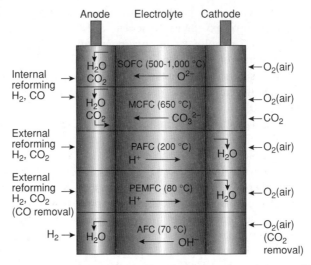

Figure 7.13 *Summary of fuel cell types (SO, solid oxide; MC, molten carbonate; PA, phosphoric acid; PEM, polymer electrolyte membrane; A, alkaline). The oxidation reaction takes place at the anode (+) and involves the liberation of electrons (e.g., $O^{2-} + H_2 = H_2O + 2e^-$ or $H_2 = 2H^+ + 2e^-$). These electrons travel around the external circuit producing electrical energy by means of the external load and arrive at the cathode (–) to participate in the reduction reaction (e.g., $1/2O_2 + 2e^- = O^{2-}$ or $1/2O_2 + 2H^+ + 2e^- = H_2O$). (Reproduced with permission from [134]. © 2001 Nature Publishing Group.)*

have a reforming unit or low capacity in the hydrogen storage tank. The direct methanol fuel cell uses a liquid methanol fuel, which is easily stored and transported and simplifies the fuel cell system [135].

Here, we will discuss both direct methanol fuel cells as well as hydrogen fuel cells, focusing on the development of sustainable HTC-based carbon catalysts either as supports or as catalysts with intrinsic properties for both methanol electrooxidation at the anode as well as well as on the ORR at the cathode.

7.3.1 Catalyst Supports in Direct Methanol Fuel Cells

Catalysis in electrochemical systems such as fuel cells takes place at the interface between the reactant, the catalyst, and the electrolyte, which is known as triple-phase boundary. Therefore, the support should possess a highly accessible porosity that facilitates contact between those three phases and also good conductivity because it has to conduct the electrons generated in the reaction. The key properties of a carbon electrocatalyst support are therefore: (i) high crystallinity (good electric conductivity), (ii) relatively high surface area for a good dispersion of the catalyst nanoparticles, (iii) open and accessible porosity, and (iv) resistance against corrosion. As shown in Chapter 6, the crystallinity of hydrothermal carbon can be increased by heat treatment at high temperature or via catalytic graphitization [136]. These processes also generate a certain porosity that can be useful for the deposition of catalyst nanoparticles. Additionally, the high concentration of oxygen groups may be useful as they can act as anchoring sites for the deposition of catalyst nanoparticles, also avoiding their agglomeration. These considerations have prompted several authors to study the performance of hydrothermal carbon as an electrocatalyst support, as described below.

Currently, most research in this area is focused on exploring new anode catalysts that can effectively enhance the methanol electrooxidation kinetics, while some activities on methanol-tolerant cathode catalysts have also been carried out.

The first to explore the use of a HTC-based carbon as an electrocatalyst support for the electrooxidation of methanol was Yang *et al.* [137]. They deposited 10 wt% platinum nanoparticles over carbon spherules obtained by HTC of sucrose at 190 °C and post-treatment at 1000 °C using two different methods: a polyol method and chemical reduction with $Na_2S_2O_4$. The carbon spherules heat treated at 1000 °C possess a BET surface area of 400 m^2 g^{-1}, which arises from the presence of micropores of 0.6–1.6 nm, and an amorphous nature, although with some improvement in the structural ordering. The authors found that the medium used for the preparation of the catalyst has a strong influence on the dispersion of the platinum nanoparticles on the spherules. Thus, whereas the nanoparticles deposited through a polyol method exhibit a size of around 5 nm, those deposited in aqueous solution tend to agglomerate and exhibit a broader particle size distribution, from 6 to 40 nm. As shown in Figure 7.14A, the nanoparticles are deposited on the surface of the carbon spherules and crystallized with a face-centered cubic (fcc) structure. This agrees with the fact that the nanoparticles are bigger than the pore size of the spherules.

The prepared catalysts exhibited lower electrochemically active surface areas of platinum (18.2 and 54 m^2 g^{-1} for the polyol and chemical reduction methods, respectively) than commercial Pt/Vulcan XC-72 (61.4 m^2 g^{-1}) due to the larger particle size (3.7 nm

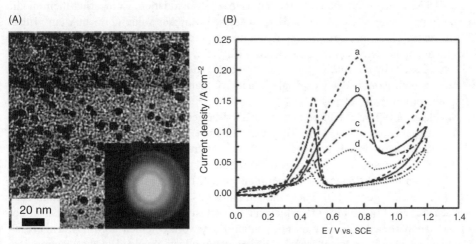

Figure 7.14 *(A) Low-magnification high-resolution TEM image of platinum nanoparticles (inset: selected area electron diffraction pattern of platinum nanoparticles). (B) CV curves of the different electrodes in 1.0 M H_2SO_4 + 1.0 M CH_3OH solution at 40 °C at a scan rate of 50 mV s^{-1}: (a) platinum catalyst prepared via polyol method, (b) commercial Pt/Vulcan XC-72, (c) platinum catalyst prepared with commercial graphitized mesocarbon microbeads (MCMB), and (d) platinum catalyst prepared via chemical reduction in aqueous solution. (Reproduced with permission from [137]. © 2005 Elsevier.)*

for Pt/Vulcan XC-72). However, the platinum utilization in the catalyst prepared through a polyol method (90.5%) was higher than that of Pt/Vulcan XC-72 (81%). The authors attributed this to the better contact of the electrolyte with the platinum particles on the monodispersed spherules. On the other hand, the low platinum utilization in the catalyst prepared via chemical reduction in aqueous solution (34.4%) is attributed to the agglomeration of the platinum nanoparticles. With regard to the electrooxidation of methanol (see Figure 7.1b), the catalyst prepared through the polyol method exhibits the highest current. Although this sample possesses a lower electrochemically active surface area of platinum than Pt/Vulcan, it has a better performance in methanol electrooxidation probably due to the higher utilization of the platinum nanoparticles as a result of a higher accessibility.

Kim *et al.* analyzed the performance of a HTC-derived graphitic carbon made in the presence of iron and post-treated at 900° as support for PtRu particles [138]. A BET surface area of 252 m^2 g^{-1} was measured for that material (SC-g). The authors also performed the HTC process in the absence of iron and under static or dynamic conditions (SC-1 and SC-2, respectively), obtaining materials with BET surface areas of 112 and 383 m^2 g^{-1}, respectively. However, in this case, the samples were composed of amorphous carbon. The PtRu nanoparticles (60 wt%) were deposited by a NaBH$_4$ reduction method. The size of the particles, determined by X-ray diffraction (XRD) analysis, was 3.5, 2.6, and 2.7 nm for SC-1, SC-2, and SC-g. The larger particle size in the former, SC-1, is due to its smaller surface area, as demonstrated by TEM images, which showed the formation of more agglomerates than in the other two materials. The performance of these catalysts towards methanol electrooxidation was analyzed by CV at room temperature in 0.5 M H$_2$SO$_4$ solution containing 2 M CH$_3$OH with a scan rate of 10 mV s^{-1} and compared to that of PtRu/Vulcan. All the supported PtRu catalysts showed an anodic peak current at 0.45–0.5 V, which is attributed to the methanol electrooxidation. Comparing their anodic peak current densities, only PtRu/SC-g possessed a higher value (21.3 mA cm^{-2}) than PtRu/Vulcan (16.4 mA cm^{-2}). This is due to its graphitic structure, which enhances its electrical conductivity. On the other hand, PtRu/SC-2 exhibited a higher catalytic activity (12.8 mA cm^{-2}) than PtRu/SC-1 (9.9 mA cm^{-2}) due to the higher metal dispersion in SC-2. However, both exhibited lower catalytic activity than PtRu/Vulcan owing to the higher graphitic ordering in Vulcan.

Sevilla *et al.* followed a different approach for the generation of graphitic structures from hydrothermal carbon [139, 140]. They used a two-step process in which the synthesized hydrothermal carbon was impregnated with nickel nitrate and subjected to a heat treatment at 900 °C. As a result, carbon nanocoils were formed, as shown in Figure 7.15a. They are highly crystalline as revealed by the well-defined (002) lattice fringes in the high-resolution TEM image in Figure 7.15b. These carbon nanocoils also exhibited relatively high BET surface areas of 114–134 m^2 g^{-1}, which can be exclusively ascribed to the external surface of the nanoparticles (i.e., they do not contain framework-confined porosity). This result implies that mass transfer resistances of reactant/products involved in the electrooxidation of methanol will be minimized. These nanocoils thus present the key properties of electrocatalyst supports: relatively high and easily accessible surface area, and high crystallinity. This was confirmed by their use as electrocatalyst supports towards the oxidation of methanol. Thus, Sevilla *et al.* deposited PtRu nanoparticles [139] or platinum nanoparticles [140] over graphitic carbon nanocoils and analyzed their behavior for methanol electrooxidation, comparing it with that of a carbon support widely used as electrocatalyst

(a)

(b)

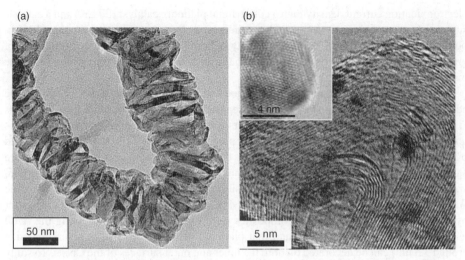

4 nm

50 nm

5 nm

Figure 7.15   *(a) TEM image of a carbon nanocoil obtained from HTC of sucrose and (b) high-resolution TEM image of a carbon nanocoil obtained from HTC of sucrose with deposited catalyst nanoparticles (dark points) (inset: detail of a PtRu nanoparticle showing the cubic structure). (Reproduced with permission from [139, 140]. © 2007 and 2009 Elsevier.)*

support (i.e., Vulcan XC-72R). It is remarkable that a high dispersion of nanoparticles (black dots in Figure 7.15b) is obtained over the nanocoils (platinum particle size = 3–3.3 nm) although their surface area is half that of Vulcan (platinum particle size = 2.6 nm). For both PtRu and platinum nanoparticles, the electrocatalysts prepared with the carbon nanocoils exhibited a higher activity than that of Vulcan, which is a consequence of the fact that nanocoils combine good electrical conductivity and accessible surface that allows the diffusional resistances of reactants/products to be minimized. Additionally, the crystalline structure confers those materials a high resistance against oxidation, which suggests that these electrocatalytic systems will have, under an oxidative environment (typical of fuel cell electrodes), a longer durability.

Joo *et al.*, on the other hand, integrated the templating route and the catalytic graphitization technique in the HTC process with the aim of synthesizing graphitic porous carbons [141]. The authors used uniform silica particles (100 nm) as a sacrificial template, sucrose as carbon precursor, and iron as graphitization catalyst. The HTC process (190 °C for 10 h under vigorous stirring) was followed by heat treatment at 900 °C. In this way, the authors synthesized a carbon material composed of large spherical pores of 100 nm, which are a faithful replica of the silica particles, exhibiting a high surface area (425 m^2 g^{-1}) and a large pore volume (0.42 cm^3 g^{-1}). Additionally, this material possesses a graphitic nature, as probed by high-resolution TEM, XRD, and Raman spectroscopy. Thus, well-defined (002) lattice fringes are observed in high-resolution TEM images, a sharp (002) diffraction peak in the XRD patterns (d_{002} = 0.335 nm and L_c = 9.8 nm), and the G′-band at around 2720 cm^{-1} in the Raman spectra, besides the D- and G-bands at 1350 and 1580 cm^{-1}, respectively. The preparation of the platinum catalyst, on the other hand, was carried out by the formaldehyde reduction method. This catalyst exhibited higher methanol electrooxidation current density than a commercial platinum catalyst (ETEK) and a catalyst supported over porous amorphous carbon synthesized by polymerization of sucrose without hydrothermal treatment. This high activity is closely related to the unique properties of graphitic carbon together with the porous characteristics of the material, which favor the mass transfer.

Wen *et al.* combined the templating technique with the HTC process, although without the addition of any graphitization catalyst, and prepared platinum catalysts with those materials [142, 143]. In one of the works, these authors used an anodic aluminum oxide (AAO) as template and glucose as hydrothermal carbon precursor, and the HTC process was followed by heat treatment at 900 °C [142]. The authors obtained CNTs with an open-ended structure and a diameter of around 200 nm (wall thickness around 10 nm), which is close to the pore size of the AAO template used. For the deposition of platinum nanoparticles (20 wt%), the AAO/CNT composites were immersed in a H_2PtCl_6 and $NaBH_4$ solution several times after pyrolysis. The Pt-CNT-Pt hybrid composites were then liberated by dissolving the AAO template with HF. The authors found that platinum nanoparticles (fcc structure and size around 3.5 nm) can be decorated on both the inner and outer surfaces of the CNTs. To study the electrocatalytic performance of these materials for methanol oxidation, a commercial carbon material (Vulcan XC-72) was also utilized as platinum catalyst support (containing about 16.7 wt% Pt). In this case, the size of the nanoparticles was about 3.7 nm and they were uniformly dispersed on the surface of the material. The electrochemically active surface area, measured in 0.5 M H_2SO_4, was around 39 m^2 g^{-1} for the Pt-CNT-Pt catalyst and around 25 m^2 g^{-1} for Pt/Vulcan XC-72. The larger electrochemically active surface area may be attributed to the good dispersion of platinum nanoparticles on the CNTs. When subjected to CV in 0.5 M H_2SO_4 solution containing 0.5 M CH_3OH, the Pt-CNT-Pt catalyst exhibits a higher mass peak current density (25.3 mA g^{-1}) than Pt/Vulcan (14.7 mA g^{-1}), which indicates a higher catalytic activity for the CNT-supported catalyst. Additionally, Pt-CNT-Pt possesses a higher tolerance than Pt/Vulcan to incompletely oxidized species accumulated on the surface of the electrode. This superior catalytic performance of Pt-CNT-Pt in the electrochemical oxidation of methanol may be due to the fact that the aligned nanochannels within the catalyst facilitate diffusion of the electrolyte and methanol, producing a better contact between them and the platinum nanoparticles and, therefore, more triple-phase boundaries are achieved.

In another publication, the authors used a different template, SBA-15, and introduced the platinum precursor directly into the autoclave, so that HTC and deposition of the platinum nanoparticles on the hydrothermal carbon took place simultaneously [143]. The samples were then carbonized at 750 °C and, finally, the template removed to obtain the Pt@C/MC catalyst. For comparison reasons, these authors also prepared Pt/CMK-3 and Pt/Vulcan XC-72 catalysts. TEM inspection of Pt@C/MC shows a good morphology replication from SBA-15 and a uniform dispersion of platinum nanoparticles on the porous carbon. Furthermore, high-resolution TEM shows that the small platinum nanoparticles (around 3–5 nm and fcc structure) are well detached from each other, which suggests that this synthetic method could effectively prevent agglomeration of platinum nanoparticles. However, replication of the porous structure of SBA-15 was not achieved, as confirmed by XRD. This could be ascribed to the fragility of the thin film of carbon, which would partially collapse during template removal. On the other hand, platinum nanoparticles were covered by a thin layer of carbon film. The BET surface area of the catalyst Pt@C/MC was 633 m^2 g^{-1} and the pore volume 0.55 cm^3 g^{-1}, the PSD exhibiting one sharp maximum at 3.5 nm and a weak broad peak at 14.5 nm. Additionally, textural mesoporosity was also observed, as well as the presence of microporosity. This catalyst exhibited no catalytic activity towards methanol oxidation and, in fact, a capacity to tolerate high concentrations of methanol. However, it had an admirable ORR activity due to the large surface of the support (mesoporous carbon) as well as the well-distributed platinum nanoparticles. On the contrary, the ORR activity was greatly impaired on the electrodes prepared with Pt/CMK-3 and Pt/Vulcan XC-72 because of methanol oxidation. The authors believe that it is the unique nanostructure of the Pt@C/MC composites obtained that endowed the nanoscale hybrid material with high catalytic activity for methanol-tolerant ORR, which is also a very important issue in methanol fuel cells. Since the platinum nanoparticles were overlaid by a film of carbon that contained micropores formed during the thermal treatment, this made it possible for oxygen to access the nanoparticles while methanol was hindered. Finally, the authors also evaluated the electrocatalyst durability through repeated CV cycles in an oxygen-saturated electrolyte consisting of 0.5 M methanol. The variation in the current density was only around 4% after 40 cycles, suggesting that the Pt@C/MC electrode has a considerable stable electrocatalytic activity for the ORR despite the existence of the well-known "poisonous" methanol in the electrolyte. Furthermore, the loss of the electrochemically active surface area of platinum would be greatly alleviated as a result of the carbon film on the surface of the nanoparticles.

In an earlier work, Wen *et al.* deposited platinum nanoparticles through chemical reduction with $NaBH_4$ over hollow carbon spheres/semispheres [144]. These hollow carbon spheres/semispheres were synthesized by HTC of glucose with the aid of sodium dodecyl sulfate at 170 °C (10 h) followed by heat treatment at 900 °C. A good dispersion of platinum nanoparticles (fcc structure and particle size = 5.7 nm) on the inner and outer surface of the hollow hemispheres was observed. This catalyst exhibited a higher activity towards methanol oxidation than those prepared with Vulcan XC-72 and microspheres prepared by HTC of glucose in the absence of sodium dodecyl sulfate. This can be attributed to the higher BET surface area, well-dispersed platinum nanoparticles, high conductivity, and the reduction of the liquid sealing effect.

Another work where the deposition of PtRu nanoparticles was also performed during the HTC process has been recently published by Marques Tusi *et al.* [145]. Thus, platinum and ruthenium salts were added to an aqueous solution of starch, the pH adjusted at around 11, then the mixture was subjected to HTC at 200 °C for 6 h and finally to a heat treatment at 900 °C. Without this high-temperature heat treatment, the samples were not active for methanol electrooxidation, which is probably due to their low electrical conductivity. The XRD analysis of the catalysts showed the coexistence of PtRu alloy with fcc structure and metallic ruthenium phase with hexagonal close-packed (hcp) structure. The average crystallite sizes determined by XRD agreed well with the values determined by TEM observation and were in the range 8–13 nm depending on the Pt/Ru atomic ratio. However, the particle size distributions were quite broad. The electrochemical oxidation of methanol was evaluated in a solution of 1 M methanol in 0.5 M H_2SO_4 at a sweep rate of 10 mV s^{-1} at room temperature. Electrooxidation of methanol started at 0.45–0.55 V versus a reversible hydrogen electrode, a value typical for PtRu carbon supported catalysts (versus 0.7–0.8 V for platinum carbon-supported catalysts), and an increase of current values was observed with the increase of ruthenium content in the catalysts.

In all the works described so far, the HTC process was carried out at a low temperature (i.e., 180–200 °C) so that an additional step was necessary for increasing the conductivity of the materials (graphitization process or heat treatment at a higher temperature, up to 1000 °C). That additional step was avoided by Xu *et al.* by performing the HTC process at a much higher temperature of 600 °C [146]. As a result, they obtained carbon microspheres of about 1500–2000 nm that, despite the high temperature used, exhibited abundant hydroxyl groups, as probed by Fourier transform infrared spectroscopy (FTIR). The authors used those carbon microspheres as support for palladium and platinum nanoparticles, deposited through chemical reduction using NaBH₄. The obtained results showed well-dispersed platinum and palladium nanoparticles on the outer surface of the carbon microspheres. The size of the nanoparticles was smaller than for the carbon black used as reference. This seems a bit striking as hydrothermal carbons normally exhibit very low surface areas, which would lead to coalescence of the metal nanoparticles. As the authors have pointed out, this is probably due to the stabilization of the nanoparticles by strong bonding interactions with the oxygen groups still present on the surface of the microspheres. As a result, the catalysts prepared with the carbon microspheres exhibited larger electroactive surface areas (double that of the catalyst with carbon black) and higher activity towards methanol/ethanol oxidation in alkaline media than the catalyst prepared with carbon black. This higher activity is probably also due to the fact that the nanoparticles are more accessible in the carbon microspheres as they are deposited over their external surface. Additionally, the micrometer-sized carbon spheres act as structure units to form pores and channels that significantly reduce the liquid-sealing effect.

A variation of the HTC process, called the solvothermal method, where ethanol is used as carbon source and solvent, was employed by Yuan *et al.* for the first time to synthesize coin-like hollow carbon (CHC) [147] and graphitized lace-like carbon (GLC) [148], which were used as supports for palladium and platinum nanoparticles, respectively. The CHC was synthesized using Mg/NiCl₂ as catalyst [147]. The hollow CHCs have a diameter of 1–3 nm and a thickness less than 154 nm. They exhibit a disordered graphitic structure with an interlayer distance of 0.34 nm in some areas, as deduced by high-resolution TEM corroborated by selected area electron diffraction and XRD patterns, as well as Raman analysis. The authors found that the formation of those CHCs takes place at temperatures of 550 °C or higher and the morphology is retained for temperatures below 700 °C. On the other hand, regardless of the high synthesis temperatures, the CHC contained 12% oxygen, which is present as hydroxyl and carbonyl groups, as deduced by infrared (IR) analysis. Prior to the deposition of palladium nanoparticles (by direct chemical reduction of palladium chloride with tannic acid), the CHCs were treated with 6 M HNO₃ solution at 30 °C for 4 h. As a result of this treatment, an increase in the surface area took place, the BET surface area of the acid treated CHCs having a value of 400 m² g⁻¹, which is 20 times higher than without the treatment. As in all the works previously described, a catalyst with Vulcan XC-72 as support was also prepared for comparison reasons. The XRD analysis of the catalyst revealed that the palladium nanoparticles deposited over the CHCs have a size of 7.4 nm – a value that agrees well with TEM observations. The electrochemical active surface area of the Pd/CHC catalyst was found to be 3 times higher than that of Pd/Vulcan, which implies a larger three-phase interface for the reaction. This is consequence of the high surface area

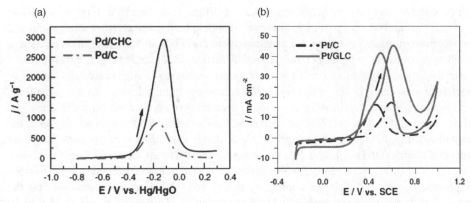

Figure 7.16 (a) Linear potential sweep curve of Pd/CHC and Pd/Vulcan in 1 M KOH + 1 M methanol (sweep rate = 5 mV s^{-1}). (b) CV of Pd/GLC and Pd/Vulcan electrocatalysts in 1 M CH$_3$OH + 0.5 M H$_2$SO$_4$ solution at 30 °C (scan rate = 5 mV s^{-1}). (Extracted with permission from [148]. © 2008 Elsevier.)

of the CHC and its unique morphology. Thus, the micrometer-sized CHCs act as structure units to form pores and channels in the catalyst layer that can significantly reduce the liquid-sealing effect, as observed previously for carbon microspheres. As a consequence, the Pd/CHC catalyst exhibited an activity towards methanol oxidation 3 times higher than Pd/Vulcan (see Figure 7.16a). Concerning GLC, magnesium was used as reducing agent [148]. The optimized conditions for GLC synthesis were found to be 12–16 h at 600–650 °C. As denoted by a sharp (002) reflection in the XRD pattern, the GLC exhibited a high crystalline order. Prior the deposition of platinum nanoparticles, GLC was activated in molten KOH. As a result, the graphitic content on the activated GLC increased, as revealed by Raman and XRD analysis, which was reflected in an increase of the electrical conductivity of the material from 186 to 236 S cm^{-1}. Furthermore, the BET surface area increased from 26 to 1710 m^2 g^{-1}. Platinum nanoparticles were deposited over this material, as well as Vulcan XC-72, by direct chemical reduction of chloroplatinic acid at 130 °C using ethylene glycol as reducing agent. Compared to Vulcan XC-72, the authors observed a better dispersion of the nanoparticles on the GLC support with a narrower particle size distribution centered at 5.2 nm – a value consistent with XRD analysis (5.6 nm). This led to a 2.6 times higher catalytic activity towards methanol oxidation (see Figure 7.16b) and 2.5 times higher electrochemically active surface area of platinum in Pt/GLC compared to Pt/Vulcan XC-72. Considering that both electrocatalysts have similar particle sizes, the results imply that a larger three-phase interface was formed on Pt/GLC due to the large surface area and unique morphology of the GLC. The same phenomenon as for CHCs would take place with the GLCs. Such a structure makes it easier for the liquid electrolyte and methanol to diffuse into the electrocatalyst layer. Therefore, a better utilization of platinum and reduced concentration polarization are obtained.

It can be concluded that the unique morphology of the HTC-derived products (spherical, coin-like, or lace-like) provides a clear advantage for their use as supports for electrocatalyst, as the particles act as structure units to form connected mesopores or channels in the electrocatalyst layer that makes it easier for the diffusion of the electrolyte and methanol towards the catalyst nanoparticles. However, it should be pointed out that some additional procedure to introduce porosity and conductivity in the HTC-derived materials is necessary, such as pyrolysis at high temperature, template mediation, or catalytic graphitization.

In the future, the diminution or even complete elimination of such scarce and expensive metals able to catalyze the electrooxidation reaction is highly desired. At the same time, CO is a very stable intermediate resulting from methanol decomposition and at the same time a strong poison for the platinum electrode, thus seriously reducing its activity. With the aim of finding alternative active catalysts, Mavrikakis *et al.* have recently applied density function theory (DFT) to investigate the structure sensitivity of methanol electrooxidation on eight transition metals [149]. In this respect, Yang *et al.* reported ruthenium-free, carbon-supported cobalt- and tungsten-containing binary and ternary platinum catalysts for the anodes of direct methanol fuel cells [150]. Maier *et al.* described the results of a high-throughput screening study for direct methanol fuel cell anode catalysts consisting of new elemental combinations with an optical high-throughput screening method, which allowed the quantitative evaluation of the electrochemical activity of catalysts [151]. It is predictable that novel catalyst supports will need to be developed for these new emerging active centers for methanol electrooxidation where HTC will also play an important role.

7.3.2   Heteroatom-Doped Carbons with Intrinsic Catalytic Activity for the ORR

Fuel cell reactions invariably involve an ORR at the cathode, which is one of the main rate-decreasing steps on platinum catalysts in terms of the water formation reaction and energy conversion efficiency in polymer electrolyte membrane fuel cells. The scarcity and cost of platinum have led to the development of alternative catalyst materials for fuel cell applications.

As a solution to increase activity towards oxygen reduction, and provide a long-term solution to platinum cost and scarcity, a variety of non-noble-metal-based catalysts have been investigated as promising cathode catalysts for fuel cells. The choice of suitable materials for this purpose is obviously restricted by all the conditions required to obtain a long lifetime under the working conditions of a fuel cell. Since Jasinski's first report on the ORR catalytic activity of metal–N_4 chelates as cobalt phthalocyanines [152], transition metal porphyrins have been thoroughly studied as attractive candidates for active and reliable catalysts for fuel cell cathodes [153, 154].

Thus, recently such nitrogen-doped carbons have played one of the most important roles for the ORR reaction in fuel cells [155]. They can either enhance the ORR for nonprecious metal catalysts [156] or show methanol-tolerant oxygen reduction [157]. Some of the carbon support used for the ORR in fuel cells showed catalytic properties themselves, in the absence of any metal. Hence, it was showed that nitrogenated CNT (NCNT) arrays might have unusually higher electrocatalytic activity for oxygen reduction than nitrogen-free CNTs [158, 159]. Vertically aligned NCNTs actually are an effective ORR electrocatalyst, even after complete removal of residual iron. These metal-free vertically aligned NCNTs catalyze ORR through a four-electron process with an enhanced electrocatalytic activity, long operation stability, and lower overpotential and tolerance to cross-over effects than platinum in alkaline electrolytes [158]. The integration of electron-accepting nitrogen atoms in the conjugated CNT plane appears to impart a relatively high positive charge density on adjacent carbon atoms. The charge density distribution, coupled with aligning the NCNTs, provides a four-electron reduction and excellent performance (a steady-state output potential of -80 mV and a current density of 4.1 mA cm^{-2} at -0.22 V, compared with -85 mV and 1.1 mA cm^{-2} at -0.20 V for a Pt/C electrode) [160]. Doping CNTs with nitrogen heteroatoms leads to active sites that promote parallel diatomic adsorption of O_2, which could effectively weaken the O–O bond. Simultaneous doping of CNTs with boron and nitrogen also proved efficient, possibly due to active centers as B–N–C moieties [161].

Concerning sustainability, most of the synthesis methods used to produce the afore-mentioned nitrogen-doped carbon materials show drawbacks in terms of the, often harsh, reaction conditions used. In order to avoid these aspects, nitrogen-doped materials have been prepared using the HTC process either directly using nitrogen-containing precursors [87], or by incorporation of amino acids [162] or proteins [85] into the HTC process. These procedures are in detail in Chapter 5.

While the HTC-derived materials have been intensively used as catalyst supports for the anode electrooxidation reaction, reports involving the use of HTC for the ORR at the cathode are far less numerous.

To the best of our knowledge, there is no report yet in the literature concerning the ORR behavior of nitrogen-doped hydrothermal carbons, except our own, which will be presented below [163].

As described in detail in Chapter 5, complementing nitrogen as a dopant, sulfur is receiv-ing increasing attention in current carbon materials research. Overall, literature reports suggest that nitrogen is the dopant of preference concerning the tunability of electronic properties of the carbon material, whereas sulfur, due to its large size, has been used more for applications where its easily polarizable lone pairs (and thus chemical reactivity) are of importance.

The synthesis of these sulfur-doped materials generally involves the pyrolysis of sulfur-containing polymer-based carbons [164–167], but also arc vaporization in the presence of sulfur-containing compounds such as thiophenes [168].

Concerning the combined incorporation of sulfur and nitrogen within the same material, only a few reports exist in the literature. Sulfur-assisted growth of CNTs by CVD of acetonitrile was shown to increase the nitrogen doping levels as well as the magnetic properties of the nanotubes [169].

Regarding sulfur-doped materials for ORR, Choi *et al.* synthesized heteroatom-doped carbon materials by the pyrolysis of amino acid/metal chloride composites. They obtained sulfur doping levels of 2.74 wt% using cysteine, and were able to show that materials containing both nitrogen and sulfur increased the material's ORR activity in acidic media, relative to undoped or purely nitrogen-doped carbons [170].

Regarding hydrothermal carbons doped with sulfur or dually doped with sulfur and nitrogen, Wohlgemuth and Titirici were the first to report on the one-pot hydrothermal synthesis of tunable dual heteroatom-doped carbon microspheres using carbohydrates and cysteine or cysteine derivatives (see Chapter 5) [171]. The addition of cysteine gives rise to pending sulfur functionalities, while addition of thienyl-cysteine results in structurally bound sulfur within the carbonaceous framework. Postpyrolysis offers an additional tool for controlling material stability and results in microporosity as well as superior conductivity relative to undoped carbon microspheres from glucose. Nitrogen doping levels of about 4 wt% and sulfur doping levels of 3–12 wt% could be achieved.

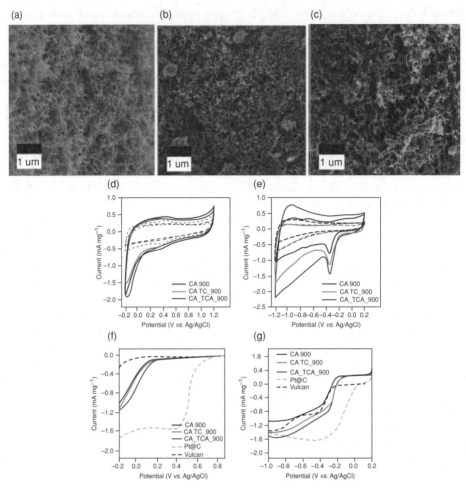

Figure 7.17   *SEM micrographs of (a) nitrogen-doped carbon aerogel (CA), (b) nitrogen- and sulfur-doped carbon aerogel obtained using thienyl-cysteine (CA-TC), and (c) nitrogen- and sulfur-doped aerogel obtained using 2-thiophene carboxaldehyde (CA-TCA). CV of doped carbon aerogels compared to Pt@C and Vulcan in (d) 0.1 M HClO₄ and (e) 0.1 M KOH. RDE polarization curves at 1600 rpm of doped carbon aerogels compared to 20 wt% Pt@C and Vulcan in (f) 0.1 M HClO₄ and (g) 0.1 M KOH. (Reproduced with permission from [148]. © 2008 Elsevier.)*

The same authors have introduced porosity in such nitrogen-, sulfur-, or dual nitrogen/sulfur-doped materials using protein gelation [86]. The morphology of the solely nitrogen-doped and the dual nitrogen/sulfur-doped materials is shown in Figure 7.17a–c. The simultaneous incorporation of nitrogen and sulfur into the HTC-derived carbons was confirmed by elemental analysis and X-ray photoelectron spectroscopy (XPS). Overall, all three materials exhibited promising structures and compositions for catalytic applications, such as high surface areas, large diameters, and continuous three-dimensionally arranged porous morphologies (and hence good mass transfer properties) leading to accessible dopant sites. However, the amorphous nature of hydrothermal carbon directly after HTC at 180 °C has the drawback of a rather low electrical conductivity. In order to convert the "organic aerogels" into "carbon aerogels" that are suitable for electrocatalytic applications, a pyrolysis step at 900 °C was added to the synthesis process. During this step, the nitrogen and sulfur groups were converted into functionalities incorporated into aromatic carbon domains while the conductivity increased up to 2657 S m^{-1}.

The effect of sulfur doping in addition to nitrogen doping (i.e., the catalytic activity of CA-TC-900 (materials prepared from glucose and thienyl-cysteine in the presence of ovalbumin) and CA-TCA 900 (materials prepared from glucose and 2-thiophene carboxaldehyde in the presence of ovalbumin)) was compared to solely nitrogen doped CA-900 (material from glucose and albumin).

CV at a scan rate of 100 mV s^{-1} and rotating disk electrode (RDE) voltammetry at a scan rate of 10 mV s^{-1} were conducted in 0.1 M KOH and 0.1 M HClO$_4$ (Figure 7.17d and e). In both cases, featureless voltammetric curves are observed for all doped carbon aerogels in the nitrogen-saturated solution. The area of the voltammograms arises from capacitive currents of the electrodes. In contrast, a well-defined cathodic peak appeared in the oxygen-saturated 0.1 M KOH solution (Figure 7.17e), clearly demonstrating the electrocatalytic activity of the doped carbon aerogels towards oxygen reduction. The areas of the voltammograms are larger in oxygen-saturated solution due to Faradaic currents (i.e., current generated by charge transfer between reacting species).

In 0.1 M HClO$_4$, Faradaic currents are similarly observed for the doped aerogels oxygen-saturated solution (Figure 7.17d). As the aerogels are less active in acidic than in alkaline media, the pronounced cathodic peaks found in 0.1 M KOH are not visible in 0.1 M HClO$_4$ at the same scan rate of 10 mV s^{-1}. The small redox peaks observed for CA-900 and CA-TC-900 in nitrogen-saturated solution may be a result of heteroatom protonation on the carbon surface. The polarization curves obtained from RDE (1600 rpm) voltammetry in oxygen-saturated 0.1 M HClO$_4$ and 0.1 M KOH are shown in Figure 7.17f and e, respectively. Compared to Pt@C, however, the aerogels are still not competitive. In 0.1 M KOH, the performance of the doped aerogels is far more comparable with that of Pt@C. Compared to Vulcan, CA-900 shows an improved onset potential of -185 mV as well as slightly improved current densities within the scanned potential range. The onset potential of the sulfur- and nitrogen-containing aerogels is more positive (both at around -130 mV), and the maximum current density is considerably higher than for Vulcan.

The current knowledge in the scientific literature regarding sulfur and ORR is limited, and sulfur is usually thought to improve platinum particle adsorption onto carbon supports and thereby the lifetime of the electrode [172]. However, the role of sulfur in metal-free catalysts and how sulfur compares to nitrogen as a dopant remain open questions.

In nitrogen-doped carbons, factors such as enhanced π-bonding, electrical conductivity, and Lewis basicity may facilitate reductive oxygen adsorption at the carbon surface [173]. Structural defects in the carbon crystal lattice, which are caused by the introduction of dopants, also result in more edge-active sites [174]. It has, however, been shown that undoped carbon materials with more edge sites did not have an improved catalytic performance [175], indicating that edge-bound heteroatoms (e.g., pyridinic nitrogen) are catalytically important. It is generally accepted that the binding state is relevant with respect to the catalytic activity of nitrogen, although there are different opinions as to which exact binding states are responsible for the good activity and two- or four-electron process selectivity. Pyridinic edge sites have been proposed as a likely candidate because edge planes facilitate oxygen chemisorption [176]. On the other hand, some reports also suggest that pyridinic nitrogen may not be an effective promoter of the four-electron ORR process. Luo *et al.* synthesized purely pyridinic nitrogen-doped carbons and found them to be selective for a two-electron reduction pathway [177]. Lui *et al.* recently proposed that graphitic nitrogen accounts for good catalytic activity. They also showed that the nitrogen content does not directly correlate with the catalyst performance – materials (nitrogen-doped mesoporous graphitic arrays) with higher nitrogen content showed lower selectivity and activity [178].

A report by Strelko *et al.* suggested that there is a critical concentration of heteroatoms in a carbon matrix that will exhibit maximum catalytic activity, and that this can be explained by the collective electronic properties and a minimal bandgap. They identified pyrrolic nitrogen as the binding state that gives rise to the smallest bandgap and thus the best electron transfer capabilities [179]. DFT calculations suggest that nitrogen is not itself the catalytically active site, but that the high electronegativity of nitrogen polarizes the C—N bond and the adjacent carbon atom therefore has a reduced energy barrier towards the ORR [174, 180].

As if for nitrogen, the opinions are much divided and there is no clear path. It seems to be dependent on electronic, morphological, and pore properties; for sulfur doping the situation is even more complicated.

Sulfur and carbon have electronegativities of 2.58 and 2.55, respectively (on the Pauling scale). Nitrogen, however, has an electronegativity of 3.04. This means that the C—S bond is not as polarized as the C—N bond, so a catalytic pathway based on a $\delta+$ adjacent carbon atom is unlikely to occur for sulfur.

Sulfur is a large atom with an atomic radius of 100 pm compared to nitrogen (65 pm) and carbon (70 pm) [181]. The disruption of the carbon connection pattern is therefore more pronounced than for nitrogen. It is therefore likely that sulfur doping will induce more strain and defect sites in the carbon material, which may facilitate charge localization and the coupled chemisorption of oxygen.

Sulfur has large, polarizable d orbitals (sulfur groups are usually soft nucleophiles). The lone pairs of sulfur can therefore easily interact with molecules in the surrounding electrolyte. This effect is expected to be much more pronounced than that for nitrogen.

Sulfur is, however, known to take part in proton transfer reactions. DFT calculations carried out by Chamorro *et al.* suggested that proton transfer in thiooxalic acid derivatives is facilitated by the high polarizability of the sulfur atom, which mediates ion-pair-like transition states during the transfer process [182]. Scheiner *et al.* carried out *ab initio* calculations and showed that the greater polarizability of SH_2 as compared to OH_2 leads to greater charge transfer between $(H_2S—H—SH_2)^+$ units than between $(H_2O—H—OH_2)^+$ units and to a larger extent of spatial regions of density charge [183].

In their publication on sulfur-doped graphene as ORR catalysts, Yang *et al.* propose that the increased spin density of sulfur compared to nitrogen or other dopants may be responsible for the increased catalytic activity [184]. This would mean that sulfur is favorable to interact with the triplet state of oxygen, as preservation of spin is a serious catalytic problem.

Given all these hypotheses existing in the literature for the solely nitrogen- and sulfur-doped carbons, in the case of dual sulfur/nitrogen-doped carbon aerogels, a synergistic mechanism between sulfur and nitrogen, whereby nitrogen activates the oxygen molecule (either directly or indirectly via the adjacent carbon atom) while sulfur facilitates the proton transfer during the reduction process, could be likely to happen. However, careful theoretical studies are necessary in order to understand the electron transfer process and the role of each individual dopant as well as the synergetic effect.

What is clear is the fact that sulfur has a clear positive effect when added to nitrogen doping on the electrocatalytic performance in the ORR in both acidic and basic media. More studies need to follow in the future.

To summarize, very little research has been done regarding the use of heteroatom-doped hydrothermal carbons as electrocatalysts for the ORR reaction. This is surprising given all the advantages offered by HTC in terms of low cost, easy production, and especially easy functionalization with any heteroatom of choice using organic chemistry (e.g., see Chapter 5 on the Maillard reaction). Given the huge amount of literature already available on the performance of heteroatom-doped hydrothermal carbons in supercapacitors, we are expecting a rapid and significant growth of research interest in this exciting and important field of research.

7.4 Photocatalysis

Efficient photocatalytic processes have the potential to yield major steps forward in tackling some of society's greatest challenges: clean energy demands (water splitting for hydrogen generation) and environmental pollution (degradation of environmental pollutants in aqueous contamination and waste water treatment, CO_2 remediation, self-cleaning activity, and air purification).

The development of effective semiconductor photocatalysts has therefore emerged into one of the most important goals in materials science. Indeed, since the first demonstration of photocatalytic water splitting on a titanium dioxide (TiO_2) electrode by Fujishima and Honda [185] the level of research in the field has grown at an exponential rate.

Energy provided by the Sun (around 1.5×10^5 TW [186]) greatly exceeds that consumed by human civilization (around 13 TW [186]). However, only a fraction of this energy is harvested by current photocatalytic materials, which typically have solar photoconversion efficiencies of below 5%.

TiO_2 has been widely used as a photocatalyst for solar energy conversion and environmental applications because of its low toxicity, abundance, high photostability, and high efficiency [187–189]. However, the application of pure TiO_2 is limited because it requires ultraviolet (UV) light, which makes up only a small fraction (less than 5%) of the total solar spectrum reaching the surface of the Earth.

Therefore, over the past few years, considerable efforts have been directed towards the improvement of the photocatalytic efficiency of TiO_2 in the visible light region [190, 191]. This has been mainly achieved by introducing various dopants into the TiO_2 structure, which can narrow its bandgap. In this respect, nonmetal doping with atoms such as with boron, fluorine, nitrogen, carbon, and sulfur has demonstrated a significant improvement of the visible light photocatalytic efficiency [189, 192, 193]. Among these, carbon doping has received particular attention [189]. An active debate regarding the fundamental nature of the nonmetal species causing the visible light absorption in such modified TiO_2 materials has continued in the community and two hypotheses have coexisted for several years: (i) the nonmetal substitutes a lattice atom (i.e., doping) and (ii) the nonmetal forms chromophoric complexes at the surface (i.e., sensitization).

TiO_2 was first doped both hydrothermally and solvothermally with carbon by Titirici and Wang [194]. They showed that the surface of nanometer-sized carbon materials can also show collective polarization modes and, therefore, these optical absorption transitions are feasible to sensitize TiO_2, showing an improved TiO_2 hole reactivity, while the electron is taken up by the carbon component. This resulted in an improved photocatalytic activity over the complete spectral range. In order to avoid carbon from doping directly into the bulk TiO_2 lattice our hybrid TiO_2/C was synthesized at low temperature under solvothermal conditions by a one-step carbonization of furfural in the presence of titanium isopropoxide, allowing for the formation and coassembly of carbon and TiO_2 into an interpenetrating $C@TiO_2$ nanoarchitecture (Figure 7.18a–c). Figure 7.18d shows the comparison between $C@TiO_2$ and some control samples towards the degradation of Methyl Orange in aqueous solution under visible light. The $C@TiO_2$ showed the highest photocatalytic activity of all compared materials in visible light ($\lambda > 420$ nm). Both the commercial P25 and nitrogen-doped P25 only showed an activity level comparable with the self-degradation of Methyl Orange under the same irradiation conditions. Complementing the chemical reactivity experiments, the photocurrent generation of $C@TiO_2$ was also investigated (Figure 7.18e). Promisingly, the $C@TiO_2$ material was indeed able to generate significant photocurrents under both UV and visible light irradiation. A typical n-type photocurrent was generated

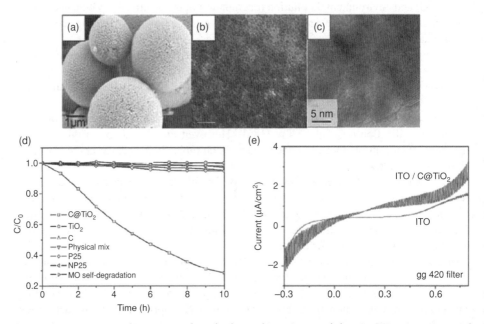

Figure 7.18 (a) SEM, (b) TEM, and (c) high-resolution TEM of the $C@TiO_2$ composite. (d) Photocatalytic degradation of Methyl Orange in the presence of $C@TiO_2$ and other samples under visible light irradiation ($\lambda > 420$ nm). (e) Photocurrent of indium tin oxide (ITO)/$C@TiO_2$ as a function of potential under chopped visible light.

with UV (λ > 320 nm), while under visible light irradiation the material exhibited both n-type and p-type currents. This was explained by the fact that UV absorption also activates pure TiO_2 bands, while the ambipolar, biphasic nature of $C@TiO_2$ is activated under visible light. This allows both electrons and holes to contribute to the charge transport, which is crucial for photoelectrochemical applications.

Following this report, Chen *et al.* also prepared a carbon-deposited TiO_2 using a very similar one-pot hydrothermal process and glucose as a carbon source [195]. This $TiO_2@C$ composite also had remarkable light absorption in the visible region. It was found that the photocatalytic activity of $TiO_2@C$ was greatly enhanced compared to noncarbon/TiO_2 under visible irradiation for the degradation of Acid Orange 7 and 2,4-dichlorophenol. The authors claim that two kinds of sensitization processes (i.e., carbon sensitization and dye sensitization) are responsible for the visible-light-induced photocatalysis of $TiO_2@C$.

Zhang *et al.* reported the fabrication of a novel mesoporous $TiO_2@C$ photocatalyst via an ethanol supercritical solvothermal method involving tetrabutyl titanate and raw rice [196]. The as-prepared $TiO_2@C$ possessed a bimodal carbon-modification effect, including carbon doping in the lattice of TiO_2 and carbon sensitizing of the surface of TiO_2. The ethanol supercritical treatment also contributed to the development of a mesoporous structure with a large surface area ($160 \, m^2 \, g^{-1}$) and high crystallinity of anatase. These materials exhibited an excellent photocatalytic performance and recyclability for phenol oxidation under visible light irradiation (above 420 nm).

Hierarchical porous $TiO_2@C$ hybrid composites with a hollow structure were successfully fabricated by Zhuang *et al.* using a one-pot low-temperature solvothermal approach in the presence of dodecylamine [197]. The growth mechanism of the hierarchical hollow spheres was demonstrated to include the condensation of a carbon source and the coinstantaneous *in situ* hydrolysis of titanium alkoxide, and the consequent assembly of $TiO_2@C$ hybrid nanoparticles on a self-conglobated template. As compared with the $TiO_2@C$ solid spheres (86%), the hierarchical $TiO_2@C$ hybrid hollow spheres exhibited enhanced photocatalytic efficiency (97%) for the visible light photooxidation of rhodamine B. Investigations demonstrated that the enhancement can be attributed to the hierarchical porous hollow structure. Moreover, the superoxide radical was detected as the main active species generated in the oxidation reaction of rhodamine B over $TiO_2@C$ photocatalysts. A corresponding mechanism was also proposed for the photocatalysis process.

Zhang *et al.* prepared core/shell nanofibers of $TiO_2@C$ embedded with silver nanoparticles ($TiO_2@C$/Ag nanofibers) combining the electrospinning technique, the hydrothermal method, and an *in situ* reduction approach [198]. The results showed that a uniform carbon layer of approximately 8 nm in thickness was formed around the electrospun TiO_2 nanofibers and small silver nanoparticles were dispersed well inside the carbon layer. The $TiO_2@C$/Ag nanofibers had remarkable light absorption in the visible region. The photocatalytic studies revealed that the $TiO_2@C$/Ag nanofibers exhibited enhanced photocatalytic efficiency of photodegradation of rhodamine B and Methyl Orange compared with the pure TiO_2 nanofibers, $TiO_2@C$ core/shell nanofibers and TiO_2/Ag nanofibers under visible light irradiation, which might be attributed to the good light absorption capability and high separation efficiency of photogenerated electron–hole pairs based on the photosynergistic effect among the three components of TiO_2, carbon and silver.

Wang *et al.* used an approach to prepare carbon-doped TiO_2 hollow spheres with a hierarchical pore structure using hydrothermal carbon spheres according to the initial procedure described by Titirici *et al.* [35]. The hydrothermal carbon spheres played a dual role as both a single hard template and the source of carbon doping. It is important to note that the resultant TiO_2 hollow spheres interconnected with each other via smaller pores to form hierarchical macroporous channels and plentiful mesopores located at the macropore walls. Furthermore, the size of macroporous channels and the thickness of macropore walls can be systematically tuned by adjusting synthesis parameters. The unique hierarchical macroporous channel structure was confirmed to dramatically enhance the photocatalytic performance of TiO_2 hollow spheres [199].

Shao *et al.* synthesized Zn_2TiO_4@C core/shell nanofibers with different thickness of carbon layers (from 2 to 8 nm) combining the electrospinning technique and HTC method [200]. The results showed that a uniform carbon layer was formed around the electrospun Zn_2TiO_4 nanofibers. By adjusting the hydrothermal fabrication parameters, the thickness of the carbon layer varied linearly with the concentration of glucose. Furthermore, the core/shell structure formed between Zn_2TiO_4 and carbon enhanced the charge separation of pure Zn_2TiO_4 under UV excitation, as evidenced by photoluminescence spectra. The photocatalytic studies revealed that the Zn_2TiO_4@C nanofibers exhibited enhanced photocatalytic efficiency for the photodegradation of rhodamine B compared with the pure Zn_2TiO_4 nanofibers under UV excitation, which might be attributed to the high separation efficiency of photogenerated electrons and holes based on the synergistic effect between carbon and Zn_2TiO_4. Notably, the Zn_2TiO_4@C nanofibers could be recycled easily by sedimentation without a decrease of the photocatalytic activity.

Another very interesting approach recently published by Ozin *et al.* is based on using carbon quantum dots (CQDs) as sensitizers for photovoltaic applications in nanocrystalline TiO_2-based solar cells [201].

CQDs are intriguing, recently discovered members of the carbon nanomaterials family alongside CNTs, fullerenes, and graphene. The general description of CQDs is that of nanometer-size particles consisting of a sp^2 hybridized graphitic core functionalized with polar carboxyl or hydroxyl groups on the surface. Initial CQD synthesis efforts focused on top-down methods such as laser ablation or electrochemical oxidation. Less expensive and more scalable solution-phase methods involve (hydro) thermal decomposition of carbon precursors and microwave-assisted pyrolysis of carbohydrates. Ozin *et al.* used the dehydration of γ-butyrolactone using concentrated sulfuric acid, resulting in a carbonaceous material dispersed in residual acidic butyrolactone. Due to the low electron contrast between CQDs and carbon-coated TEM grids, acquiring high-quality TEM images was a challenge and therefore atomic force microscopy (AFM) imaging was performed to obtain a more reliable nanoparticle size distribution (Figure 7.19a and b). The average height of CQD nanoparticles was found to be 9–6 nm by AFM with significant numbers of larger particles in the range of 20–30 nm, some of which were also visible by TEM. Photoluminescence spectra of aqueous CQDs showing the excitation-dependent nature of the photoluminescence are shown in Figure 7.19c. A large red-shift in the photoluminescence maximum from approximately 470 to 580 nm was observed upon an increase of the excitation wavelength from 350 to 500 nm, most likely attributed to optical selection of differently sized CQDs in the colloidal dispersion. The absolute photoluminescence quantum yield of our CQD dispersions was determined to be 0.5%. This value is likely related to efficient nonradiative competing pathways involving vibrational relaxation enabled by high-frequency core and surface modes.

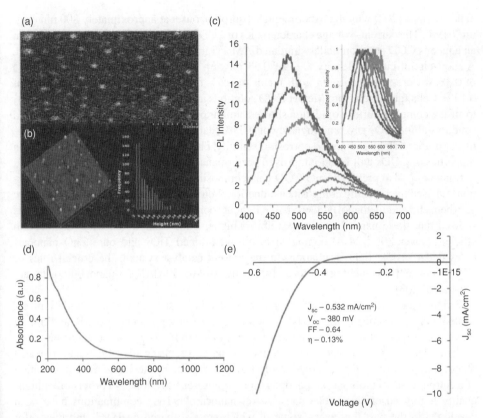

Figure 7.19 (a) TEM image of CQDs drop cast on a carbon-coated TEM grid. (b) AFM image and particle height distribution of a CQD thin film spin-coated onto a silicon wafer. (c) Photoluminescence spectra of purified CQDs at different excitation wavelengths from 350 to 500 nm increasing at 25-nm intervals (inset: normalized photoluminescence spectra at the corresponding excitation wavelengths). (d) UV/Vis/near-IR absorption spectrum of CQD thin film on quartz substrate. (e) Current–voltage characteristics of an aqueous CQD-sensitized solar cell. (Reproduced with permission from [201]. © 2011 Royal Society of Chemistry.)

The broad absorption spectra, ease of synthesis from a single-molecule precursor, and solution-processable nature of CQDs makes them potentially valuable as sensitizers for nanocrystalline TiO$_2$-based solar cell applications. Dye-sensitized solar cells have the capacity to be a useful solution to the impending energy challenge facing our planet due to their low cost, environmental friendliness, and nearly angle-independent performance under diffuse light. The current state-of-the-art ruthenium-complex sensitizers offer good performance, but rely on expensive, scarce ruthenium and are often time-consuming to synthesize. To explore whether these issues can be surmounted, the CQDs synthesized here were checked as alternative sensitizers for TiO$_2$ solar cells.

The absorption spectrum of a thin film of CQDs is shown in Figure 7.19d, displaying broad absorption throughout the visible region. An electronic bandgap of approximately 1.1 eV for an indirect transition and 3.1 eV for a direct transition was calculated using

Tauc relations [202] with the experimental absorption onset at approximately 800 nm (Figure 7.19d). The current–voltage characteristics of a CQD-sensitized solar cell prepared from an aqueous CQD solution under simulated AM 1.5 irradiation are shown in Figure 7.19d. A short-circuit current density (J_{sc}) of 0.53 mA cm^{-2} and an open-circuit voltage (V_{oc}) of 0.38 V were produced with a fill factor of 0.64, for a power conversion efficiency of 0.13%. Cells fabricated from MeOH CQD solutions and aqueous CQDs refluxed in HNO$_3$ to ensure complete surface oxidation showed similar performance, whereas a nonsensitized nanocrystalline TiO$_2$ gave an efficiency of 0.03% (data not shown). It is hoped that refinements in the surface chemistry, corrosion stability, charge transport, and charge injection properties of CQDs will lead to improved performance.

Kang *et al.* also prepared ZnO/CQD nanocomposites by a one-step hydrothermal reaction and used the resulting composite as superior photocatalysts for the degradation of toxic gas (benzene and methanol) under visible light at room temperature. The presented results showed that these nanocomposites exhibited higher photocatalytic activity (degradation efficiency over 80%, 24 h) compared to nitrogen-doped TiO$_2$ and pure ZnO nanoparticles under visible light irradiation. In the present catalyst system, the crucial roles of CQDs in the enhancement of photocatalytic activity of the ZnO/CQDs nanocomposites are illustrated [203].

Various nanocarbon/TiO$_2$ (and other semiconductor) systems have been widely investigated and are promising materials for future high-activity photocatalysts. In addition to providing a high-surface-area support and immobilization for TiO$_2$ photocatalyst particles, the presence of the carbonaceous material and nanostructuring may facilitate enhanced photocatalytic activity through one or all of the three primary mechanisms: minimization of electron–hole recombination, bandgap tuning/photosensitization, and provision of high-quality highly adsorptive active sites. Novel nanocarbon/TiO$_2$ combinations have been developed in the past few years, some of which are also based on HTC, and they offer opportunities for the design of new photocatalytic systems. The primary challenge to further exploitation of synergistic effects lies in a better understanding of the mechanisms of enhancement, in parallel with control and understanding of synthesis.

7.5   Gas Storage

7.5.1   CO$_2$ Capture Using HTC-Based Carbons

The mitigation of CO$_2$ emissions is a crucial issue as this gas is the main anthropogenic contributor to climate change. Among the possible strategies for CO$_2$ abatement, capture and storage have attracted keen interest. In this regard, the use of solid sorbents to capture CO$_2$ by means of pressure, temperature, or vacuum swing adsorption systems constitutes a promising alternative [204]. To accomplish this objective, the sorbents need to satisfy important conditions: (i) low cost and high availability, (ii) large CO$_2$ uptake, (iii) high sorption rate, (iv) good selectivity between CO$_2$ and other competing gases (i.e., nitrogen), and (v) easy regeneration. However, the development of a solid sorbent that satisfies all these conditions has so far proved to be complex. Taking into account the potential scale involved in the production of porous carbons for CO$_2$ capture, the use of renewable sources

Table 7.1 *Textural properties of HTC-based activated carbons derived from eucalyptus sawdust. (Adapted with permission from [205]. © 2011 Royal Society of Chemistry.)*

Activation temperature (°C)	KOH/hydrothermal carbon = 2		KOH/hydrothermal carbon = 4	
	S_{BET} (m^2 g^{-1})^a	V_p (cm^3 g^{-1})^a	S_{BET} (m^2 g^{-1})^a	V_p (cm^3 g^{-1})^a
600	1260 (98)	0.62 (89)	2370 (86)	1.15 (79)
650	1380 (98)	0.67 (91)	—	—
700	1390 (98)	0.69 (90)	2250 (93)	1.03 (88)
800	1940 (95)	0.97 (85)	2850 (95)	1.35 (91)

*The percentage of surface area and pore volume corresponding to micropores is indicated in parentheses.

for fabricating these materials would seem highly desirable. A good option is the use of biomass or biomass-derived products as precursors for the production of carbon sorbents for CO_2 capture. In this respect, low-cost sustainable porous carbons such as those derived from HTC-derived materials would constitute a good alternative.

The CO_2 capture performance of HTC-based activated carbons was recently investigated by Sevilla and Fuertes [205]. These authors analyzed the use as CO_2 adsorbents of porous carbons obtained by chemical activation of several HTC-derived materials prepared from starch, cellulose, and eucalyptus sawdust. The HTC of these materials was carried out at temperatures in the range of 230–250 °C for 2 h. The chemical activation was performed with KOH at temperatures in the range of 600–850 °C. They observed that the textural properties of the activated carbons derived from sawdust are similar to those obtained from starch or cellulose. This result demonstrates that an inexpensive and widely available biomass subproduct such as sawdust constitutes an excellent precursor for the preparation of chemically activated carbons via HTC-derived materials. Table 7.1 lists the textural properties (i.e., surface area and pore volume) of the sawdust-based HTC-derived activated carbons produced at several reaction temperatures by using two KOH/hydrothermal carbon mass ratios. They exhibit BET surface areas between 1260 and 1940 m^2 g^{-1}. As indicated, both the surface area and the pore volumes are mostly associated to micropores (less than 2 nm).

The CO_2 adsorption uptake for several HTC-based activated carbons is listed in Table 7.2 for a pressure of 1 bar and three adsorption temperatures (0, 25, and 50 °C). It can be seen that the CO_2 capture capacities of the HTC-based porous carbons prepared from starch, cellulose, or eucalyptus sawdust are quite substantial and similar (about 5.5–5.8 mmol CO_2 g^{-1}, 243–256 mg CO_2 g^{-1}). This result is coherent with the fact that the pore characteristics of these materials are analogous. The sawdust-based HTC-derived activated carbons prepared by using KOH/hydrothermal carbon = 4 exhibit similar CO_2 uptakes irrespective of the activation temperature. Indeed, the capture capacities at adsorption temperatures of 0 and 25 °C are in the 5.2–5.8 and 2.9–3.5 mmol CO_2 g^{-1} ranges, respectively. Interestingly, the HTC-sawdust-based porous carbons prepared under mild activation conditions (KOH/hydrothermal carbon = 2) exhibit better CO_2 capture capacities than under KOH/hydrothermal carbon = 4. Thus, at room temperature (25 °C), CO_2 adsorption capacities up to 6.6 and 4.8 mmol CO_2 g^{-1} were obtained for the samples prepared with KOH/hydrothermal carbon = 2 and reaction temperatures of 700 and 600 °C,

Table 7.2   *CO$_2$ capture capacities of the porous carbons at different adsorption temperatures and 1 atm. (Adapted with permission from [205]. © 2011 Royal Society of Chemistry.)*

		Chemical activation	CO$_2$ uptake (mmol g^{-1} (mg g^{-1}))		
HTC precursor	*T* (°C)	KOH/hydrothermal carbon	0 °C	25 °C	50 °C
Starch	700	4	5.6 (247)	3.5 (152)	2.2 (196)
Cellulose	700	4	5.8 (256)	3.5 (155)	1.8 (79)
Eucalyptus sawdust	600	4	5.2 (230)	2.9 (128)	—
	700	4	5.5 (243)	2.9 (128)	1.8 (79)
	800	4	5.2 (227)	3.0 (130)	—
	600	2	6.1 (270)	4.8 (212)	3.6 (158)
	650	2	6.0 (262)	4.7 (206)	3.3 (145)
	700	2	6.6 (288)	4.3 (190)	2.6 (116)
	800	2	5.8 (255)	3.9 (170)	3.1 (136)

respectively. These outstanding CO$_2$ adsorption uptakes are ascribed to the fact that a large fraction of the porosity of the mildly activated HTC-derived samples corresponds to narrow micropores, which have strong adsorption potentials that enhance their filling by the CO$_2$ molecules.

For practical applications, in addition to a high CO$_2$ adsorption capacity, sorbents need to show fast adsorption kinetics, a high selectivity towards CO$_2$, and must also be easy to regenerate. Sevilla and Fuertes [205] examined the performance of sawdust-based HTC-derived activated carbons in relation to these prerequisites. As shown in Figure 7.20a, the CO$_2$ adsorption is very fast, around 95% of CO$_2$ uptake occurring in a span of 2 min.

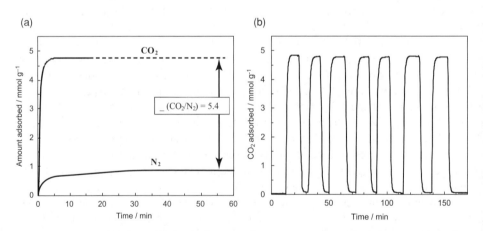

Figure 7.20   *(a) Adsorption kinetics of CO$_2$ and nitrogen at 25 °C and (b) CO$_2$ adsorption/ desorption cycles obtained at 25 °C (CO$_2$ concentration: 100%). The carbon sample used in this experiment was prepared from a sawdust-derived hydrothermal carbon activated at 600 °C with KOH/hydrothermal carbon = 2. (Reproduced with permission from [201]. © 2011 Royal Society of Chemistry.)*

By contrast, nitrogen adsorption is slower, around 60 min being needed for the maximum adsorption uptake to take place. Then, the $[CO_2/N_2]$ selectivity measured under equilibrium conditions is 5.4. Easy regeneration is another critical property that must be considered when designing CO_2 sorbents. In this respect, for the HTC-based carbons more than 95% of CO_2 is desorbed within 3 min under these conditions. This is illustrated in Figure 7.20b, where the adsorption/desorption cycles are represented. The experiments were repeated 7 times and no noticeable changes were observed in the desorption kinetics or CO_2 uptake.

Another interesting approach in relation to the use of HTC-derived materials for CO_2 capture is based on their functionalization with amine groups, which exhibit a high affinity to CO_2. In this respect, Titirici *et al.* [206] reported CO_2 capture by means of an amine-rich HTC-derived product. This material was prepared by a two-step process: (i) HTC of glucose in the presence of small amounts of acrylic acid and (ii) functionalization of carboxylic-rich HTC products with triethylamine. This aminated HTC-derived material shows high CO_2 capture capacities (up to 4.3 mmol CO_2 g^{-1} at -20 °C). More importantly, these materials exhibit a very high $[CO_2/N_2]$ selectivity at low (-20 °C) and high (70 °C) temperatures, up to 110 at 70 °C.

Zhang *et al.* prepared nitrogen-containing porous carbon from an ocean pollutant, *Enteromorpha prolifera*, via HTC followed by KOH activation [207]. Carbons contained as much as 2.6% nitrogen in their as-prepared state. The inorganic minerals contained in the carbon matrix contributed to the development of mesoporosity and macroporosity, functioning as an *in situ* hard template. The carbon manifested high CO_2 capacity and facile regeneration at room temperature. The CO_2 sorption performance was investigated in the range of 0–75 °C. The dynamic uptake of CO_2 was 61.4 and 105 mg g^{-1} at 25 and 0 °C, respectively, using 15% CO_2 (v/v) in nitrogen. Meanwhile, regeneration under argon at 25 °C recovered 89% of the carbon's initial uptake after eight cycles.

Sevilla *et al.* recently prepared highly porous nitrogen-doped carbon materials with apparent surface areas in the 1300–2400 m^2 g^{-1} range and pore volumes up to 1.2 cm^3 g^{-1} from mixtures of algae and glucose. The porosity of these materials was made up of uniform micropores, most of them having sizes below 1 nm. Moreover, they had nitrogen contents in the range of 1.1–4.7 wt%, this heteroatom being mainly in pyridone-type structures. These microporous carbons present unprecedented large CO_2 capture capacities, up to 7.4 mmol g^{-1} (1 bar, 0 °C). The importance of the pore size on the CO_2 capture capacity of microporous carbon materials was clearly demonstrated. Indeed, a good correlation between the CO_2 capture capacity at subatmospheric pressure and the volume of narrow micropores was observed, pointing out the unimportance of the pyridone-type structures present in the structure of these materials [208].

Recently, a variety of microporous polymers have been reported to have high $[CO_2/N_2]$ selectivity, which is a very interesting feature in the field of CO_2 capture [209–213]. As the as-prepared HTC-based carbons are microporous and possess polymer-like behavior, it was interesting to investigate their ability as selective CO_2 sorption materials. To compare the adsorption ability for CO_2 and nitrogen, the nitrogen uptake of xylose-derived HTC-derived material (HTC-X) at 273 K was measured (Figure 7.21).

The uptake of around 1.3 cm^3 g^{-1} at standard temperature and pressure is small as compared to the CO_2 uptake of 26.0 cm^3 g^{-1} under the same conditions. This can be calculated to an apparent $[CO_2/N_2]$ selectivity of 20 at 273 K, 1 bar, and 1 : 1 gas composition. This selectivity is not extraordinarily high, but it definitely represents a competitive result. The

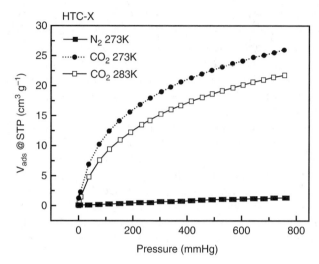

Figure 7.21   *Nitrogen and CO₂ sorption isotherms of HTC-X, measured at 273 K and CO₂ sorption isotherm measured at 283 K. (Adapted with permission from [205]. © 2011 Royal Society of Chemistry.)*

selectivity range is comparable to that of some microporous polymers, which show comparable CO_2 uptakes [209, 212]. The fact that the HTC-derived materials are easily made out of biomass by simple autoclave treatment below 200 °C is an extra advantage as compared with the complicated and more expensive synthesis of porous polymers.

7.5.2   Hydrogen Storage Using HTC-Based Activated Carbons

Hydrogen storage is currently one of the main obstacles towards the commercial use of hydrogen in fuel cell systems. Among the large variety of materials investigated as carriers for hydrogen storage, porous carbons have generated great attention. The interest in porous carbons lies in the fact that they are relatively cheap and accessible on a commercial scale, and, in addition, their pore structure can be easily designed, which is important to optimize the hydrogen storage capacity. In this sense, porous carbons with very large BET surface areas (above 2000 m² g⁻¹) and a porosity made up by micropores of around 1 nm have demonstrated large capacities to adsorb hydrogen [215].

Carbons with these characteristics have been prepared by chemical activation of a variety of precursors (i.e., coal, petroleum pitch, polymers, biomass products, etc.). Recently, Sevilla *et al.* reported the use as hydrogen adsorbents of porous carbons produced by chemical activation (KOH) of HTC-derived materials obtained by HTC of glucose, starch, furfural, cellulose, and sawdust [216]. The main textural parameters for the HTC-based activated carbons, as well as their hydrogen storage capacity at 1 and 20 and at −196 °C, are given in Table 7.3. Independently on the type of precursor selected to produce the HTC-derived material, all the activated samples exhibited similar textural properties: (i) large BET surface areas around 2200 m² g⁻¹, (ii) pore volume of around 1 cm³ g⁻¹, and (iii) a porosity made up mainly by micropores (below 2 nm). Their hydrogen storage capacity

Table 7.3 *Hydrogen uptake capacity and textural properties of chemically activated carbons produced from HTC-derived materials. (Adapted with permission from [216]. © 2011 Royal Society of Chemistry.)*

Precursor[a]	BET surface area (m² g⁻¹)	Pore volume (cm³ g⁻¹)	Micropore volume (cm³ g⁻¹)	H₂ uptake (wt%)[b]
Furfural	2180	1.03	0.94	5.4 (2.5)
Glucose	2120	1.00	0.91	5.3 (2.4)
Starch	2190	1.01	0.92	5.4 (2.4)
Cellulose	2370	1.08	0.96	5.6 (2.5)
Eucalyptus sawdust	2250	1.03	0.91	5.6 (2.5)

[a]The HTC-derived products were obtained at a temperature of 230 (glucose and starch) and 250 °C (furfural, cellulose, and eucalyptus sawdust). For the chemical activation of HTC-derived materials, a temperature of 700 °C and a KOH/hydrothermal carbon mass ratio of 4 were used.
[b]Hydrogen uptake capacity measured at −196 °C and 20 bar. The hydrogen uptake capacity measured at −196 °C and 1 bar is given in parentheses.

is around 2.5 wt% at 1 bar and in the 5.3–5.6 wt% range at 20 bar. These hydrogen uptakes are in most cases superior to those obtained for other activated carbons with large surface area under similar conditions.

7.6 Adsorption of Pollutants from Water

The removal of heavy metals or organic pollutants from water is of special concern due to their recalcitrance and persistence in the environment. Various methods for the treatment of water have been extensively studied. These technologies include chemical precipitation, ion-exchange, membrane filtration, coagulation/flocculation, flotation, electrochemical methods, and adsorption. In general, most of the technologies are often expensive and mostly effective in removing one single metal or organic contaminant. However, in practice, wastewater contains both heavy metal and organic pollutants, and in this respect adsorption is a viable and a cost-effective option since it can efficiently remove both metals and organics.

Among other possible applications indicated above, HTC-derived materials rich in functional groups are promising candidates for use as cheap, sustainable, and effective sorption materials for the removal of heavy metals or organic pollutants. The possibility of water purification achieved with HTC-derived materials with an overall low cost base is definitely an attractive alternative, especially for the developing world.

7.6.1 Removal of Heavy Metals

Water pollution by heavy metals is known to be a serious worldwide environmental problem with significant impact on living organisms and the environment due to their high toxicity and nonbiodegradability, tending to bioaccumulate through the food chain. Chromium, nickel, cadmium, zinc, and lead are some of the heavy metals commonly associated with water pollution. The concentration in drinking water of any of them must be in the range of parts per million. Adsorption is the most widely used method to remove heavy metals owing to its simplicity and low cost – activated carbons being normally used due to their large surface area and pore volume, as well as high chemical and thermal stability.

However, more economical sorbents as well as "greener" and easier synthesis routes are desirable, and in that sense hydrothermal carbons are great candidates. Even though they do not exhibit porosity, they possess abundant oxygen functionalities that can bind the metal ions through an ion-exchange mechanism.

Titirici and Demir-Cakan described the first use of hydrothermal carbon as sorption materials in which they concurrently demonstrated the possibility of tuning the surface functionally of carbon spheres [125]. In their strategy, the main precursor was a cheap water-soluble carbohydrate (glucose), while an organic monomer (i.e., acrylic acid) was required in very small amounts in order to provide the functionality.

This method offers a facile, cheap, and general route towards the production of carbonaceous materials with a variety of functionalities. A small amount of vinyl organic monomers addition into the HTC of D-glucose gave rise to a new type of hybrid between carbon and polymer latex. The vinyl organic monomers are partially replaced by the controlled dehydration products of carbohydrates, which are then copolymerized or undergo cycloaddition reaction with functional comonomers (see Chapter 5).

Depending on the monomer used, this present synthetic strategy can be extended to various compositions, yielding materials that combine the surface properties of the polymers with the structural, mechanical, thermal, and electric properties of the carbon framework.

In the adsorption study, acrylic acid was employed, resulting in carbon spheres rich in carboxylic groups. Acrylic acid plays a double role in this process: (i) it donates the functionality and (ii) it reacts with the compounds formed in the previous step of HTC via cycloaddition reactions, stabilizing the small droplets and leading to assembly into macroporous materials (Figure 7.22a).

The synthesized materials were investigated in adsorption experiments for the removal of heavy metals from aqueous solutions. The adsorption capacity was as high as 351.4 mg g^{-1} for Pb(II) and 88.8 mg g^{-1} for Cd(II), which is well beyond ordinary sorption capacities, proving the efficiency of the materials to bind and buffer ions, or more specifically to remove heavy metal pollutants (Figure 7.22b and c).

In a similar manner, the same group studied the impact of vinyl imidazole monomer in the HTC [126]. The successful incorporation of imidazole groups in the resulting materials was confirmed by elemental analysis, FTIR, zeta potential measurement, and ^{13}C solid-state nuclear magnetic resonance (NMR). Zeta potential measurements confirm the present of the imidazole ring anchored to the carbon materials. Thus, the composite has a negative zeta potential over all the pH range while the imidazole-containing materials show positive values at an acidic pH due to the protonation of the nitrogen atom linked to the carbon. The isoelectric point increases from pH 2 up to pH 6 depending on the amounts of monomer used, clearly demonstrating the basic character of the materials. The synthesized carbonaceous materials containing imidazole groups were used as an adsorbent for removal of hexavalent chromium from the water at pH 4. A colorimetric method was used to measure the concentrations of the Cr(VI). The pink-colored complex, formed from 1,5-diphenylcarbazide and Cr(VI) in acidic solution, was spectrophotometrically analyzed at 540 nm. Adsorption isotherms and sorption capacities of the bare hydrothermal carbon spheres without imidazole groups show the lowest binding capacity around 30 mg g^{-1}.

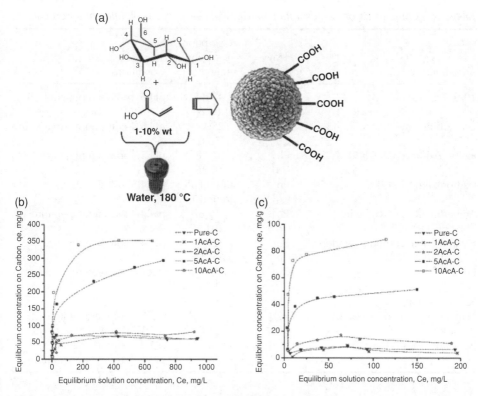

Figure 7.22   *(a) Schematic representation of carboxylic acid functionalization techniques using glucose and small amounts of acrylic acid. (b) Pb(II) and (c) Cd(II) adsorption isotherms on the obtained carbonaceous material (1–10 represents the wt% of acrylic acid added in respect to the total glucose concentration). (Reproduced with permission from [214]. © 2012 American Chemical Society.)*

However, the capacity increased upon increasing the functionality and reached 70 mg g^{-1} Cr(VI) binding capacity. It has been proved that, when Cr(VI) comes in contact with organic substances or reducing agents, especially in an acidic medium, it is spontaneously reduced to Cr(III) [217]. The total chromium species in the equilibrium solution was also analyzed by inductively coupled plasma techniques and it was observed that there is no difference in the concentration of the final liquids detected by colorimetric methods. In this respect, it is concluded that the removal of Cr(VI) by imidazole-functionalized materials from the water took place via anionic adsorption and not via reduction of hexavalent chromium to trivalent chromium (R.D.-C. and M.-M.T., unpublished results).

Chen *et al.* also studied the adsorption of palladium and cadmium metals with a material prepared by HTC of glucose post-treated at 300 °C in air [218]. The authors found that, whereas the temperature of the post-treatment in air had a great influence on the amount of carboxylic groups introduced in the HTC-derived materials, the time had little effect. For example, the content of carboxylic groups at 350 °C was 5 times higher than at 100 °C, while at 300 °C the content increased from 3.42 mmol g^{-1} for 1 h of heating to 3.70 mmol g^{-1} for

Table 7.4   *Comparison of the adsorption capacity of heavy metals of different adsorbents.*

Adsorbent	Heavy metal	Adsorption capacity (mg g^{-1})	Conditions
10AcA-C [45]	Pb^{2+}	351.4	pH 6; room temperature
	Cd^{2+}	88.8	
Leonardite (low-rank coal) [45]	Pb^{2+}	250.7	pH 5.4–5.6; room temperature
	Cd^{2+}	50.6	
HNO$_3$ oxidized CNT [45]	Pb^{2+}	97.08	pH 5; room temperature
	Cd^{2+}	10.86	
Carbon aerogel [45]	Pb^{2+}	around 35	pH 4.5; 37 °C
	Cd^{2+}	around 15	
Amberlite IR-120 synthetic sulfonated resin [45]	Pb^{2+}	19.6	pH 4–8; room temperature
	Cd^{2+}	201.1	
Hydrothermal carbon treated in air at 300 °C [46]	Pb^{2+}	326.1	pH 4; room temperature
	Cd^{2+}	150.7	
H300 [47]	Cu^{2+}	4.46	Room temperature
P700 [47]		2.75	
Sawdust carbon [47]	Cu^{2+}	5.73	30 °C
Nanoarchitectured activated carbon [48]	CrO$_4$$^{2-}$	20.9	pH 8; room temperature
	Fe^{3+}	83.8	pH 1.8; room temperature
Activated carbon [48]	CrO$_4$$^{2-}$	3.71	pH 8; room temperature
	Fe^{3+}	50.9	pH 1.8; room temperature

5 h. In this way, the adsorption capacity of the sample treated in air at 300 °C was 326.1 ± 3.0 mg g^{-1} for Pb^{2+} and 150.7 ± 2.7 mg g^{-1} for Cd^{2+} (Table 7.4), values 3 and 30 times higher than that of as-synthesized hydrothermal carbon.

Removal of radioactive uranium was also investigated using HTC technology from lignocellulosic biomass [219]. Most precisely, biochar was produced from switchgrass at 300 °C in subcritical water, and characterized using XRD, FTIR spectroscopy, SEM, and thermogravimetric analysis. The physiochemical properties indicated that the biochar could serve as an excellent adsorbent to remove uranium from groundwater. A batch adsorption experiment at the natural pH of biochar indicated a H-type isotherm. The adsorption data were fitted using a Langmuir isotherm model and the sorption capacity was estimated to be around 2.12 mg U g^{-1} biochar. The adsorption process was highly dependent on the pH of the system. An increase towards neutral pH resulted in the maximum adsorption of around 4 mg U g^{-1}. These results indicated that biochar could be used as an effective adsorbent for U(VI), as a reactive barrier medium.

The purpose of removal of uranium was also studied by a salicylideneimine-functionalized hydrothermal carbon [220]. The resulting adsorption material was obtained via HTC at mild temperature (573.15 K), amination, and grafting with salicylaldehyde in sequence. Adsorption behaviors of the extractant on U(VI) were investigated by varying the pH values of the solution, adsorbent amounts, contact times, initial metal concentrations, temperatures, and ionic strengths. An optimum adsorption capacity of 1.10 mmol g^{-1} (261 mg g^{-1}) for U(VI) was obtained at pH 4.3. The adsorption process fitted a pseudo-second-order model and Langmuir isotherm. Thermodynamic parameters ($H = +8.81$ kJ mol^{-1}, $S = +110$ J K^{-1} mol^{-1}, $G = -23.0$ kJ mol^{-1}) indicated the adsorption process was endothermic and spontaneous. Results from batch adsorption test in simulated nuclear industrial effluent, containing Cs$^+$, Sr^{2+}, Ba^{2+}, Mn^{2+}, Co^{2+}, Ni^{2+}, Zn^{2+}, La^{3+}, Ce^{3+}, Nd^{3+}, Sm^{3+}, and Gd^{3+}, showed the adsorbent could separate U(VI) from those competitive ions with high selectivity.

Another study compared two types of biochars prepared from hydrothermal liquefaction of pinewood and rice husk for lead removal from aqueous solution [221]. The results indicated that the investigated biochars were effective for lead removal with capacities of 4.25 and 2.40 mg g^{-1} for pine wood- and rice husk-derived hydrothermal carbons, respectively. The adsorption equilibrium was achieved in around 5 h, whilst higher temperatures favored the removal capacity, implying that the adsorption was an endothermic process.

Liu *et al.* used also a hydrothermal carbon for the removal of heavy metals (i.e., copper) [222]. In this case, the hydrothermal carbon was prepared from pinewood hydrothermally treated at 300 °C for 20 min (denoted H300). For comparison purposes, a pyrolytic char was also prepared by pyrolysis of pinewood at 700 °C for 2 h (denoted P700). The elemental analysis clearly shows the higher oxygen content of the hydrothermal carbon (34.8%) in comparison to the pyrolytic char (3.8%). This was further confirmed by Boehm titration and FTIR analysis. From a textural point of view, both materials exhibited a low porosity development, with BET surface areas of 21 and 29 m^2 g^{-1} for the hydrothermal carbon and the pyrolytic char, respectively. In spite of that, the hydrothermal carbon had an estimated maximum adsorption capacity of Cu^{2+} of 4.46 versus 2.75 mg g^{-1} for the pyrolytic carbon (Table 7.4). The adsorption of Cu^{2+} took place through an ion-exchange mechanism with the H$^+$ of the oxygen-containing groups, which was manifested by a decrease of the pH of the solution.

Xu *et al.* functionalized the surface of an activated carbon (obtained by calcining bulk activated carbon at 400 °C for 4 h) with glucose-derived carbon nanospheres (obtained at 180–190 °C and 3–6 h) [223]. The analysis of the as-synthesized composite by SEM and TEM showed that the carbon nanospheres (around 100 nm) were well-dispersed on the surface and pores of the activated carbon host, and they were anchored onto or embedded in the host. This was translated into a large decrease of the BET surface area and pore volume, which reached values of 222 m^2 g^{-1} and 0.147 cm^3 g^{-1}, respectively. However, it also translated into the increase of the amount of oxygen-containing functional groups with respect to the activated carbon, as evidenced by temperature programmed desorption, FTIR and XPS analysis. As a result, the composite material showed much better adsorption capability of CrO$_4$$^{2-}$ and Fe^{3+} than the initial activated carbon. Thus, the estimated maximum adsorption capacity of CrO$_4$$^{2-}$ and Fe^{3+} for this composite were 180 µmol g^{-1} (0.81 µmol m^{-2}) and 1501 µmol g^{-1} (6.76 µmol m^{-2}) respectively, whereas for the activated carbon was 32 mmol g^{-1} (0.030 µmol m^{-2}) and 912 µmol g^{-1} (0.84 µmol m^{-2}) respectively.

An attapulgite clay (ATP)@C nanocomposite adsorbent was used for removal of Cr(VI) and Pb(II) ions [224]. The composite was synthesized by a one-pot HTC process under mild conditions using two materials: ATP, which is a magnesium aluminum silicate that is abundant in nature, and glucose. Compared to carbon-based materials, this ATP@C nanocomposite exhibits a high adsorption ability for Cr(VI) and Pb(II) ions with maximum adsorption capacities of 177.74 and 263.83 mg g^{-1}, respectively.

Based on all these reports, a comparison of different sorption capacities for various heavy metals using different HTC-derived adsorbents is given in Table 7.4.

Ni *et al.* have prepared honeycomb-like Ni@C composite nanostructures via a two-step solution route [225]. Homogeneous nickel nanospheres with an average diameter of 100 nm were first obtained via a dimethylformamide/water mixed solvothermal route; then honeycomb-like Ni@C composite nanostructures were prepared through the HTC of glucose solutions with suitable amounts of nickel nanospheres. The products were characterized by XRD, SEM, TEM, electron diffraction, energy-dispersive X-ray spectroscopy, and Raman spectroscopy. Furthermore, the honeycomb-like Ni@C composite nanostructures presented good capacities for selective adsorption of Pb^{2+}, Cd^{2+}, and Cu^{2+} ions in water. The order of adsorbing capacities for the three ions was $Pb^{2+} > Cu^{2+} > Cd^{2+}$, and the removal capacities were calculated to be 21.45, 14.3, and 6.43 mg g^{-1}, respectively. Owing to the presence of magnetic cores, the present adsorbents can be easily separated from solutions under an external magnetic field and can be reused via washing with deionized water.

Magnetic carbonaceous nanoparticles were also investigated by Lee *et al.* through a two-step solution-phase thermal synthesis [226]. Fe_3O_4 magnetic nanoparticles (MNPs) with a size less than 100 nm were first generated from $FeCl_3$ in a solvothermal reaction. The size could be significantly reduced to approximately 30 nm when 1,6-hexanediamine was employed in the reaction solution to functionalize the surface of MNPs with amine. Both the plain and amine-functionalized MNPs (MH) were encapsulated in the carbonaceous shell by hydrothermal treatment in 0.5M glucose solution. The saturation magnetization of MH decreased significantly from 70 to 25 emu g^{-1} after the carbonaceous shell was formed. The surface charge of MNPs and MNP@C particles was studied by measuring their zeta potentials at pH ranging from 2 to 12. All the particles had their zeta potential decrease with pH. MH particles possesses the highest point of zero charge (PZC) at pH about 6.5, which decreased to 2 after HTC treatment. The plain particle had a slightly lower PZC of pH 6.0 and reduced to 3.5 after carbonization treatment. The reduced PZC of MNP@C particles is due to the presence of the carbonaceous shell since the carbonaceous surface is known to be hydrophilic, and mainly consists of acidic carboxyl and carbonyl groups. The carbonaceous shell not only can protect the MNPs from the corrosive environment, but also possesses a high adsorption capacity towards Pb(II). The adsorption isotherm at room temperature was well-fitted by a Langmuir model with a maximum adsorption capacity of 123 mg g^{-1}.

Xu *et al.* reported the synthesis and activation of colloidal carbon nanospheres (400–500 nm in diameter) via hydrothermal treatment of glucose solution and used as adsorbents for Ag(I) ions from aqueous solutions [226]. The surface of nonporous carbon nanospheres after being activated by NaOH was enriched with —OH and —COO$^-$ functional groups. Despite the low surface area (less than 15 m^2 g^{-1}), the activated carbon nanospheres exhibited a high adsorption capacity of 152 mg Ag g^{-1}. Under batch conditions, all Ag(I) ions can be completely absorbed in less than 6 min with the initial Ag(I) concentrations lower than 2 ppm. This was attributed to the minimum mass transfer resistance as Ag(I) ions were all deposited and reduced as Ag^0 nanoparticles on the external surface of carbon nanospheres. The kinetic data was fitted to the pseudo-second-order kinetics model. The NaOH activated carbon nanospheres reported could represent a low-cost adsorbent nanomaterials for removal of trace Ag(I) ions for drinking water production.

Hydrothermal carbon materials can also be employed as a membrane structure. and this was first reported by Yu *et al.* performing the filtration and separation of nanoparticles with different sizes from solution [227]. For this study, initially hydrothermal CNFs were synthesized [228] by dispersion of tellurium nanowires in glucose solution followed by hydrothermal treatment of the mixed solution at 160 °C. This resulted in Te@C nanocables and pure CNFs were obtained after removal of the tellurium cores by chemical etching. Then, those free-standing CNF membranes were fabricated through a solvent-evaporation-induced self-assembly process. A wool-like homogeneous suspension was obtained by vigorous magnetic stirring of the CNFs in ethanol for several hours. After casting the suspension onto a Teflon substrate and drying at ambient temperature, a brown paper-like material was formed, which could be easily detached from the substrate without cracking. The fabricated CNF membranes are very flexible and mechanically robust enough for filtration under a high applied pressure without any damage. These CNF membranes have very narrow PSDs and show size-selective rejection properties. The cut-off sizes of these free-standing membranes could be controlled precisely by carefully regulating the diameters of the CNFs. The performance of the CNF membrane was tested by filtering different size particles from solution. The separation of gold nanoparticles of two different sizes has been successfully achieved. It is believed that the size-dependent separation ability of these membranes also makes them suitable for separation of a wide range of other materials on the nanometer or micrometer scale, such as polymers, viruses, bacteria, microorganisms, and so on. The results suggest two unique advantages of the CNF membranes: high flux and high selectivity. Although there are commercially available membranes with a range of pore sizes, their filtration rate is usually low when the cut-off size is below 100 nm because of their intrinsic structures.

7.6.2 Removal of Organic Pollutants

HTC-derived materials were also studied for organic pollutants removals. Hydrothermal CNFs functionalized by β-cyclodextrins (β-CD) fabricated as a membrane was investigated for molecular filtration [229]. β-CDs are important for the removal of organic pollutants to form inclusion complexes in aqueous media with a wide variety of organic substrates, including many organic pollutants. First, CNFs were synthesized thorough a method described elsewhere [228]. Then, those CNFs were functionalized by β-CD and the free-standing CNF-β-CD membrane was prepared by a simple filtration process. The membrane shows a capability to function as an ideal molecular filter through complexation of phenolphthalein molecules with the β-CD molecules grafted on the CNFs. It was considered that as a typical dye pollutant, fuchsin acid can also be effectively removed from the solution through such a membrane.

Xing *et al.* compared pyrolysis and HTC-derived biochars for the adsorption of bisphenol A, 17α-ethinyl estradiol, and phenanthrene from water [230]. The thermal biochars were produced from feedstocks of poultry litter and wheat straw, respectively, through pyrolysis at 400 °C. The HTC-derived biochars were prepared from poultry litter and swine solids at 300 °C. XRD and solid-state NMR spectroscopy results suggested that HTC-derived biochars consisted of more amorphous aliphatic-C, possibly being responsible for their high sorption capacity of phenol. This study demonstrated that hydrothermal biochars could absorb a wider spectrum of both polar and nonpolar organic contaminants than thermally produced biochars, suggesting that the investigated HTC-derived biochar is a potential sorbent for agricultural and environmental applications.

Another interesting example of the use of HTC-derived materials is to remove dye pollutants from water [231]. Orange-II ($C_{16}H_{11}N_2NaO_4S$), an anionic dye, was chosen as the model pollutant in water. Rattle-type C/γ-Al_2O_3 particles were prepared by using glucose and aluminum nitrate via a one-pot hydrothermal synthesis at 180 °C for 24 h followed by calcinations at 450 °C for 2 h. The microstructure, morphology, and chemical composition of the resulting materials were characterized by XRD, energy dispersive X-ray spectroscopy, SEM, TEM, and nitrogen adsorption/desorption techniques. These rattle-type spheres are composed of a porous Al_2O_3 shell (thickness around 80 nm) and a solid carbon core (diameter around 200 nm) with variable space between the core and shell. Furthermore, adsorption experiments indicate that the resulting C/Al_2O_3 particles are powerful adsorbents for the removal of Orange-II dye from water with a maximum adsorption capacity of 210 mg g^{-1}. At the pH studied, Orange-II is expected to be adsorbed on the surface of Al_2O_3 due to electrostatic attraction between the Al_2O_3 shell and sulfonic acid groups of the Orange-II molecules. The fast Orange-II uptake is attributed to the combination of several factors, such as high specific surface area, large pore volume, unique rattle-type structure, and high adsorption affinity of the carbon core. Moreover, keeping pH in the range between 5 and 6 facilitated electrostatic interactions because the Al_2O_3 surface was positively charged ($pI \sim 9.0$). The adsorption kinetics of Orange-II on the C/Al_2O_3 samples studied follows the pseudo-second-order kinetic model. The equilibrium adsorption data are well represented by the Langmuir isotherm equation.

The results described above clearly indicate that for adsorption of both heavy metals as well as organic pollutants, the low surface area of HTC products does not constitute a problem, as is the case for supercapacitors, but it is their functional groups that are the main driving force for the adsorption of heavy metals.

7.7   HTC-Derived Materials in Sensor Applications

The emerging applications of carbon nanomaterials in electrochemical sensors has led to production of CNTs, crystalline diamond, and diamond-like carbons on a large scale. It will soon be possible to take advantage of the demanding properties of novel carbon-derived materials to develop a myriad of new applications for chemical sensing. Among these novel carbon materials, HTC-derived materials will also play an important role as they are cheap, have chemical functionalities, and can be easily hybridized with inorganic nanoparticles.

7.7.1   Chemical Sensors

Marken *et al.* [232] recently reported the use of electrochemical properties of HTC-derived materials with positive surface charge obtained in the form of a nanoaggregate via mild hydrothermal treatment of chitosan following the procedure initially reported by Zhao *et al.* [87]. The hydrothermal carbon was deposited on ITO as a film and employed to bind redox-active anions. The extremely low electrical conductivity of the hydrothermal nanocarbon limited the accuracy of specific binding capacity measurements and therefore a nanocomposite approach was investigated next. Negatively charged Emperor 2000 carbon nanoparticles were mixed in equal weight amounts with hydrothermal nanocarbon (180 °C) and, due to the strong interaction of positive and negative carbon, robust composite films are formed on ITO electrodes upon solvent evaporation. The redox system chosen here for a survey of anion binding ability was the reduction of indigo carmine – a hydrophobic double negatively charged dye molecule. When dipped into 1 mM indigo carmine in 0.1 M phosphate buffer pH 2, rinsed, and transferred into 0.1 M phosphate buffer pH 2 solution, a well-defined reversible reduction response is observed. With the negatively charged carbon nanoparticle additive, the hydrothermal nanocarbon deposit becomes electrically conducting and the peak currents for indigo carmine are enhanced. A comparison of indigo carmine binding for hydrothermal nanocarbons produced at 180, 200, and 230 °C results in specific anion-binding capacities of around 14, 70, and 20 C g^{-1}, suggesting highest efficiency for materials produced at 200 °C. Given a typical nitrogen content of 6.6%, the availability of nitrogen as surface ammonium appears to be around 15%.

The honeycomb-like Ni@C composite nanostructures described in Section 7.6 for the removal of Pb^{2+}, Cd^{2+}, and Cu^{2+} also showed also good electrochemical response in 0.1 M NaOH, and could promote the oxidation of glucose [225]. This property was used in sensors for the detection of glucose (detection limit = 0.9×10^{-6} M).

Sun *et al.* reported a simple, economical, and green preparative strategy towards water-soluble, fluorescent carbon nanoparticles with a quantum yield of around 6.9% using the HTC of pomelo peel wastes as precursor [233]. The resulting nanoparticles were used as probes for a fluorescent Hg^{2+} detection, based on Hg^{2+}-induced fluorescence quenching of carbon nanoparticles. This sensing system exhibited excellent sensitivity and selectivity toward Hg^{2+}, and a detection limit as low as 0.23 nM. The practical use of this newly designed system for Hg^{2+} determination in lake water samples was successfully demonstrated.

The same authors also prepared fluorescent nitrogen-doped carbon nanoparticles using the HTC of grass as an effective fluorescent sensing platform for label-free detection of Cu(II) Ions. The application of this method to detect Cu^{2+} ions in real water samples was also demonstrated successfully [234].

Li *et al.* reported on the production of a C@Ag nanocomposite using the HTC of glucose in the presence of AgNO$_3$ according to the publication of Sun *et al.* [235]. The C@Ag composite were used for the modification of glassy carbon electrodes and used for the detection of tryptophan [236]. Tryptophan is a vital constituent of proteins, and is indispensable in human nutrition for establishing and maintaining a positive nitrogen balance. In particular, the improper metabolization of tryptophan can create a waste product in the brain, causing hallucinations and delusions. Therefore, it is of great significance to develop a simple, accurate, rapid, and inexpensive method for the determination of tryptophan. In recent decades, electrochemical techniques have been the most attractive methods for the determination of amino acids due to their favorable properties, such as high sensitivity, accuracy, easy operation, and low cost. Silver has the highest electrical conductivity among all metals and is probably the most important material in plasmonics. However, silver nanostructures are considered to be toxic and instable. It is well established that applying a coating of different material can eliminate the reactivity and toxicity of nano-silver. The carbon shell can not only protect the silver core, but also contribute to the enhanced substrate accessibility and tryptophan/substrate interactions, while the nano-silver core can display good electrocatalytic activity to tryptophan at the same time. Under the optimum experimental conditions the oxidation peak current was linearly dependent on the tryptophan concentration in the range of 1.0×10^{-7} to 1.0×10^{-4} M with a detection limit of 4.0×10^{-8} M (S/N = 3). In addition, the proposed electrode was applied for the determination of tryptophan concentrations in real samples and satisfactory results were obtained. The technique offers enhanced sensitivity and may trigger the possibilities of the use of Ag@C nanocomposites for diverse applications in biosensor and electroanalysis.

Using a similar methodology, Zhu *et al.* fabricated monodisperse colloidal carbon nanospheres from glucose solution and gold nanoparticles using a rapid microwave/hydrothermal approach. The resulting gold nanoparticle/colloidal carbon nanosphere hybrid materials have been characterized, and used as a promising template for biomolecule immobilization and biosensor fabrication because of their satisfactory chemical stability and the good biocompatibility of gold nanoparticles. In one particular publication, as an example, it was demonstrated that the as-prepared gold nanoparticle/C hybrid material can be conjugated with horseradish peroxidase-labeled antibody (HRP-Ab2) to fabricate HRP-Ab2-gold nanoparticle/C bioconjugates, which can then be used as a label for the sensitive detection of protein [237]. The amperometric immunosensor fabricated on a CNT-modified glass carbon electrode was very effective for antibody immobilization. The approach provided a linear response range between 0.01 and 250 ng ml^{-1} with a detection limit of 5.6 pg ml^{-1}. The developed assay method was versatile, offered enhanced performance, and could be easily extended to the detection of other proteins as well as DNA analysis.

7.7.2 Gas Sensors

As previously mentioned, hydrothermal carbon spheres produced in the presence of various metal precursors can act as sacrificial templates from the production of a large variety of metal oxide hollow spheres [35]. Such porous inorganic metal oxides can also be used as gas sensors. When the ionic oxygen species (O^{2-}, O^-) fixed on the surface of the metal oxide react with reducing gas molecules, the trapped electrons release back the crystal grains, the potential barriers at the crystal grain boundary decrease and the resistance is dramatically increased [238]. For the porous hollow structures, the ionic oxygen can absorb onto both inner and outer surfaces, and the detected gas molecules can penetrate through and react with them freely, thus increasing the sensitivity.

Thus, Li *et al.* reported a WO_3 hollow sphere structure that showed certain selective responses to organic gases [239]. They found that only with porous morphologies and small nanocrystals in the shell wall did the WO_3 hollow spheres exhibit high sensitivity to organic gases at intermediate temperatures. The sensors were fabricated using thin films prepared from a powder suspension of as-synthesized WO_3 hollow spheres. They investigated the response of various gases, including alcohol, acetone, CS_2, NH_3, H_2S, benzene, petroleum ether, acetonitrile, and so forth, using the HW-30 (Hanwei Electronics) commercial gas-sensing measurement system. Table 7.5 summarizes all of the response results. With increasing concentration of the gases, the sensitivity of the sensors sharply increased. The sensors were more sensitive to alcohol and acetone than to other organics. They showed much higher responses to H_2S and acetone than to alcohol, CS_2, and NH_3. At low concentrations, the sensor sensitivities to H_2S, acetone, alcohol, CS_2, and NH_3 were in descending sequence. The sensors had no response to benzene, petroleum ether, and acetonitrile when the gas species were at low concentrations. Only when the concentration of those gases was

Table 7.5 *Response results (ppm) to various gases. (Reproduced with permission from [239]. © 2004 American Chemical Society.)*

Sensor	10	50	100	200	500	1000	2000	5000
H_2S	21.8	52.9	67.2					
NH_3		1.5	1.7	2.14	2.89	3.15	3.57	4.38
Alcohol		2.09	2.46	3.16	6.14	7.79	9.35	12.4
Acetone		3.53	4.56	6.04	13.5	16	18.06	23.1
CS_2		1.56	1.83	2.54	5.06	7.2	11.02	14.3
Benzene					2.56	3.06	3.43	4.06
CH_3CN					3.18	4.07	4.85	5.73
Petroleum ether					2.72	2.92	3.67	4.91

quite high (up to 500 ppm) did the sensors give a small response. The sensing mechanism of semiconducting oxide sensors is usually believed to be the surface conduction modulation by the absorbed gas molecules: (i) the electrical properties of the semiconducting oxides showed dramatic changes with or without the adsorption of gas molecules (mechanism of sensitivity), and (ii) with different kinds of adsorbing gas molecules, the electrical properties would show different changes (mechanism of selectivity).

Similarly, using glucose-derived carbon spheres as a template, Liu *et al.* [158, 240] synthesized porous In_2O_3 hollow nanospheres that had a satisfactory response for ethanol, methanol, and other organic gases even at a very low concentration. In addition, the sensors had good recovery ability. After every measurement, the response curves returned back to the baseline very quickly.

Wu *et al.* reported on a hollow SnO_2 sphere structure that was very sensitive to $(C_2H_5)_3N$ and ethanol (7.1 for 1 ppb $(C_2H_5)_3N$ at 150 °C and 2.7 ppb for ethanol at 250 °C, respectively) with a short response time, while the sensor based on SnO_2 nanoparticles alone showed much less response [238].

Here, we have only provided a few examples where again HTC biochar carbons proved to have excellent properties for the production of various sensing devices. There is a lot of progress in the literature on using CNTs in sensing devices [241]. As HTC is much cheaper, biomass based, and easy to make, together with its easy chemical modification, it is strongly believed that it will play an important role in sensing in the future.

7.8 Bioapplications

Recent years have witnessed unprecedented growth in research and applications in the area of nanoscience and nanotechnology. There is increasing optimism that nanotechnology, as applied to medicine, will bring significant advances in the diagnosis and disease treatment, more specifically in fields such as drug delivery, diagnostics and bioimaging, and the production of biocompatible materials [242]. However, many challenges must be overcome if the application of nanotechnology is to realize the anticipated improved understanding of the pathophysiological basis of disease, bring more sophisticated diagnostic opportunities, and yield improved therapies.

Among the nanomaterials utilized for the above-mentioned applications, CNTs and fullerenes have been studied in detail, although recent studies show that they exert adverse effects and toxicity [243, 244]. Therefore, new and nontoxic alternatives are still necessary in this field.

Recently, scientists have also started exploring favorable applications of HTC-based materials in medical applications such as drug delivery and bioimaging.

7.8.1   Drug Delivery

An ideal carrier for a targeted drug delivery system should have three prerequisites: (i) they themselves have target effects, (ii) they have sufficiently strong adsorptive effects for drugs to ensure they can transport the drugs to the effect-relevant sites, and (iii) they can release the drugs from them in the effect-relevant sites and only in this way can the treatment effects develop. However, the requirement for new drug delivery systems is to improve the pharmacological profiles while decreasing the toxicological effects of the delivered drugs.

The transporting capabilities of CNTs combined with appropriate surface modifications and their unique physicochemical properties show great promise to meet the three prerequisites [245]. As previously mentioned, there are a lot of issues related to the biosafety of CNTs [246, 247]. In addition they are expensive, difficult to make, and based on fossil-derived precursors. Their functionalization plays the most important role in drug delivery; however, the functionalization of CNTs is not a trivial task [243]. As described in Chapter 5, the functionalization of HTC-derived materials is straightforward. The HTC-derived materials should possess low toxicity as they are derived from biomass. In addition, they are water dispersible and easy to upscale. They can be shaped either as hollow spheres [19], aerogels [86], or nanotubes [248]. All these characteristics makes them very promising for applications in drug delivery.

Before going into such drug delivery applications using HTC, it is crucial to study the toxicity of these carbon materials. Thus, Stanisheysky *et al.* studied the toxicity of hydrothermal carbons of various sizes (from 10 to 500 nm) *in vitro* with a variety of cultured cell lines and found them nontoxic [249]. A size-dependent effect of hydrothermal carbon sphere addition on cell function has been observed. For example, hydrothermal carbon spheres can, in some cases, substantially increase interleukin-12 production by bone marrow dendritic cells. It has been further demonstrated that hydrothermal carbon spheres can be modified with fluorescent dye molecules or loaded with anticancer drugs for bioimaging or therapeutic purposes, respectively. The results of these tests and the strategies for particle preparation and functionalization for biomedical applications have been discussed.

Kundu and Selvi demonstrated that glucose-derived HTC-based carbon nanospheres are an emerging class of intracellular carriers. The surfaces of these spheres are highly functionalized and do not need any further modification. In addition, the intrinsic fluorescence property of carbon nanospheres helps in tracking their cellular localization without any additional fluorescent tags. The spheres were found to target the nucleus of mammalian cells, causing no toxicity. Interestingly, the *in vivo* experiments showed that these nanospheres have an important ability to cross the blood–brain barrier and localize in the brain, besides being localized in the liver and the spleen. There is also evidence showing that they are continuously being removed from these tissues over time. Furthermore, these nanospheres were used as a carrier for the membrane-impermeable molecule *N*-(4-chloro-3-trifluoromethylphenyl)-2-ethoxybenzamide) (CTPB) – the only known small-molecule activator of histone acetyltransferase (HAT). Biochemical analyses such as Western blotting, immunohistochemistry, and gene expression analysis showed the induction of the hyperacetylation of HAT p300 (autoacetylation) as well as histones both *in vitro* and *in vivo*, and the activation of HAT-dependent transcription upon CTPB delivery. These results establish an alternative path for the activation of gene expression mediated by the induction of HAT activity instead of histone deacetylase inhibition [250].

Using a combination of HTC and templating of mesoporous silica spheres, Shi *et al.* prepared mesoporous carbon nanoparticles (MCNs) with a size below 200 nm and good water dispersibilty [251]. They used MCM-48-type mesoporous silica as a hard template previously functionalized with amino groups in order to provide effective electrostatic attraction between positively charged $-NH_3^+$ ions on the pore surface of the solid template and negatively charged carbonaceous polysaccharide. Considering their biomedical application potential, the uptake of the synthesized MCNs labeled with rhodamine B (RhB@MCNs) by MCF-7 (human breast adenocarcinoma) cell lines was examined via TEM and confocal microscopy. These results suggest that those rhodamine molecules were actually transported into the MCF-7 cells by MCN carriers. This further confirms that MCNs could be effectively internalized into living cells. In addition, the authors have also evaluated the biocompatibility of the synthesized MCNs by incubating the MCF-7 cells with different concentrations of MCN suspensions in growth media. The results showed that no growth inhibition was present when the MCN concentration was below 5 μg ml^{-1} even after 72 h of incubation. This and other experiments not only with cancer cells but also with fibroblast cells showed that the cells survived with a concentration of 160 μg ml^{-1} and incubation time as long as 72 h, indicating the biocompatibility of HTC-derived materials *in vitro*. As a preliminary application study, the authors employed these well-dispersed and uniform MCNs as hydrophobic anticancer drug carriers and took the representative chemotherapeutic drug camptothecin (CPT) as an example. It was found that drug-loading capacity was as high as 17%, which is considerably higher than the reported loading amount using MSNs as carriers. The drug-release profile is of great importance in applying the synthesized systems to practical drug delivery. MCN delivery systems demonstrate no initial burst release, but a sustained release feature with around 80% CPT being released for as long as 5 days, as compared with the 99.8% free CPT released from the dialysis bag in 18 h. The sustained release characteristics are probably attributed to the slow drug diffusion from the mesochannels to the media solution. It should be noted that this investigation employed dimethylsulfoxide as a release medium while no CPT was released when phosphate-buffered saline solution was used as a release medium, probably because free CPT is poorly soluble in the aqueous solution. This means that CPT will remain inside the mesochannels of MCNs in aqueous environment, but release in the hydrophobic regions of the cell compartments once internalized by the cancer cells. These properties are vitally important for drug delivery systems because they allow the delivery vehicles to reach the cancer cells without significant leakage of drug from the delivery system on their paths through the physiological environment. To test the therapeutic effect of the synthesized CPT@MCNs, the authors studied the cytotoxicity of the synthesized samples against MCF-7 cell lines. Growth inhibition and killing of cancer cells were obviously observed when the cells were treated with either the suspension of CPT@MCNs or a dimethylsulfoxide solution of free CPT. In addition, the synthesized CPT@MCNs demonstrate a dose-dependent cytotoxic effect against MCF-7 cells and cytotoxicity is almost equivalent to the free drugs. Thus, the synthesized MCNs served as effective carriers for CPT delivery and subsequent release in cancer cells, and efficiently inhibited the growth of the cancer cell lines with minimal leakage into hydrophilic dispersion media.

In a different publication, Cui *et al.* used a hydrothermal method to synthesize in one step magnetic Fe_3O_4@C nanotubes (50–100 nm in diameter and several micrometers in length) via a hydrothermal rolling of grapheme oxide sheets. Thus, the authors synthesized Fe_3O_4@C nanotubes with a high saturation magnetization of 24.5 emu g^{-1}, good dispensability in water, and no significant toxicity on healthy human gastric cells (GES-1). Moreover, these magnetic nanotubes exhibited a strong affinity to adriamycin with a high adsorption capacity (101.3 μg mg^{-1}) at room temperature [252].

Nanocapsules are very attractive morphologies for drug/gene delivery due to their hollow and porous structures, and their facile surface functionalization. The inner void can take up a large amount of drug while the open-ended pores serve as gates that control their release. Furthermore, hollow structures can be differently functionalized between the inner and the outer surface, thus providing multifunctionality throughout sequential surface modifications. Most importantly, unlike spherical nanoparticles, the hollow structures can isolate the drug/gene payload from the environment and they can transport the payload safely into the cell without hydrolytic degradation of biological payload or aggregation of nanomaterials caused by many hydrophobic drug molecules during delivery into the cell [253].

There are many possibilities to produce hydrothermal carbon hollow structures either by using various sacrificial templates such as alumina membranes [248], latex and silica nanoparticles [19], or selenium cores [254]. However, despite the promising potential, the drug delivery capabilities of such hollow carbon nanostructures remained unexplored.

As mentioned previously, hydrothermal carbon nanospheres can be used as sacrificial templates to produce a wide variety of hollow metal oxide nanostructures [35] with many applications in drug delivery [255]. We will only mention here only a few of these examples where the HTC-derived materials mediated the formation of such inorganic hollow structures.

Hanagata and Kaskel prepared rattle-type Fe_3O_4@SiO_2 hollow mesoporous spheres with different particle sizes, different mesoporous shell thicknesses, and different levels of Fe_3O_4 content using hydrothermal carbon spheres as templates. The effects of particle size and concentration of Fe_3O_4@SiO_2 hollow mesoporous spheres on cell uptake and their *in vitro* cytotoxicity to HeLa cells was evaluated. The spheres exhibited relatively fast cell uptake. Concentrations of up to 150 μg ml^{-1} showed no cytotoxicity, whereas a concentration of 200 μg ml^{-1} showed a small amount of cytotoxicity after 48 h of incubation. Doxorubicin hydrochloride, an anticancer drug, was loaded into the Fe_3O_4@SiO_2 hollow mesoporous spheres and the doxorubicin-loaded spheres exhibited a higher cytotoxicity than free doxorubicin. These results indicate the potential of Fe_3O_4@SiO_2 hollow spheres for drug loading and delivery into cancer cells to induce cell death [256]. The situation might have been different if the authors had used directly the carbon spheres loaded with magnetic F_3O_4 which can also be synthesized as a rattle-type structure.

Using the same HTC-directed template route, You *et al.* prepared well-dispersed Lu_2O_3 hollow spheres. The as-obtained Lu_2O_3:Eu^{3+} hollow spheres exhibited red emission under UV excitation, and the Lu_2O_3:Er^{3+} and Lu_2O_3:Yb^{3+},Er^{3+} samples showed green and red emissions under 980-nm light excitation, respectively. The experimental result and analysis revealed that the two-photon process may responsible for the upconversion luminescence of Yb^{3+}- and Er^{3+}-doped samples. The potential of such materials for drug delivery is discussed as a potential future application [257].

Rare-earth-doped gadolinium oxide (Gd_2O_3) hollow nanospheres have been successfully prepared on a large scale via a template-directed method using hydrothermal carbon spheres as sacrificial templates [258]. SEM and TEM images revealed that these hollow-structured nanosphere spheres have mesoporous shells that are composed of a large amount of uniform nanoparticles. By doping the rare-earth ions (Yb/Er) into the Gd_2O_3 host matrix, these nanoparticles emitted bright multicolored upconversion emissions that can be fine-tuned from green to red by adjusting the codoped Yb/Er ratio under 980-nm near-IR laser excitation. The possibility of using these upconversion nanoparticles for optical imaging *in vivo* has been demonstrated. It was also shown that these Gd_2O_3 nanospheres brightened the T_1-weighted images and enhanced the r_1 relaxivity of water protons, which suggested that they could act as T_1 contrast agents for magnetic resonance imaging. Moreover, these hollow spheres can be used as drug delivery host carriers and their drug storage/release properties were investigated using ibuprofen as the model drug. As a result, the so-prepared nanoscaled Gd_2O_3 hollow spheres bearing upconversion luminescence, magnetic resonance imaging, and drug delivery capabilities could be potentially employed for simultaneous magnetic resonance/fluorescent imaging and therapeutic applications.

7.8.2 Bioimaging

The development of targeted contrast agents such as fluorescent probes has made it possible to selectively view specific biological events and processes in both living and nonviable systems with improved detection limits, imaging modalities, and engineered biomarker functionality. These contrast agents have become a mainstay in modern medicinal and biological research. The fabrication of luminescent-engineered nanoparticles (with multifunctional features) is expected to be integral to the development of next-generation therapeutic, diagnosis, and imaging technologies. Here, we present a few examples where HTC technology has proved a powerful technique to produce small luminescent carbon nanoparticles.

Semiconductor quantum dots are the most widely used fluorescent nanoparticles; they have strong fluorescence, high photostability, broad excitation spectra, long fluorescence lifetime, and narrow and tunable emission spectra. However, the release of heavy metals results in cytotoxicity and their environmental concerns limit their application [259]. Therefore, it is an urgent challenge to search for environmentally friendly fluorescent nanomaterials.

Fluorescent carbon nanoparticles were first produced by Sun *et al.* [260] and were shown to exhibit bright photoluminescence in the visible spectrum with either one or two photoexcitations. Carbon nanoparticles have received increasing attention due to their high photostability, tunable excitation and emission wavelength, excellent biocompatibility, and environmental friendliness [261–264].

Yu *et al.* synthesized biocompatible and green luminescent monodisperse Ag/phenol formaldehyde resin core/shell spheres with controllable sizes, in the range of 180–1000 nm, and interesting architectures (centric, eccentric, and coenocytic core/shell spheres) via

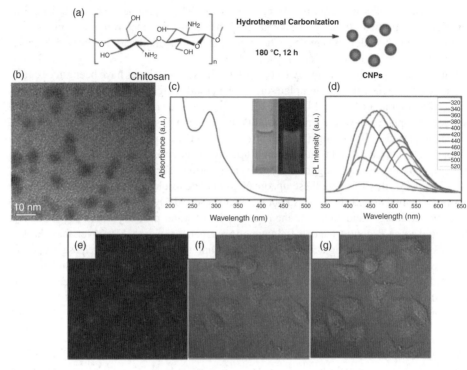

Figure 7.23 (a) Schematic illustration of the preparation procedure of carbon nanoparticles by HTC of chitosan. (b) High-resolution TEM image of carbon nanoparticles. (c) Absorption spectra of the carbon nanoparticles (inset: photograph of the samples excited by daylight and a 365-nm UV lamp). (d) Emission spectra of the carbon nanoparticles at different excitation wavelengths as indicated. (e) Confocal fluorescence microphotograph of A549 cells labeled with the carbon nanoparticles at $37\,^{\circ}C$ for 24 h (λ_{ex}: 405 nm). (f) Bright-field microphotograph of the cells. (g) Overlay image of (e) and (f).

a facile one-step hydrothermal approach. These spheres can be used as bioimaging labels for human lung cancer H1299 cells. The results demonstrate that the nanoparticles can be internalized into cells and exhibit no cytotoxic effects, showing that such novel biocompatible core/shell structures can potentially be used as *in vivo* bioimaging labels. This facile one-pot polymerization and encapsulation technique may provide a useful tool to synthesize other core/shell particles that have potential application in biotechnology [265].

Liu *et al.*, inspired by a publication from Titirici *et al.* [87], explored the one-step synthetic method to produce highly amino-functionalized fluorescent carbon nanoparticles by HTC of chitosan at a mild temperature (180 °C) [266]. The carbonization and functionalization occurs through the dehydration of the chitosan, which leads to the formation of the fluorescent carbon nanoparticles (see Figure 7.23a). High-resolution TEM images (Figure 7.23b) clearly revealed that the carbon nanoparticles were monodisperse and had a narrow size distribution of 4–7 nm in diameter. The optical properties of the carbon nanoparticles showed a strong UV/Vis absorption feature centered at 288 nm. The inset photograph of the carbon nanoparticles solution is yellowish, transparent, and clear under daylight, and exhibited strong blue luminescence under UV excitation (Figure 7.23c). The fluorescence spectra are given in Figure 7.23d. The phenomenon of λ_{ex}-dependent λ_{em} is similar to that reported by Sun *et al.* and other groups; they concluded that surface passivation could produce defect sites on the surface of carbon nanoparticles and the fluorescence emission arises from the radiative recombination of the excitons trapped by the defects. Thus, the formation of carbon nanoparticles and their surface functionalization takes place simultaneously during the HTC process. The abundant functional groups, such as carboxyl acids and amines, can introduce different defects on the surface, acting as excitation energy traps and leading to the different photoluminescence properties. The carbon nanoparticles showed excellent photostability as the fluorescence intensity did not change, even after continuous excitation under a 150-W Xe lamp for several hours. The photoluminescence quantum yield of the upper brown carbon nanoparticles was as high as 43% when excited at 360 nm, which is much higher than most of the carbon nanoparticles reported so far. To assess the prospects of carbon nanoparticles as a bioimaging material, A549 human lung adenocarcinoma cells were used to evaluate the cytocompatibility of the carbon nanoparticles. The cell viability of the carbon nanoparticles was determined by a methylthiazol terazolium assay [267].

Assays of cell viability studies suggested that the carbon nanoparticles demonstrate low cytotoxicity and were not found to pose any significant toxic effects. This result concluded that carbon nanoparticles can be used in a high concentration for imaging or other biomedical applications. The carbon nanoparticles were introduced into the A549 cells to show their bioimaging capabilities using a confocal microscopy test *in vitro*. The results showed that the photoluminescent spots were observed only in the cell membrane and cytoplasmic area of the cell, but were very weak at the central region corresponding to the nucleus, which indicated that the carbon nanoparticles easily penetrated into the cell, but did not enter the nuclei (Figure 7.23e–g).

It was thus demonstrated that it is possible to synthesize amino-functionalized fluorescent carbon nanoparticles as novel bioimaging agents by hydrothermal treatment of chitosan. No external surface passivation agent or further modification is necessary in their preparation. This green, cheap, and convenient process represents a potential advance for large-scale production.

Highly fluorescent crystalline carbon nanoparticles can also be synthesized by one-step microwave irradiation of carbohydrates with phosphoric acid at 100 W for 3 min 40 s. This method is also very simple, rapid, and economical, and hence can be used for large-scale applications. The average particle sizes are 3–10 nm and they emit bright green fluorescence under UV irradiation. Therefore, the particles can be used as a unique material for bioimaging as well as drug delivery. To further increase the fluorescence property of the synthetic carbon nanoparticles we simply functionalized them by using different organic dyes, such as fluorescein, rhodamine B, and α-naphthylamine; the maximum fluorescence intensity was observed for the particles functionalized with fluorescein. It is very interesting to note that all of those compounds show maximum fluorescence intensity at 225 nm excitation wavelength and for any excitation wavelength the peak positions are exactly at same the position as that of carbon nanoparticles themselves, which is completely different from the individual precursors (dyes). All of the above compounds, including carbon nanoparticles, have also been successfully introduced into the erythrocyte-enriched fraction of healthy human blood cells with minimum cytotoxicity [268].

Qu *et al.* reported a microwave method to convert carbohydrates into fluorescent CQDs without an additional passivation step. The secret is the presence of a tiny amount of an inorganic ion during the microwave treatment. Additionally, the photoluminescence intensity of the as-synthesized carbon dots does not change at the physiological and pathological pH range of 4.5–9.5, and shows no photobleaching. Furthermore, these carbon dots can enter into cells and can be used for photoluminescence-based cell-imaging applications [269].

Combining their green synthetic method, favorable optical properties, low cytotoxicity, and ease of labeling, such carbon nanoparticles will provide promise for applications in biological labeling, disease diagnosis, and biosensors in the future.

Although only in their incipient phase, we hope that throughout this section dedicated to the bioapplications of HTC-derived materials, researchers working in the field have become aware of the potential such materials hold for future drug delivery as well as bioimaging, either as carbon capsules, carbon nanostructures decorated with MNPs, or fluorescent CQDs. They offer several advantages compared with the classical nanomaterials in terms of low-cost precursors, easy synthesis and processing, and plenty of functional groups.

7.9   Conclusions and Perspectives

The aim of this chapter was to show that HTC-derived materials have already been used in a plethora of very important applications ranging from renewable energy to catalysis, photocatalysis, gas storage, water purification, sensing, and medicine.

The HTC-derived materials have multiple advantages when compared with CNTs, fullerenes, and even graphene materials in terms of easy preparation, natural abundance of the precursors, CO_2 neutrality, upscaling, low-temperature processes, water compatibility, and functional surface groups, to mention only a few. The wide range of applications speaks for itself. The HTC-derived materials can be applied with the same or even better performance in many important technological fields of the twenty-first century. Among these, CO_2 capture from plant material into carbon solids and soil enrichment is discussed in Chapter 8.

So far, mainly pure carbohydrates have been employed as precursors for the synthesis of such highly functional materials. Whilst still comparatively inexpensive and sustainable, the production of such high-performance materials will be even more valuable if entirely based on (waste) biomass. Furthermore, we predict that heteroatom-doped carbon materials will continue to play an important role, especially in fields such as adsorption, electrocatalysis, and energy storage. HTC-based hybrid materials will continue to develop and expand into new and exciting nature-inspired self-assembled materials with future applications in catalysis, photocatalysis, and sensors.

In particular, in the field of biofuels production using the catalytic liquefaction of biomass [127], hydrothermal carbons and hybrids with a basic or acid character could play a specific role for various catalytic reactions when used as solid catalysts. It would of course be of great advantage to use biomass-derived carbons to catalyze the production of chemicals or biofuels out of biomass itself, providing a complete synergistic and holistic approach that could easily be integrated into future biorefinery schemes.

Another field where low-cost porous HTC-derived materials could be of great importance is chromatographic separations. (i) The HTC-derived materials can be shaped either as porous monoliths or porous spheres (see Chapter 2). (ii) They possess surface functional groups just like silica-based stationary phases. This offers them flexibility for various applications in either hydrophilic interaction liquid chromatography [270] or reversed-phase liquid chromatography [271]. (iii) When postcarbonized at higher temperatures the carbons will become increasingly condensed, aromatized, and hydrophobic, leading to stationary-phase-like properties more commonly associated with porous graphitic carbon columns [272]. The advantage of the HTC approach is the ability to moderate "π" stacking/delocalization characteristics via carbonization temperature selection – a material feature believed to determine the retention strength of such polar analytes. Such stationary phases are known for the efficient separations of polar analytes, including sugars and phenols also derived from biomass. It will be extremely interesting to produce biomass-derived stationary phases for the separation of biomass components – a field of extreme importance for the production of biofuels.

Whilst the (re)naissance of HTC technology is now effectively into its tenth year, and many further developments and applications are to be still discovered, the authors hope to have provided a suitable introduction to the reader to the extremely broad scope and applicability of these exciting biomass-derived materials.

References

(1) Kroto, H.W., Heath, J.R., O'Brien, S.C. *et al.* (1985) *Nature*, **318**, 162–163.

(2) Iijima, S. (1991) *Nature*, **354**, 56–58.

(3) (a) Su, D.S. and Schloegl, R., *ChemSusChem*, **3**, 136–168; (b) Radovic, L.R. and RodriguezReinoso, F. (1997) in *Chemistry and Physics of Carbon*, vol. **25** (ed. P.A. Thrower), Dekker, New York.

(4) Lu, A.H., Hao, G.P., Sun, Q. *et al.* (2012) *Macromolecular Chemistry and Physics*, **213**, 1107–1131.

(5) (a) Nogueira, C.A. and Delmas, F. (1999) *Hydrometallurgy*, **52**, 267–287; (b) Shukla, A.K., Venugopalan, S., and Hariprakash, B. (2001) *Journal of Power Sources*, **100**, 125–148.

(6) (a) Kuriyama, N., Sakai, T., Miyamura, H. *et al.* (1993) *Journal of Alloys and Compounds*, **202**, 183–197; (b) Reilly, J.J., Adzic, G.D., Johnson, J.R. *et al.* (1999) *Journal of Alloys and Compounds*, **293**, 569–582; (c) Kleperis, J., Wojcik, G., Czerwinski, A. *et al.* (2001) *Journal of Solid State Electrochemistry*, **5**, 229–249; (d) Sakai, T., Yuasa, A., Ishikawa, H. *et al.* (1991) *Journal of the Less-Common Metals*, **172**, 1194–1204.

(7) (a) Li, H., Wang, Z.X., Chen, L.Q., and Huang, X.J. (2009) *Advanced Materials*, **21**, 4593–4607; (b) Bruce, P.G., Scrosati, B., and Tarascon, J.-M. (2008) *Angewandte Chemie International Edition*, **47**, 2930–2946; (c) Endo, M., Kim, C., Nishimura, K. *et al.* (2000) *Carbon*, **38**, 183–197; (d) Scrosati, B. and Garche, J. (2010) *Journal of Power Sources*, **195**, 2419–2430; (e) Tarascon, J.M. and Armand, M. (2001) *Nature*, **414**, 359–367; (f) Bates, J.B., Dudney, N.J., Neudecker, B. *et al.* (2000) *Solid State Ionics*, **135**, 33–45.

(8) (a) Park, C.K., Kakirde, A., Ebner, W. *et al.* (2001) *Journal of Power Sources*, **97–8**, 775–778; (b) Zaghib, K., Simoneau, M., Armand, M., and Gauthier, M. (1999) *Journal of Power Sources*, **81–82**, 300–305.

(9) (a) Song, J.Y., Wang, Y.Y., and Wan, C.C. (1999) *Journal of Power Sources*, **77**, 183–197; (b) Croce, F., Appetecchi, G.B., Persi, L., and Scrosati, B. (1998) *Nature*, **394**, 456–458.

(10) Jayaprakash, N., Shen, J., Moganty, S.S. *et al.* (2011) *Angewandte Chemie International Edition*, **50**, 5904–5908; (b) Su, Y.-S. and Manthiram, A. (2012) *Electrochimica Acta*, **77**, 272–278; (c) Ji, X. and Nazar, L.F. (2010) *Journal of Materials Chemistry*, **20**, 9821–9826; (d) Hagen, M., Doerfler, S., Althues, H. *et al.* (2012) *Journal of Power Sources*, **213**, 239–248; (e) Mikhaylik, Y.V. and Akridge, J.R. (2004) *Journal of the Electrochemical Society*, **151**, A1969–A1976; (f) Song, M.S., Han, S.C., Kim, H.S. *et al.* (2004) *Journal of the Electrochemical Society*, **151**, A791–A795.

(11) Abraham, K.M., Rauh, R.D., and Brummer, S.B. (1978) *Electrochimica Acta*, **23**, 501–507.

(12) (a) Tao, Z.L., Xu, L.N., Gou, X.L. *et al.* (2004) *Chemical Communications*, 2080–2081; (b) Aurbach, D., Weissman, I., Gofer, Y., and Levi, E. (2003) *Chemical Record*, **3**, 61–73; (c) Morita, M., Yoshimoto, N., Yakushiji, S., and Ishikawa, M. (2001) *Electrochemical and Solid State Letters*, **4**, A177–A179; (d) Levi, E., Levi, M.D., Chasid, O., and Aurbach, D. (2009) *Journal of Electroceramics*, **22**, 13–19; (e) Aurbach, D., Gofer, Y., Lu, Z. *et al.* (2001) *Journal of Power Sources*, **97–98**, 28–32; (f) Aurbach, D., Lu, Z., Schechter, A. *et al.* (2000) *Nature*, **407**, 724–727.

(13) (a) Lee, J.-S., Kim, S.T., Cao, R. *et al.* (2011) *Advanced Energy Materials*, **1**, 34–50; (b) Laoire, C.O., Mukerjee, S., Abraham, K.M. *et al.* (2009) *Journal of Physical Chemistry C*, **113**, 20127–20134; (c) Girishkumar, G., McCloskey, B., Luntz, A.C. *et al.* (2010) *Journal of Physical Chemistry Letters*, **1**, 2193–2203; (d) Blurton, K.F. and Sammells, A.F. (1979) *Journal of Power Sources*, **4**, 263–279; (e) Xiao, J., Wang, D., Xu, W. *et al.* (2010) *Journal of the Electrochemical Society*, **157**, A487–A492; (f) Kowaluk, I., Read, J., and Salomon, M. (2007) *Pure and Applied Chemistry*, **79**, 851–860.

(14) Winter, M. and Besenhard, J.O. (1998) in *Lithium Ion Battery: Fundamentals and Performance* (eds M. Wakihara and O. Yamamoto), Wiley-VCH Verlag GmbH, Weinheim.

(15) Kaskhedikar, N.A. and Maier, J. (2009) *Advanced Materials*, **21**, 2664–2680.

(16) (a) Liu, H., Fu, L.J., Zhang, H.P. *et al.* (2006) *Electrochemical and Solid State Letters*, **9**, A529–A533; (b) Liu, C. and Cheng, H.M. (2005) *Journal of Physics D – Applied Physics*, **38**, R231–R252.

(17) Wang, Q., Li, H., Chen, L., and Huang, X. (2001) *Carbon*, **39**, 2211–2214.

(18) Hu, J., Li, H., and Huang, X.J. (2005) *Solid State Ionics*, **176**, 1151–1159.

(19) White, R.J., Tauer, K., Antonietti, M., and Titirici, M.-M. (2010) *Journal of the American Chemical Society*, **132**, 17360–17363.

(20) Tang, K., White, R.J., Mu, X. *et al.* (2012) *ChemSusChem*, **5**, 400–403.

(21) Subramanian, V., Zhu, H., and Wei, B. (2006) *Journal of Physical Chemistry B*, **110**, 7178–7183.

(22) Wu, X.-L., Chen, L.-L., Xin, S. *et al.* (2010) *ChemSusChem*, **3**, 703–707.

(23) Olson, C.L., Nelson, J., and Islam, M.S. (2006) *Journal of Physical Chemistry B*, **110**, 9995–10001; (b) Södergren, S., Siegbahn, H., Rensmo, H. *et al.* (1997) *Journal of Physical Chemistry B*, **101**, 3087–3090; (c) Wagemaker, M., Kearley, G.J., van Well, A.A. *et al.* (2002) *Journal of the American Chemical Society*, **125**, 840–848.

(24) Zhu, G.-N., Wang, Y.-G., and Xia, Y.-Y. (2012) *Energy & Environmental Science*, **5**, 6652–6667.

(25) Yang, Z., Choi, D., Kerisit, S. *et al.* (2009) *Journal of Power Sources*, **192**, 588–598.

(26) Park, S.J., Kim, Y.J., and Lee, H. (2011) *Journal of Power Sources*, **196**, 5133–5137.

(27) Xuefei, G., Chengyang, W., Mingming, C. *et al.* (2012) *Journal of Power Sources*, **214**, 107–112112.

(28) Beaulieu, L.Y., Hewitt, K.C., Turner, R.L. *et al.* (2003) *Journal of the Electrochemical Society*, **150**, A149–A156; (b) Huggins, R.A. (1999) *Journal of Power Sources*, **81**, 13–19.

(29) Hu, Y.-S., Demir-Cakan, R., Titirici *et al.* (2008) *Angewandte Chemie International Edition*, **47**, 1645–1649.

(30) Demir-Cakan, R., Hu, Y.-S., Antonietti, M. *et al.* (2008) *Chemistry of Materials*, **20**, 1227–1229.

(31) Kim, C., Noh, M., Choi, M. *et al.* (2005) *Chemistry of Materials*, **17**, 3297–3301.

(32) Yang, H.X., Qian, J.F., Chen, Z.X. *et al.* (2007) *Journal of Physical Chemistry C*, **111**, 14067–14071.

(33) Ping, W., Ning, D., Hui, Z. *et al.* (2011) *Nanoscale*, **3**, 746–750.

(34) Xiaoyuan, Y., Siyuan, Y., Baohua, Z. *et al.* (2011) *Journal of Materials Chemistry*, **21**, 12295–12302.

(35) Titirici, M.-M., Antonietti, M., and Thomas, A. (2006) *Chemistry of Materials*, **18**, 3808–3812.

(36) Lou, X.W., Chen, J.S., Chen, P., and Archer, L.A. (2009) *Chemistry of Materials*, **21**, 2868–2874.

(37) Poizot, P., Laruelle, S., Grugeon, S. *et al.* (2000) *Nature*, **407**, 496–499.

(38) Badway, F., Cosandey, F., Pereira, N., and Amatucci, G.G. (2003) *Journal of the Electrochemical Society*, **150**, A1318–A1327.

(39) Poizot, P., Laruelle, S., Grugeon, S., and Tarascon, J.-M. (2002) *Journal of the Electrochemical Society*, **149**, A1212–A1217.

(40) Huang, X.H., Tu, J.P., Zhang, C.Q. *et al.* (2007) *Electrochimica Acta*, **52**, 4177–4181.

(41) Jiang, J., Zhu, J.H., Ding, R.M. *et al.* (2011) *Journal of Materials Chemistry*, **21**, 15969–15974.

(42) Qi, Y., Du, N., Zhang, H. *et al.* (2011) *Journal of Power Sources*, **196**, 10234–10239.

(43) Ellis, B.L., Lee, K.T., and Nazar, L.F. (2010) *Chemistry of Materials*, **22**, 691–714.

(44) Nanjundaswamy, K.S., Padhi, A.K., Goodenough, J.B. *et al.* (1996) *Solid State Ionics*, **92**, 1–10.

(45) Padhi, A.K., Nanjundaswamy, K.S., Masquelier, C. *et al.* (1997) *Journal of the Electrochemical Society*, **144**, 1609–1613.

(46) Padhi, A.K., Nanjundaswamy, K.S., and Goodenough, J.B. (1997) *Journal of the Electrochemical Society*, **144**, 1188–1194.

(47) MacNeil, D.D., Lu, Z., Chen, Z., and Dahn, J.R. (2002) *Journal of Power Sources*, **108**, 8–14.

(48) Delacourt, C., Wurm, C., Laffont, L. *et al.* (2006) *Solid State Ionics*, **177**, 333–341.

(49) Chung, S.-Y., Bloking, J.T., and Chiang, Y.-M. (2002) *Nature Materials*, **1**, 123–128.

(50) Popovic, J., Demir-Cakan, R., Tornow, J. *et al.* (2011) *Small*, **7**, 1127–1135.

(51) Yoon, S., Liao, C., Sun, X.-G. *et al.* (2012) *Journal of Materials Chemistry*, **22**, 4611–4614.

(52) Dominko, R. (2008) *Journal of Power Sources*, **184**, 462–468.

(53) Aravindan, V., Karthikeyan, K., Lee, J.W. *et al.* (2011) *Journal of Physics D – Applied Physics*, **44**.

(54) Ellis, B.L., Makahnouk, W.R.M., Makimura, Y. *et al.* (2007) *Nature Materials*, **6**, 749–753.

(55) Kim, S.-W., Seo, D.-H., Ma, X. *et al.* (2012) *Advanced Energy Materials*, **2**, 710–721.

(56) Chevrier, V.L. and Ceder, G. (2011) *Journal of the Electrochemical Society*, **158**, A1011–A1014.

(57) Palomares, V., Serras, P., Villaluenga, I. *et al.* (2012) *Energy & Environmental Science*, **5**, 5884–5901.

(58) Lee, K.T., Ramesh, T.N., Nan, F. *et al.* (2011) *Chemistry of Materials*, **23**, 3593–3600.

(59) Jian, Z., Zhao, L., Pan, H. *et al.* (2012) *Electrochemistry Communications*, **14**, 86–89.

(60) Sangster, J. (2007) *Journal of Phase Equilibria and Diffusion*, **28**, 571–579.

(61) Xin, X., Obrovac, M.N., and Dahn, J.R. (2011) *Electrochemical and Solid-State Letters*, **14**, A130–A133.

(62) (a) Stevens, D.A. and Dahn, J.R. (2000) *Journal of the Electrochemical Society*, **147**, 1271–1273; (b) Stevens, D.A. and Dahn, J.R. (2001) *Journal of the Electrochemical Society*, **148**, A803–A811.

(63) Alcantara, R., Jimenez-Mateos, J.M., Lavela, P., and Tirado, J.L. (2001) *Electrochemistry Communications*, **3**, 639–642.

(64) Wenzel, S., Hara, T., Janek, J., and Adelhelm, P. (2011) *Energy & Environmental Science*, **4**, 3342–3345.

(65) Komaba, S., Murata, W., Ishikawa, T. *et al.* (2011) *Advanced Functional Materials*, **21**, 3859–3867.

(66) Qian, J., Chen, Y., Wu, L. *et al.* (2012) *Chemical Communications*, **48**, 7070–7072.

(67) Ellis, B.L. and Nazar, L.F. (2012) *Current Opinion in Solid State & Materials Science*, **16**, 168–177; (b) Cao, Y., Xiao, L., Sushko, M.L. *et al.* (2012) *Nano Letters*, **12**, 3783–3787.

(68) Kun, T., Lijun, F., White, R.J. *et al.* (2012) *Advanced Energy Materials*, **2**, 873–877.

(69) Yan, H., Huang, X., Li, H., and Chen, L. (1998) *Solid State Ionics*, **11**, 113.

(70) Bruce, P.G., Freunberger, S.A., Hardwick, L.J., and Tarascon, J.M. (2012) *Nature Materials*, **11**, 19–29.

(71) Rauh, R.D., Abraham, K.M., Pearson, G.F. *et al.* (1979) *Journal of the Electrochemical Society*, **126**, 523–527.

(72) Yamin, H., Gorenshtein, A., Penciner, J. *et al.* (1988) *Journal of the Electrochemical Society*, **135**, 1045–1048.

(73) Ji, X., Lee, K.T., and Nazar, L.F. (2009) *Nature Materials*, **8**, 500–506.

(74) Jun, S., Joo, S.H., Ryoo, R. *et al.* (2000) *Journal of the American Chemical Society*, **122**, 10712–10713.

(75) Zhi, L., Hu, Y.S., El Hamaoui, B., *et al.* (2008) *Advanced Materials*, **20**, 1727; (b) Liang, C., Dudney, N.J., and Howe, J.Y. (2009) *Chemistry of Materials*, **21**, 4724–4730.

(76) Xiulei, J., Evers, S., Black, R., and Nazar, L.F. (2011) *Nature Communications*, **2**, 325–327.

(77) Brun, N., Sakaushi, K., Yu, L. *et al.* (2013) *Physical Chemistry Chemical Physics*, **15**, 6080–6087.

(78) Conway, B.E. (1999) *Electrochemical Supercapacitors: Scientific Fundamentals and Technological Applications*, Kluwer, New York.

(79) Bleda-Martínez, M.J., Maciá-Agulló, J.A., Lozano-Castelló, D. *et al.* (2005) *Carbon*, **43**, 2677–2684.

(80) Lota, G., Grzyb, B., Machnikowska, H. *et al.* (2005) *Chemical Physics Letters*, **404**, 53–58.

(81) Li, W., Chen, D., Li, Z. *et al.* (2007) *Carbon*, **45**, 1757–1763.

(82) Seredych, M., Hulicova-Jurcakova, D., Lu, G.Q., and Bandosz, T.J. (2008) *Carbon*, **46**, 1475–1488.

(83) Sevilla, M. and Fuertes, A.B. (2009) *Chemistry – A European Journal*, **15**, 4195–4203.

(84) Sevilla, M. and Fuertes, A.B. (2009) *Carbon*, **47**, 2281–2289.

(85) Baccile, N., Antonietti, M., and Titirici, M.-M. (2010) *ChemSusChem*, **3**, 246–253.

(86) White, R.J., Yoshizawa, N., Antonietti, M., and Titirici, M.-M. (2011) *Green Chemistry*, **13**, 2428–2434.

(87) Zhao, L., Baccile, N., Gross, S. *et al.* (2010) *Carbon*, **48**, 3778–3787.

(88) Wei, L., Sevilla, M., Fuertes, A.B. *et al.* (2011) *Advanced Energy Materials*, **1**, 356–361.

(89) Wang, L., Guo, Y., Zou, B. *et al.* (2011) *Bioresource Technology*, **102**, 1947–1950.

(90) Lozano-Castelló, D., Cazorla-Amorós, D., and Linares-Solano, A. (2002) *Fuel Processing Technology*, **77–78**, 325–330.

(91) Molina-Sabio, M. and Rodrı—guez-Reinoso, F. (2004) *Colloids and Surfaces A: Physicochemical and Engineering Aspects*, **241**, 15–25.

(92) Raymundo-Piñero, E., Kierzek, K., Machnikowski, J., and Béguin, F. (2006) *Carbon*, **44**, 2498–2507.

(93) Largeot, C., Portet, C., Chmiola, J. *et al.* (2008) *Journal of the American Chemical Society*, **130**, 2730–2371.

(94) Zhao, L., Fan, L.-Z., Zhou, M.-Q. *et al.* (2010) *Advanced Materials*, **22**, 5202–5206.

(95) Drage, T.C., Arenillas, A., Smith, K.M. *et al.* (2007) *Fuel*, **86**, 22–31.

(96) Sevilla, M., Mokaya, R., and Fuertes, A.B. (2011) *Energy & Environmental Science*, **4**, 2930–2936.

(97) Frackowiak, E., Lota, G., Machnikowski, J. *et al.* (2006) *Electrochimica Acta*, **51**, 2209–2214.

(98) (a) Zhu, H., Wang, X., Yang, F., and Yang, X. (2011) *Advanced Materials*, **23**, 2745–2748; (b) Zhou, H., Zhu, S., Hibino, M. *et al.* (2002) *Advanced Materials*, **15**, 2107–2111.

(99) Raymundo-Piñero, E., Leroux, F., and Béguin, F. (2006) *Advanced Materials*, **18**, 1877–1882.

(100) Lee, H.Y. and Goodenough, J.B. (1999) *Journal of Solid State Chemistry*, **144**, 220–223.

(101) Chen, S., Zhu, J., Wu, X. *et al.* (2010) *ACS Nano*, **4**, 2822–2830.

(102) Zhang, J.T., Ma, J.Z., Jiang, J.W., and Zhao, X.S. (2010) *Journal of Materials Research*, **25**, 1476–1484.

(103) Dupont, J., de Souza, R.F., and Suarez, P.A.Z. (2002) *Chemical Reviews*, **102**, 3667–3691.

(104) Rolison, D.R. (2003) *Science*, **299**, 1698–1701.

(105) Serp, P., Corrias, M., and Kalck, P. (2003) *Applied Catalysis A: General*, **253**, 337–358.

(106) Yang, Y., Chiang, K., and Burke, N. (2011) *Catalysis Today*, **178**, 197–205.

(107) Makowski, P., Demir-Cakan, R., Antonietti, M. *et al.* (2008) *Chemical Communications*, 999–1001.

(108) White, R.J., Luque, R., Budarin, V.L. *et al.* (2009) *Chemical Society Reviews*, **38**, 481–494.

(109) Astruc, D., Lu, F., and Aranzaes, J.R. (2005) *Angewandte Chemie International Edition*, **44**, 7852–7872.

(110) Fechete, I., Wang, Y., and Vedrine, J.C. (2012) *Catalysis Today*, **189**, 2–27.

(111) Wu, S.Y., Ding, Y.S., Zhang, X.M. *et al.* (2008) *Journal of Solid State Chemistry*, **181**, 2171–2177.

(112) Yu, G., Sun, B., Pei, Y. *et al.* (2010) *Journal of the American Chemical Society*, **132**, 935–937.

(113) Soorholtz, M., White, R.J., Zimmermann, T., *et al.* (2013) *Chemical Communications*, **49**, 240–242.

(114) Periana, R.A., Taube, D.J., Gamble, S. *et al.* (1998) *Science*, **280**, 560–564.

(115) (a) Arakawa, H., Aresta, M., Armor, J.N. *et al.* (2001) *Chemical Reviews*, **101**, 953–996; (b) Lunsford, J.H. (2000) *Catalysis Today*, **63**, 165–174; (c) Crabtree, R.H. (2001) *Journal of the Chemical Society, Dalton Transactions*, 2437–2450; (d) Alvarez-Galvan, M.C., Mota, N., Ojeda, M. *et al.* (2011) *Catalysis Today*, **171**, 15–23.

(116) (a) Shilov, A.E. and Shul'pin, G.B. (1997) *Chemical Reviews*, **97**, 2879–2932; (b) Shilov, A.E. and Shul'pin, G.B. (2000) *Activation and Catalytic Reactions of Saturated Hydrocarbons in the Presence of Metal Complexes*, Kluwer, Dordrecht; (c) Labinger, J.A. and Bercaw, J.E. (2002) *Nature*, **417**, 507–514.

(117) (a) Wolf, D. (1998) *Angewandte Chemie International Edition*, **37**, 3351–3353; (b) Conley, B.L., Tenn, W.J., Young, K.J.H. *et al.* (2006) in *Activation of Small Molecules: Organometallic and Bioinorganic Perspectives* (ed W.B. Tolman), Wiley-VCH Verlag GmbH, Weinheim.

(118) Palkovits, R., Antonietti, M., Kuhn, P. *et al.* (2009) *Angewandte Chemie. International Edition*, **48**, 6909–6912.

(119) Qian, H.S., Antonietti, M., and Yu, S.H. (2007) *Advanced Functional Materials*, **17**, 637–643.

(120) Xiang, X., Bai, L., and Li, F. (2010) *AIChE Journal*, **56**, 2934–2945.

(121) Ming, J., Liu, R.X., Liang, G.F. *et al.* (2011) *Journal of Materials Chemistry*, **21**, 10929–10934.

(122) Sanchez, C., Julian, B., Belleville, P., and Popall, M. (2005) *Journal of Materials Chemistry*, **15**, 3559–3592.

(123) Sanchez, C., Boissière, C., Grosso, D. *et al.* (2008) *Chemistry of Materials*, **20**, 682–737.

(124) Kaper, H., Grandjean, A., Weidenthaler, C. *et al.* (2012) *Chemistry – A European Journal*, **18**, 4099–4106.

(125) Demir-Cakan, R., Baccile, N., Antonietti, M., and Titirici, M.-M. (2009) *Chemistry of Materials*, **21**, 484–490.

(126) Demir-Cakan, R., Makowski, P., Antonietti, M. *et al.* (2010) *Catalysis Today*, **150**, 115–118.

(127) Chheda, J.N., Huber, G.W., and Dumesic, J.A. (2007) *Angewandte Chemie International Edition*, **46**, 7164–7183.

(128) Melero, J.A., Iglesias, J., and Morales, G. (2009) *Green Chemistry*, **11**, 1285–1308.

(129) Macia-Agullo, J.A., Sevilla, M., Diez, M.A., and Fuertes, A.B. (2010) *ChemSusChem*, **3**, 1352–1354.

(130) Huiquan, X., Yingxue, G., Xuezheng, L., and Chenze, Q. (2010) *Journal of Solid State Chemistry*, **183**, 1721–1725.

(131) Berge, N.D., Ro, K.S., Mao, J. *et al.* (2011) *Environmental Science & Technology*, **45**, 5696–5703.

(132) White, R.J., Budarin, V., Luque, R. *et al.* (2009) *Chemical Society Reviews*, **38**, 3401–3418.

(133) (a) Budarin, V.L., Clark, J.H., Luque, R., and Macquarrie, D.J. (2007) *Chemical Communications*, 634–636; (b) Luque, R., Herrero-Davila, L., Campelo, J.M. *et al.* (2008) *Energy & Environmental Science*, **1**, 542–564; (c) Budarin, V.L., Clark, J.H., Luque, R. *et al.* (2008) *Green Chemistry*, **10**, 382–387; (d) Gronnow, M.J., Luque, R., Macquarrie, D.J., and Clark, J.H. (2005) *Green Chemistry*, **7**, 552–557.

(134) Steele, B.C.H. and Heinzel, A. (2001) *Nature*, **414**, 345–352.

(135) Liu, H.S., Song, C.J., Zhang, L. *et al.* (2006) *Journal of Power Sources*, **155**, 95–110.

(136) Sevilla, M. and Fuertes, A.B. (2010) *Chemical Physics Letters*, **490**, 63–68.

(137) Yang, R., Qiu, X., Zhang, H. *et al.* (2005) *Carbon*, **43**, 11–16.

(138) Kim, P., Joo, J., Kim, W. *et al.* (2006) *Catalysis Letters*, **112**, 213–218.

(139) Sevilla, M., Lota, G., and Fuertes, A.B. (2007) *Journal of Power Sources*, **171**, 546–551.

(140) Sevilla, M., Sanchís, C., Valdés-Solís, T. *et al.* (2009) *Electrochimica Acta*, **54**, 2234–2238.

(141) Joo, J.B., Kim, Y.J., Kim, W. *et al.* (2008) *Catalysis Communications*, **10**, 267–271.

(142) Wen, Z., Wang, Q., and Li, J. (2008) *Advanced Functional Materials*, **18**, 959–964.

(143) Wen, Z., Liu, J., and Li, J. (2008) *Advanced Materials*, **20**, 743–747.

(144) Wen, Z., Wang, Q., Zhang, Q., and Li, J. (2007) *Electrochemistry Communications*, **9**, 1867–1872.

(145) Marques Tusi, M., Soares de Oliveira Polanco, N., Brandalise, M. *et al.* (2011) *International Journal of Electrochemcal Science*, **6**, 484.

(146) Xu, C., Cheng, L., Shen, P., and Liu, Y. (2007) *Electrochemistry Communications*, **9**, 997–1001.

(147) Yuan, D., Xu, C., Liu, Y. *et al.* (2007) *Electrochemistry Communications*, **9**, 2473–2478.
(148) Yuan, D., Tan, S., Liu, Y. *et al.* (2008) *Carbon*, **46**, 531–536.
(149) Ferrin, P. and Mavrikakis, M. (2009) *Journal of the American Chemical Society*, **131**, 14381–14389.
(150) Zeng, J. and Lee, J.Y. (2007) *International Journal of Hydrogen Energy*, **32**, 4389–4396.
(151) Welsch, F.G., Stowe, K., and Maier, W.F. (2011) *ACS Combinatorial Science*, **13**, 518–529.
(152) Jasinski, R. (1964) *Nature*, **201**, 1212–1213.
(153) Wang, B. (2005) *Journal of Power Sources*, **152**, 1–15.
(154) Koslowski, U.I., Abs-Wurmbach, I., Fiechter, S., and Bogdanoff, P. (2008) *Journal of Physical Chemistry C*, **112**, 15356–15366.
(155) Wang, H., Maiyalagan, T., and Wang, X. (2012) *Acs Catalysis*, **2**, 781–794; (b) Su, D.S., Zhang, J., Frank, B. *et al.* (2010) *ChemSusChem*, **3**, 169–180; (c) Zhou, Y., Neyerlin, K., Olson, T.S. *et al.* (2010) *Energy & Environmental Science*, **3**, 1437–1446; (d) Paraknowitsch, J.P. and Thomas, A. (2012) *Macromolecular Chemistry and Physics*, **213**, 1132–1145.
(156) Onodera, T., Suzuki, S., Mizukami, T., and Kanzaki, H. (2011) *Journal of Power Sources*, **196**, 7994–7999.
(157) Liu, S.-H. and Wu, J.-R. (2011) *International Journal of Hydrogen Energy*, **36**, 87–93.
(158) Gong, K., Du, F., Xia, Z. *et al.* (2009) *Science*, **323**, 760–764.
(159) Bron, M., Fiechter, S., Hilgendorff, M., and Bogdanoff, P. (2002) *Journal of Applied Electrochemistry*, **32**, 211–216.
(160) Joo, S.H., Choi, S.J., Oh, I. *et al.* (2001) *Nature*, **412**, 169–172.
(161) Ozaki, J.-i., Anahara, T., Kimura, N., and Oya, A. (2006) *Carbon*, **44**, 3358–3361.
(162) Baccile, N., Laurent, G., Coelho, C. *et al.* (2011) *Journal of Physical Chemistry C*, **115**, 8976–8982.
(163) Wohlgemuth, S.-A., White, R.J., Willinger, M.-G. *et al.* (2012) *Green Chemistry*, **14**, 1515–1523.
(164) Sevilla, M. and Fuertes, A.B. (2012) *Microporous and Mesoporous Materials*, **158**, 318–323.
(165) Petit, C., Peterson, G.W., Mahle, J., and Bandosz, T.J. (2010) *Carbon*, **48**, 1779–1787.
(166) Seredych, M., Khine, M., and Bandosz, T.J. (2011) *ChemSusChem*, **4**, 139–147.
(167) Paraknowitsch, J.P., Thomas, A., and Schmidt, J. (2011) *Chemical Communications*, **47**, 8283–8285.
(168) Glenis, S., Nelson, A.J., and Labes, M.M. (1999) *Journal of Applied Physics*, **86**, 4464–4466.
(169) Cui, T., Lv, R., Huang, Z.-h. *et al.* (2011) *Nanoscale Research Letters*, **6**, 77.
(170) Choi, C.H., Park, S.H., and Woo, S.I. (2011) *Green Chemistry*, **13**, 406–412.
(171) Wohlgemuth, S.-A., Vilela, F., Titirici, M.-M., and Antonietti, M. (2012) *Green Chemistry*, **14**, 741–749.
(172) Baker, W.S., Long, J.W., Stroud, R.M., and Rolison, D.R. (2004), *Journal of Non-Crystalline Solids*, **350**, 80–87.

(173) Shao, Y., Sui, J., Yin, G., and Gao, Y. (2008) *Applied Catalysis B: Environmental*, **79**, 89–99.

(174) Jin, H., Zhang, H., Zhong, H., and Zhang, J. (2011) *Energy & Environmental Science*, **4**, 3389–3394.

(175) Xu, Y., Tomita, A., and Kyotani, T. (2005) *Chemistry of Materials*, **17**, 2940–2945.

(176) Matter, P.H., Zhang, L., and Ozkan, U.S. (2006) *Journal of Catalysis*, **239**, 83–96.

(177) Luo, Z., Lim, S., Tian, Z. *et al.* (2011) *Journal of Materials Chemistry*, **21**, 8038–8044.

(178) Liu, R., Wu, D., Feng, X., and Müllen, K. (2010) *Angewandte Chemie International Edition*, **49**, 2565–2569.

(179) Strelko, V.V., Kuts, V.S., and Thrower, P.A. (2000) *Carbon*, **38**, 1499–1503.

(180) Sidik, R.A., Anderson, A.B., Subramanian, N.P. *et al.* (2006) *Journal of Physical Chemistry B*, **110**, 1787–1793.

(181) Slater, J.C. (1964) *The Journal of Chemical Physics*, **41**, 3199–3204.

(182) Chamorro, E., Toro-Labbe, A., and Fuentealba, P. (2002) *Journal of Physical Chemistry A*, **106**, 3891–3898.

(183) Scheiner, S. and Bigham, L.D. (1985) *The Journal of Chemical Physics*, **82**, 3316–3321.

(184) Yang, Z., Yao, Z., Li, G. *et al.* (2011) *ACS Nano*, **6**, 205–211.

(185) Fujishima, A. and Honda, K. (1972) *Nature*, **238**, 37–38.

(186) Crabtree, G.W. and Lewis, N.S. (2007) *Physics Today*, **60**, 37–42.

(187) Asahi, R., Morikawa, T., Ohwaki, T. *et al.* (2001) *Science*, **293**, 269–271.

(188) Fox, M.A. and Dulay, M.T. (1993) *Chemical Reviews*, **93**, 341–357.

(189) Khan, S.U.M., Al-Shahry, M., and Ingler, W.B. (2002) *Science*, **297**, 2243–2245.

(190) Vinodgopal, K., Wynkoop, D.E., and Kamat, P.V. (1996) *Environmental Science & Technology*, **30**, 1660–1666.

(191) Sakthivel, S., Janczarek, M., and Kisch, H. (2004) *Journal of Physical Chemistry B*, **108**, 19384–19387.

(192) In, S., Orlov, A., Berg, R. *et al.* (2007) *Journal of the American Chemical Society*, **129**, 13790–13791.

(193) Ohno, T., Tsubota, T., Nishijima, K., and Miyamoto, Z. (2004) *Chemistry Letters*, **33**, 750–751.

(194) Zhao, L., Chen, X., Wang, X. *et al.* (2010) *Advanced Materials*, **22**, 3317–3321.

(195) Zhong, J., Chen, F., and Zhang, J. (2009) *Journal of Physical Chemistry C*, **114**, 933–939.

(196) Zhang, Y., Zhang, P., Huo, Y. *et al.* (2012) *Applied Catalysis B: Environmental*, **115–116**, 236–244.

(197) Zhuang, J., Tian, Q., Zhou, H. *et al.* (2012) *Journal of Materials Chemistry*, **22**, 7036–7042.

(198) Zhang, P., Shao, C., Zhang, Z. *et al.* (2011) *Journal of Materials Chemistry*, **21**, 17746–17753.

(199) Shi, J.-W., Zong, X., Wu, X. *et al.* (2012) *ChemCatChem*, **4**, 488–491.

(200) Zhang, P., Shao, C., Zhang, M. *et al.* (2012) *Journal of Hazardous Materials*, **229–230**, 265–272.

(201) Mirtchev, P., Henderson, E.J., Soheilnia, N. *et al.* (2012) *Journal of Materials Chemistry*, **22**, 1265–1269.

(202) Tauc, J., Grigorovici, R., and Vancu, A. (1966) *Physica Status Solidi (b)*, **15**, 627–637.

(203) Yu, H., Zhang, H., Huang, H. *et al.* (2012) *New Journal of Chemistry*, **36**, 1031–1035.

(204) Hedin, N., Chen, L., and Laaksonen, A. (2010) *Nanoscale*, **2**, 1819–1841.

(205) Sevilla, M. and Fuertes, A.B. (2011) *Energy & Environmental Science*, **4**, 1765–1771.

(206) Zhao, L., Bacsik, Z., Hedin, N. *et al.* (2010) *ChemSusChem*, **3**, 840–845.

(207) Zhang, Z., Wang, K., Atkinson, J.D., *et al.* (2012) *Journal of Hazardous Materials*, **229–230**, 183–191.

(208) Sevilla, M., Falco, C., Titirici, M.-M., and Fuertes, A.B. (2012) *RSC Advances*, **2**, 12792–12797.

(209) Ritter, N., Senkovska, I., Kaskel, S., and Weber, J. (2011) *Macromolecules*, **44**, 2025–2033.

(210) Du, N., Robertson, G.P., Pinnau, I., and Guiver, M.D. (2009) *Macromolecules*, **42**, 6023–6030.

(211) Liu, Q., Mace, A., Bacsik, Z. *et al.* (2010) *Chemical Communications*, **46**, 4502–4504.

(212) Jeromenok, J., Böhlmann, W., Antonietti, M., and Weber, J. (2011) *Macromolecular Rapid Communications*, **32**, 1846–1851.

(213) Dybtsev, D.N., Chun, H., Yoon, S.H. *et al.* (2003) *Journal of the American Chemical Society*, **126**, 32–33.

(214) Yu, L., Falco, C., Weber, J. *et al.* (2012) *Langmuir*, **28**, 12373–12383.

(215) Yürüm, Y., Taralp, A., and Veziroglu, T.N. (2009) *International Journal of Hydrogen Energy*, **34**, 3784–3798.

(216) Sevilla, M., Fuertes, A.B., and Mokaya, R. (2011) *Energy & Environmental Science*, **4**, 1400–1410.

(217) (a) Park, D., Yun, Y.-S., and Park, J.M. (2004) *Environmental Science & Technology*, **38**, 4860–4864; (b) Park, D., Yun, Y.-S., Cho, H.Y., and Park, J.M. (2004) *Industrial & Engineering Chemistry Research*, **43**, 8226–8232.

(218) Chen, Z., Ma, L., Li, S. *et al.* (2011) *Applied Surface Science*, **257**, 8686–8691.

(219) Kumar, S., Loganathan, V.A., Gupta, R.B., and Barnett, M.O. (2011) *Journal of Environmental Management*, **92**, 2504–2512.

(220) Wang, H., Ma, L., Cao, K. *et al.* (2012) *Journal of Hazardous Materials*, **229–230**, 321–330.

(221) Liu, Z. and Zhang, F.-S. (2009) *Journal of Hazardous Materials*, **167**, 933–939.

(222) Liu, Z., Zhang, F.-S., and Wu, J. (2010) *Fuel*, **89**, 510–514.

(223) Xu, Y.-J., Weinberg, G., Liu, X. *et al.* (2008) *Advanced Functional Materials*, **18**, 3613–3619.

(224) Chen, L.-F., Liang, H.-W., Lu, Y. *et al.* (2011) *Langmuir*, **27**, 8998–9004.

(225) Ni, Y., Jin, L., Zhang, L., and Hong, J. (2010) *Journal of Materials Chemistry*, **20**, 6430–6436.

(226) Nata, I.F., Salim, G.W., and Lee, C.-K. (2010) *Journal of Hazardous Materials*, **183**, 853–858.

(227) Liang, H.-W., Wang, L., Chen, P.-Y. *et al.* (2010) *Advanced Materials*, **22**, 4691–4695.

(228) Qian, H.-S., Yu, S.-H., Luo, L.-B. *et al.* (2006) *Chemistry of Materials*, **18**, 2102–2108.

(229) Chen, P., Liang, H.-W., Lv, X.-H. *et al.* (2011) *ACS Nano*, **5**, 5928–5935.

(230) Sun, K., Ro, K., Guo, M. *et al.* (2011) *Bioresource Technology*, **102**, 5757–5763.

(231) Zhou, J., Tang, C., Cheng, B. *et al.* (2012) *ACS Applied Materials & Interfaces*, **4**, 2174–2179.

(232) Xia, F.J., Pan, M., Mu, S.C. *et al.* (2012) *Electroanalysis*, **24**, 1703–1708.

(233) Lu, W.B., Qin, X.Y., Liu, S. *et al.* (2012) *Analytical Chemistry*, **84**, 5351–5357.

(234) Liu, S., Tian, J., Wang, L. *et al.* (2012) *Advanced Materials*, **24**, 2037–2041.

(235) Li, S. (2005) *Langmuir*, **21**, 6019–6024.

(236) Mao, S.X., Li, W.F., Long, Y.M. *et al.* (2012) *Analytica Chimica Acta*, **738**, 35–40.

(237) Cui, R., Liu, C., Shen, J. *et al.* (2008) *Advanced Functional Materials*, **18**, 2197–2204.

(238) Caihong, W., Chu, X., and Wu, M. (2007) *Sensors and Actuators B: Chemical*, **120**, 508–513.

(239) Li, X.-L., Lou, T.-J., Sun, X.-M., and Li, Y.-D. (2004) *Inorganic Chemistry*, **43**, 5442–5449.

(240) Guo, Z., Liu, J., Jia, Y. *et al.* (2008) *Nanotechnology*, **19**, 345704.

(241) Sinha, N., Ma, J.Z., and Yeow, J.T.W. (2006) *Journal of Nanoscience and Nanotechnology*, **6**, 573–590.

(242) De Jong, W.H. and Borm, P.J.A. (2008) *International Journal of Nanomedicine*, **3**, 133–149.

(243) Menard-Moyon, C., Venturelli, E., Fabbro, C. *et al.* (2010) *Expert Opinion on Drug Discovery*, **5**, 691–707.

(244) Zhang, W., Zhang, Z., and Zhang, Y. (2011) *Nanoscale Research Letters*, **6**.

(245) Vashist, S.K., Zheng, D., Pastorin, G. *et al.* (2011) *Carbon*, **49**, 4077–4097.

(246) Tang, A.C.L., Hwang, G.-L., Tsai, S.-J. *et al.* (2012) *Plos One*, **7**.

(247) Yamashita, T., Yamashita, K., Nabeshi, H. *et al.* (2012) *Materials*, **5**, 350–363.

(248) Kubo, S., Tan, I., White, R.J. *et al.* (2010) *Chemistry of Materials*, **22**, 6590–6597.

(249) Stanisheysky, A.V., Styres, C., Yockell-Lelievre, H., and Yusuf, N. (2011) *Journal of Nanoscience and Nanotechnology*, **11**, 8705–8711.

(250) Selvi, B.R., Jagadeesan, D., Suma, B.S. *et al.* (2008) *Nano Letters*, **8**, 3182–3188.

(251) Gu, J., Su, S., Li, Y. *et al.* (2011) *Chemical Communications*, **47**, 2101–2103.

(252) Gao, G., Wu, H., Zhang, Y. *et al.* (2011) *Journal of Materials Chemistry*, **21**, 12224–12227.

(253) Son, S.J., Bai, X., and Lee, S.B. (2007) *Drug Discovery Today*, **12**, 650–656.

(254) Yu, J.C., Hu, X., Li, Q. *et al.* (2006) *Chemistry – A European Journal*, **12**, 548–552.

(255) (a) Ley, G.R. and Knapp, C.P. (2002) US Patent 6,361,780; (b) Janib, S.M., Moses, A.S. and MacKay, J.A. (2010) *Advanced Drug Delivery Reviews*, **62**, 1052–1063.

(256) Zhu, Y.F., Ikoma, T., Hanagata, N., and Kaskel, S. (2010) *Small*, **6**, 471–478.

(257) Guang, J., Cuimiao, Z., Liyong, W. *et al.* (2011) *Journal of Alloys and Compounds*, **509**, 6418–6422.

(258) Tian, G., Gu, Z.J., Liu, X.X. *et al.* (2011) *Journal of Physical Chemistry C*, **115**, 23790–23796.

(259) Peng, Z.A. and Peng, X. (2000) *Journal of the American Chemical Society*, **123**, 183–184.

(260) Sun, Y.-P., Zhou, B., Lin, Y. *et al.* (2006) *Journal of the American Chemical Society*, **128**, 7756–7757.

(261) Cao, L., Wang, X., Meziani, M.J. *et al.* (2007) *Journal of the American Chemical Society*, **129**, 11318–11319.

(262) Yang, S.-T., Cao, L., Luo, P.G. *et al.* (2009) *Journal of the American Chemical Society*, **131**, 11308–11309.

(263) Zhou, J., Booker, C., Li, R. *et al.* (2007) *Journal of the American Chemical Society*, **129**, 744–745.

(264) Ray, S.C., Saha, A., Jana, N.R., and Sarkar, R. (2009) *Journal of Physical Chemistry C*, **113**, 18546–18551.

(265) Guo, S.R., Gong, J.Y., Jiang, P. *et al.* (2008) *Advanced Functional Materials*, **18**, 872–879.

(266) Yang, Y.H., Cui, J.H., Zheng, M.T. *et al.* (2012) *Chemical Communications*, **48**, 380–382.

(267) Li, H.-W., Li, Y., Dang, Y.-Q. *et al.* (2009) *Chemical Communications*, 4453–4455.

(268) Chandra, S., Das, P., Bag, S. *et al.* (2011) *Nanoscale*, **3**, 1533–1540.

(269) Wang, X.H., Qu, K.G., Xu, B.L. *et al.* (2011) *Journal of Materials Chemistry*, **21**, 2445–2450.

(270) Hemstrom, P. and Irgum, K. (2006) *Journal of Separation Science*, **29**, 1784–1821.

(271) Claessens, H.A. and van Straten, M.A. (2004) *Journal of Chromatography A*, **1060**, 23–41.

(272) Knox, J.H., Kaur, B., and Millward, G.R. (1986) *Journal of Chromatography*, **352**, 3–25.

8

Environmental Applications of Hydrothermal Carbonization Technology: Biochar Production, Carbon Sequestration, and Waste Conversion

Nicole D. Berge[1], Claudia Kammann[2], Kyoung Ro[3], and Judy Libra[4]

[1]*Department of Civil and Environmental Engineering, University of South Carolina, USA*
[2]*Department of Plant Ecology, Justus-Liebig-University Giessen, Germany*
[3]*USDA-ARS, Coastal Plains Soil, Water, and Plant Research Center, USA*
[4]*Department of Technology Assessment and Substance Cycles, Leibniz Institute for Agricultural Engineering Potsdam-Bornim (ATB), Germany*

8.1 Introduction

The motivation for the development and use of hydrothermal carbonization (HTC) in the last decade came primarily from the desire to create sustainable carbon nanomaterials/nanostructures (e.g., [1–6]) for use in applications ranging from hydrogen storage to chemical adsorption (e.g., [7, 8]). These were already pointed out in Chapter 7.

The potential environmental benefits of this process [9–13] were quickly recognized, and research was expanded to include the carbonization of biomass and organic waste streams for environmental applications. The carbon-rich material, often referred to as hydrochar,

Sustainable Carbon Materials from Hydrothermal Processes, First Edition. Edited by Maria-Magdalena Titirici.
© 2013 John Wiley & Sons, Ltd. Published 2013 by John Wiley & Sons, Ltd.

resulting from the carbonization of these feedstocks in the presence of water may be used in a variety of environmental applications, such as a soil amendment [11] and as an adsorbent in environmental remediation processes [14]. HTC has also shown promise as a sustainable waste stream conversion technique, ultimately converting waste materials to value-added products, while promoting integration of carbon in the solid phase (e.g., [9, 11, 12, 15]).

Waste streams, such as municipal solid waste (MSW), animal wastes (e.g., pig, cow, poultry), and human wastes (e.g., wastewater), may be sustainably converted to a stable, sterile, carbon-rich, high-energy-density material via HTC (e.g., [9–12]). Utilization of HTC as a conversion technique may substantially reduce fugitive greenhouse gas emissions from MSW landfills, composting facilities, animal manure pits, and wastewater treatment plants by promoting carbon integration within the hydrochar. In addition, the resulting hydrochar may serve as a source of renewable solid fuel. Another advantage associated with using HTC as a waste management tool is that it results in considerable waste volume and mass reduction, requiring less ultimate storage/disposal space.

Perhaps one of the most advantageous aspects of using HTC for waste stream conversion is the generation of value-added products. The ability to recover and reuse waste materials promotes the desired waste management hierarchy (e.g., reuse, recycle) prevalent in many countries. As illustrated in Figure 8.1, hydrochar resulting from the carbonization of wastes has the potential to be used as an adsorbent in environmental remediation applications, a novel carbon material, a storable solid fuel for energy generation (via cocombustion or use in carbon fuel cells), and possibly as a soil amendment.

Using hydrochar as a soil amendment is a topic of recent exploration. There has been a recent surge in exploring land application of biochar (terminology commonly used to denote char application in soils) to increase soil fertility, while providing a long-term carbon sink

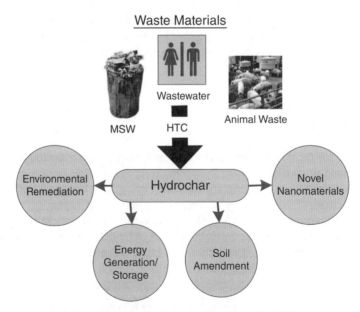

Figure 8.1 *Potential uses of hydrochar resulting from the HTC of waste streams.*

(e.g., [16]). The source of the biochar in the majority of reported studies, however, has been from the dry pyrolysis of different types of biomass. Hydrochar also has the potential to serve as a valuable soil amendment. Land application of hydrochar may increase the carbon content of degraded soils, ultimately improving soil fertility [11]. Research regarding the application of hydrochar in soils is still in its infancy [11].

Utilization of HTC in environmentally relevant applications is fairly new, and there is much additional research required to fully explore the potential and subsequent use of the process in such applications. This chapter describes the current knowledge associated with environmentally relevant HTC applications. The use of HTC as a waste management tool is first examined, followed by an overview of char applications, with a focus on char as a soil amendment. In addition, the current commercial status and research needs required to implement carbonization at the field-scale are discussed.

8.2 Waste Conversion to Useful Products

HTC of waste streams is an attractive alternative waste management strategy for biomass residuals such as agricultural residues and municipal wastes. In HTC, the feedstock is heated in subcritical water to between 150 and 250 °C at autogenic pressures for reaction times ranging from 1 to 20 h. The feedstock is decomposed by reaction mechanisms similar to those in dry pyrolysis, which include hydrolysis, dehydration, decarboxylation, aromatization, and recondensation [11]. In contrast to dry pyrolysis, however, HTC produces higher solid (i.e., hydrochar) yields, more water-soluble organic compounds, and mainly CO_2 as a major gaseous product. Furthermore, the chemical structure of the solid products more closely resemble natural coals than pyrolysis-derived chars [11, 17].

Since, during HTC, the majority of the carbon remains integrated within the solid material, successful carbonization of wastes could substantially reduce fugitive greenhouse gas emissions from current management/treatment processes. Approximately 36% of the estimated anthropogenic CH_4 emitted in the United States in 2008 was related to waste management activities (including landfills and composting [18, 19]). Waste management-related activities also accounted for approximately 8% of US N_2O emissions (a much more potent greenhouse gas). Berge [15] reports that carbon emissions resulting from HTC of MSW may be lower than those expected from landfilling (over a range gas collection efficiencies), composting, or incinerating (over a range of waste conversion efficiencies) the same waste materials.

There are many additional advantages associated with carbonizing waste streams. HTC of waste materials may require less solids processing/treatment (such as chemical or mechanical dewatering of biosolids). Ramke *et al.* [12], for example, evaluated the dewaterability of sewage sludge prior to and following carbonization, and concluded that carbonization significantly improved the dewaterability of the solids. Another significant potential advantage associated with this process is that emerging compounds, such as pharmaceuticals, personal care products, and endocrine-disrupting compounds, which currently pose significant environmental concerns in landfills (e.g., [20–22]), animal wastes (e.g., [23, 24]), and wastewater (e.g., [25, 26]), may be thermally degraded or transformed during carbonization.

The conversion of wastes to a valuable resource adds to the attractiveness of this approach. The hydrochar may be used in applications such as environmental remediation, feedstock for energy generation (via cocombustion or use in carbon fuel cells), and soil augmentation. To date, recovering waste materials (or transformed waste materials) as an energy source or other value-added product has been limited. Many of the studies that have evaluated the conversion of waste streams via HTC have been focused on creating an energy-rich hydrochar material. Use of the hydrochar as a solid fuel has been explored by Berge *et al.* [9], Ramke *et al.* [12], and Hwang *et al.* [27]. These studies report the hydrochar has characteristics similar to lignite coal. Work evaluating the burning of the char to produce energy has also been conducted. The benefits include a significant reduction in coal ignition temperature through the blending of hydrochar with coal [28], as well as a homogenized and stabilized fuel. Converting wastes (some components classified as renewable) to a source of non-fossil-fuel-derived energy has many significant potential environmental advantages.

To date, there have been a limited number of studies evaluating the carbonization of waste materials. The purpose of this section is to review the current state of knowledge associated with using HTC as a waste conversion tool, including associated environmental implications, hydrochar characteristics, and research needs.

8.2.1   Conversion of Municipal Solid Waste

8.2.1.1   MSW Characteristics

MSW is broadly defined as wastes originating from residential (i.e., product packaging, newspapers, magazines, food waste, grass clippings, yard waste, recyclables), institutional (i.e., schools, prisons), and commercial sources (i.e., restaurants). Construction and demolition debris and combustion ash are not generally characterized as MSW.

Waste properties that are important during HTC include the chemical composition, volatile fraction, combustible fraction, moisture content, particle size, and energy content. Typical ranges associated with these parameters are presented in Table 8.1 and are highly dependent on waste composition. MSW is generally comprised of a mixture of organics, inorganics, putrescibles, combustibles, and recyclables [29, 30]. Typical unsorted

Table 8.1   *Typical unsorted MSW properties relevant to carbonization [29, 30].*

Waste property	Value
Elemental analysis (%, dry ash free)	
Carbon	27–55
Hydrogen	3–9
Oxygen	22–44
Nitrogen	0.4–1.8
Sulfur	0.04–0.18
Volatile fraction (%, dry weight)	40–60
Moisture content (%, wet weight)	15–40
Noncombustible fraction (%, weight)	10–30
Particle size[a]	average: 18–20 cm range: 0.02–60 cm
Energy content (MJ $kg_{dry weight}^{-1}$)[a]	2–14

[a]Neglecting bulky items.

Table 8.2   *Comparison of MSW composition in the United States, European Union, and developing countries. (Adapted with permission from [37]. © 2009 Elsevier.)*

	Percentage of total composition						
	Paper	Textiles	Plastics	Glass	Metals	Organics	Other
United States	35	5	11	5	8	30	6
European Union	26	2	8	6	5	28	25
Developing countries[a]	13	2	11	3	5	54	12

[a]Based on average values from 19 developing countries; see source for specific countries included in the analysis.

MSW consists of 40–50% cellulose, 9–12% hemicellulose, and 10–15% lignin [31]. The biodegradable fraction of unsorted MSW varies and is dependent on waste composition. Approximately only 45% of typical unsorted MSW in the United States is biodegradable [29]. The percent biodegradability is typically larger in developing counties (see Table 8.2). Approximately 90% of the biodegradable fraction is typically comprised of cellulose plus hemicellulose [31]. In general, waste is fairly combustible (70–90% by weight) and contains a large fraction of volatile solids (40–60% by weight). The heterogeneous nature of MSW (in terms of composition, chemical properties, and particle size), particularly unsorted MSW, complicates its use as a feedstock for thermochemical conversion processes, potentially requiring the waste be processed (i.e., shredded, sorted) prior to introduction to such processes to minimize operation and maintenance issues, and to result in a consistent conversion product (e.g., oil, gas) [7, 29, 30, 32–34].

An attractive feature of carbonizing MSW for future use is that it is a continuously generated feedstock (always abundant) containing appreciable masses of renewable resources (i.e., food, paper, wood products [35]). Waste generation rates and composition vary with location and have been shown to correlate with average income, ranging from 0.1 (low-income countries) to more than 0.8 metric tons per person-year (high-income countries) [36]). In 2007, approximately 0.745, 0.522, and 0.548 metric tons per person-year of waste were generated in the United States, European Union, and Malaysia, respectively [35, 37, 38]. Variations in waste generation rates and composition can be attributed to several different factors, including country gross national product, population (and population density), collection frequency, affluence, cost of disposal, legislation, and public attitudes [29, 30, 36, 37, 39, 40]. Table 8.2 presents a comparison between typical waste composition from developed (United States and European Union) and developing countries. It should also be noted that current waste management practices also differ between countries. In the United States, the majority of waste discarded resides in MSW landfills (54% [35]), while the majority of waste discarded in Japan is incinerated (75% [41]) and in Germany is recycled or incinerated (70% and 30%, respectively [42]).

8.2.1.2   Carbonization Studies

MSW has been successfully used as a feedstock in several high-temperature, wet and dry conversion processes, such as pyrolysis, incineration, and gasification [34, 43–49]. Fewer studies evaluating the conversion of this waste stream via HTC have been conducted.

In comparison to other, more traditional thermal conversion processes, HTC occurs at comparatively lower temperatures, is simpler (e.g., compared to fluidized-bed gasification), requires the presence of water, and the main process product is a carbon-rich, high energy-density char. Oily and tar products may also result, depending on the feedstock and reaction temperature and time. Gaseous oxidation products, particularly CO_2, are limited during HTC because unlike combustion, exposure to oxygen is limited to that initially present in the reactor headspace and any dissolved oxygen in the water or initial feedstock.

Table 8.3 contains a summary of investigations evaluating the HTC of typical components of MSW and/or mixed MSW. Information, and in many cases the lack of, in Table 8.3 illustrates the current need for additional in-depth experimental studies to fully evaluate the use of HTC in this manner. What is apparent from the studies conducted is that the majority of carbon does remain in the solid material, despite variations in solids concentration, temperature, and reaction times. All studies report a fraction of carbon is transferred to the liquid-phase, and a smaller fraction of carbon is transferred to the gas phase (all studies report the majority of this gas is CO_2).

Ramke *et al.* [12] carbonized several different waste materials, most notably organics and green waste (grass cuttings). Results demonstrated that carbonization is feasible. Similarly to that observed when carbonizing pure substances, the majority of carbon remained within the hydrochar material (75–80%, see Table 8.3). Hwang *et al.* [27] also explored the HTC of components of MSW, specifically shredded paper and dog food, at different temperatures (Table 8.3). They determined that at all temperatures the majority of carbon remained within the solid-phase (greater than 80%). It was also determined that as the temperature increased, less carbon remained within the solid and solid yields decreased. Carbon balances from experiments conducted by Berge *et al.* [9] reveal results similar to those reported by others.

In all studies, an appreciable fraction of carbon is transferred to the liquid. When considering the use of this technique as a waste management tool, management/treatment of the liquid is necessary. Ramke *et al.* [12] measured gross organics in the liquid phase and reported that the BOD_5 ranges from 10 000 to 14 000 mg l^{-1} and the COD ranges from 14 000 to 70 000 mg l^{-1}. Experiments on the liquid-phase conducted by Ramke *et al.* [12] suggest that the organics in the liquid are degradable (COD approached 90% removal). Berge *et al.* [9] also identified several different organics in the leachate stream (using gross characterization and identification via GC-MS). BOD and COD results were similar to that reported by Ramke *et al.* [12]. Hwang *et al.* [27] tracked chloride and reported that the majority of the chloride remained within the char (at the tested temperatures).

Carbon in the gas phase has also been measured. As seen in many of the studies (Table 8.3), a small fraction of carbon is transferred to the gas phase. Ramke *et al.* [12] report that CO_2 is the predominant gas and represents 70–90% of the total gas produced, while trace hydrocarbons represented the balance. Berge *et al.* [9] conducted a more detailed analysis of the gas-phase components. They identified, in addition to CO_2, several trace gases, such as CH_4, hydrogen, and propene. Of environmental concern is the detection of furans. Additional studies are required to understand the furan concentrations and their associated implications (see Berge *et al.* [9]).

The carbon fractionation reported by these carbonization studies suggests that the hydrochar produced via MSW carbonization may serve as a significant carbon sink. Berge *et al.* [9] report that the carbon sequestered during carbonization of paper, food waste, and mixed MSW is greater than that currently achieved when landfilling the materials.

Table 8.3 Studies evaluating MSW conversion via HTC.

Feedstock	HTC conditions			Char characteristics							Liquid		Gas		Source
	Temperature (°C)	Reaction time (h)	Solids content (% weight)	% Carbon[a]	Volatile matter (%)	Fixed carbon (%)	Ash (%)	Energy content (KJ kg^{-1})	Yield (%)	^{13}C NMR	% Carbon[a]	Composition information	% Carbon[a]	Composition information	
Organics	180	12	>20	74.9	NR	NR	NR	15 000	NR	no	19	BOD; TOC; COD; some composition information	6.1	not provided	[12]
Green waste (grass cuttings)	180	12	>20	75.3	NR	NR	NR	NR	NR	no	19.7	BOD; TOC; COD; some composition information	5	not provided	[12]
Paper	234–294	NR	25	80–90	25–50	~15	<5	~20 000–25 000	NR	no	<5	TOC; chloride	<10	CO$_2$, CO, CH$_4$ only	[27]
Dog food	234–294	NR	25	80–90	35–55	~15	~10	25 000	NR	no	5–10	TOC; chloride	10–20	CO$_2$, CO, CH$_4$ only	[27]
Paper	250	20	20	48.8	52.8	19.8	24.2	23 900	29.2	yes	37.4	BOD; COD; TOC; composition via GC-/MS	10.8	composition via GC-MS	[9]
Food waste	250	20	20	75	53.5	29.7	11.2	29 100	43.8	yes	20.1	BOD; COD; TOC; composition via GC-MS	9	composition via GC-MS	[9]
Mixed MSW	250	20	20	74.8	33.6	14.6	46	20 000	63.2	yes	27.7	BOD; COD; TOC; composition via GC-MS	8.3	composition via GC-MS	[9]
Rabbit food	200–300	50–150 s	NR	NR	NR	NR	NR	NR	30–45	no	NR	yes; glucose, acetic acid	NR	NR	[50]
Japanese MSW	220	0.5	50	NR	70.9	12.3	16.8	16 400	NR	no	NR	no	NR	none	[51]
Indian MSW	220	0.5	50	NR	63.6	17.5	18.9	17 900	NR	no	NR	no	NR	none	[51]
Chinese MSW	220	0.5	50	NR	57.9	27.1	15.0	24 900	NR	no	NR	no	NR	none	[51]

[a]Percentage of initially present carbon.

NMR, nuclear magnetic resonance; NR, not reported; BOD, biochemical oxygen demand; TOC, total organic carbon; COD, chemical oxygen demand; GC-MS, gas chromatography-mass spectrometry.

Barlaz [31] developed carbon storage factors (CSFs: mass of carbon remaining in the solid following biological decomposition in a landfill/dry mass of feedstock) as a means to compare the mass of carbon remaining (stored) within solid material following biological decomposition in landfills. The estimated CSFs from hydrothermally carbonized office paper, food, and model mixed MSW are greater than those reported for landfilling of the same materials. This suggests that more carbon remains stored within the solid material following carbonization than if the materials had been landfilled, providing evidence that carbonization may be a promising process for mitigating carbon emissions associated with management of MSW. Berge [15] reports that retention of carbon in the char creates a carbon sink far surpassing that of current waste management processes (e.g., landfilling, composting, and incineration).

A few other studies evaluating the hydrothermal treatment of MSW have been conducted for purposes other than those described previously. Goto *et al.* [50] carbonized rabbit food at temperatures ranging from 200 to 300 °C. The objective of the experiments was to determine yields of smaller organic molecules, such as glucose and acetic acid. They measured glucose yields as high as 33%, up to 2.6% for acetic acid, and 3.2% yield of lactic acid. Onwudili and Williams [52] studied the catalytic (base catalyst: sodium hydroxide) gasification of refuse-derived fuel (material composed of MSW with recyclables removed) at temperatures ranging from 330 to 375 °C and placed emphasis on understanding how the produced gas composition varies with catalyst addition and temperature. The carbon content of the residues resulting from their studies ranged from 48 to 72%. Onwudili and Williams [52] also gasified glucose, starch, and cellulose for comparison, and reported differences in char properties and gas production.

8.2.2   Conversion of Animal Manure

"Confined animal feeding operations" in the United States and many other countries over the last few decades have undergone extensive expansions and consolidations [53]. This shift of animal production agriculture toward fewer, but larger operations has created environmental concerns in recycling and disposing of surplus animal manure because traditional manure management systems may not adequately dispose/recycle the surplus animal manure and pose potential environmental threat [54–56]. HTC of animal manure may provide a viable alternative for managing surplus animal manure and simultaneously produce value-added hydrochar.

8.2.2.1   Animal Manure Characteristics

Animal manures have widely different physical and chemical properties (Table 8.4). The dry manures such as chicken litter and feedlot manures have a moisture content ranging from 10.2 to 20.3%, whereas wet manures such as swine and dairy manures can have a moisture content greater than 96%. Except for the paved-surface feedlots, most poultry and feedlot operations collect a mixture containing manure, bedding, waste feed, and underlying soil. These mixtures generally have high ash contents. Furthermore, both volatile matter and fixed carbon content of soil-containing manure decrease, negatively affecting the higher heating value (HHV). Dairy and swine feeding operations typically produce a dilute solids waste stream comprised primarily of discharged wash water, but also manure, urine, and undigested feed. Discharged swine and dairy manure characteristics are highly dependent

Table 8.4   *Characteristics of animal manures.*

Parameters	Dry manures			Wet manures	
	Chicken litter [59]	Soil-surfaced feedlot manure [60]	Paved-surfaced feedlot manure [60]	Pit-recharge liquid swine manure	Flushed dairy manure
Proximate analyses					
Moisture contents (%)	10.2	19.8	20.3	98.0 [61]	96.2 [62]
Volatile matter (%$_{db}$)	51.1	33.8	64.6	68.7	83.8
Ash (%$_{db}$)	30.6	58.7	20.2	31.3^a	16.2^a
Fixed C (%$_{db}$)	8.2	7.5	15.2	NA	NA
HHV (MJ kg^{-1})	13.0	7.9	16.8	17.2	18.2
Ultimate analyses					
C (%$_{db}$)	34.4	21.7	43.1	45.7 [63]	44.7 [64]
H (%$_{db}$)	4.1	2.62	5.22	6.45	5.85
N (%$_{db}$)	3.27	1.94	3.11	3.45	2.05
S (%$_{db}$)	0.81	0.42	0.67	0.38	0.31
O (%$_{db}$)	23.4	14.6	27.7	31.4	38.2

^{a}Reported as "fixed solid," which is a combination of ash and fixed carbon.

on the type of manure handling and collection system (flush or pit-recharge) and the amount of added water [57]. Compared to the poultry- and cattle-based manures, dairy and swine manures have higher volatile matter and greater HHV. In a preliminary wet gasification analysis, Ro *et al.* [58] reported that swine manure among the five major types of animal manure would produce product gases with the highest energy per kilogram of dry matter [58].

8.2.2.2   Animal Manure Carbonization

The chemical structure of the solid products from HTC resembles natural coals more closely than chars from dry pyrolysis [11, 17]. Pyrochars also usually have a higher carbon and lower hydrogen content than hydrochars (likely due to temperature differences). In order to investigate how the structures differ, Cao *et al.* [65] compared two chars made from carbonizing swine manure via the two types of pyrolysis using ^{13}C solid-state NMR spectroscopy. Raw swine manure was converted into hydrochar (designated as "HTC-swine A") by anaerobically heating the manure solution (20% solid) to 250 °C under its autogenic pressure for 20 h and washed with acetone to remove mobile compounds adsorbed on hydrochar. Pyrochar was produced by pyrolyzing dried swine manure at 620 °C for 2 h with a heating rate of 13 °C min^{-1} [59].

As shown in Table 8.5, fixed carbon and ash contents of pyrochar increased dramatically from that of raw swine manure. While carbon contents in both hydrochar and pyrochar increased slightly, both hydrogen and oxygen contents of the pyrochar decreased substantially, indicating the increase in aromaticity due to pyrolysis. The NMR spectra of these chars and raw swine manure clearly showed the increase in aromaticity of the pyrochar

Table 8.5   Proximate and ultimate analyses of raw swine manure and its chars. (Reproduced with permission from [65]. © 2011 American Chemical Society.)

Parameters	Raw swine manure solid	HTC-swine A[a]	Pyrochar
Proximate analyses			
Moisture content (%)	12.8 ± 0.3	3.4 ± 0.8	3.4 ± 0.1
Volatile matter (%$_{db}$)[b,c]	60.6 ± 1.1	59.1 ± 1.5	14.1 ± 2.5
Ash (%$_{db}$)	18.5 ± 0.2	27.8 ± 0.3	44.7 ± 1.2
Fixed C (%$_{db}$)[d]	8.1 ± 0.6	13.1 ± 1.3	41.2 ± 1.3
HHV (MJ kg^{-1})	19.5 ± 0.2	NA	18.3 ± 0.4
Ultimate analyses[e]			
C (%$_{db}$)	47.3 ± 0.2	49.5 ± 2.8	50.7 ± 0.6
H (%$_{db}$)	5.9 ± 0.1	5.7 ± 0.0	1.9 ± 0.3
N (%$_{db}$)	4.58 ± 0.13	1.92 ± 0.95	3.26 ± 0.08
O (%$_{db}$)	20.1 ± 0.4	16.5 ± 6.0	<0.01
S (%$_{db}$)	0.93 ± 0.04	NA	0.66 ± 0.01
P (mg g^{-1} dry matter)[f]	23.7 ± 0.8	47.7	71.5 ± 1.3

[a]Hydrochar washed with acetone.
[b]ASTM D3175-07.
[c]Calculated as 100 − volatile matter − ash.
[d]ASTM D3172.
[e]ASTM D3176-02.
[f]USEPA Method 3052.

(Figure 8.2). Figure 8.2 also shows the NMR spectra of HTC-swine W, HTC-AW-swine A, and HTC-AC-swine A, which are water-washed hydrochar, acid-prewashed hydrochar, and acid-catalyzed hydrochar, respectively. When the raw swine manure was carbonized via HTC, the signals from aromatic or olefinic carbons around 130 ppm increased, indicating the increase in aromaticity, while the signals from COO/N—C=O functionalities decreased simultaneously. Proteins or peptides are one of the major constituents of swine manure along with carbohydrates as indicated by significant signals in the regions of 48–112 and 165–190 ppm. Furthermore, there was a substantial increase in mobile —(CH$_2$)$_n$— groups in HTC-swine W char, as indicated by the dominant band around 30 ppm in its dipolar-dephased spectrum. It suggested that the acetone wash removed some mobile —(CH$_2$)$_n$— groups in the hydrochar.

In contrast to hydrochar, the pyrochar spectra (Figure 8.2f, l, and r) showed that the pyrochar was predominantly aromatic, with only very small peaks in the alkyl region (0–48 ppm). The dipolar-dephased spectrum of the pyrochar showed a very pronounced signal from nonprotonated aromatic carbons with signals from the mobile —(CH$_2$)$_n$— and CCH$_3$ components. The absence of carbohydrates in pyrochar was evident with the complete disappearance of the signals assigned to anomeric O—C—O carbons around 103 ppm in the chemical shift anisotropy (CSA)-filtered spectrum.

The ^{1}H–^{13}C long-range recoupled dipolar dephasing experiments infer the information about the size of aromatic fused rings. The larger the ^{1}H–^{13}C distance, the slower the dephasing of the ^{13}C signal and the larger the aromatic cluster size. Figure 8.3 shows the dephasing times of HTC-swine A ranging from 0.29 to 0.86 ms, while those of pyrochar

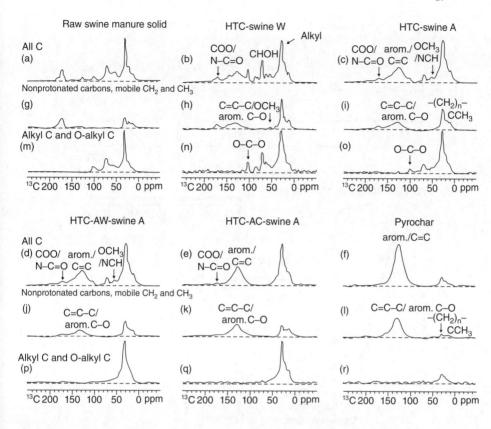

Figure 8.2 ^{13}C NMR spectral editing for identification of functional groups in raw swine manure solid (a, g, and m), HTC-swine W (b, h, and n), HTC-swine A (c, i, and o), HTC-AW-swine A (d, j, and p), HTC-AC-swine A (e, k, and q), and pyrochar (f, l, and r). (a–f) Unselective cross-polarization (CP)/total sideband suppression (TOSS) spectra. (g–l) Corresponding dipolar-dephased CP/TOSS spectra. (m–r) Selection of sp^3-hybridized carbon signals by a ^{13}C CSA filter. (Reproduced with permission from [65]. © 2011 American Chemical Society.)

ranging from 0.29 to 1.43 ms. Comparison of the dephasing curves of HTC-swine A, pyrochar, and lignin suggested the presence of fused or more substituted aromatic rings in both hydrochar and pyrochar than lignin, but a more condensed character of aromatics in pyrochar than hydrochar (Figure 8.3c).

The quantitative structural compositions of the carbons of the swine chars are shown in Table 8.6. These chars were produced from different pyrolysis and postprocessing conditions based on ^{13}C direct polarization/magic-angle spinning NMR experiments. The dominant structural component in the raw swine manure was alkyl (62.7%), followed by O-alkyl (12.6% excluding O—CH$_3$), COO/N—C=O (11.0%), and NCH (5.7%) groups. These functionalities decreased as the swine manure was carbonized, but aromatic components increased. Raw swine manure underwent substantially deeper carbonization during

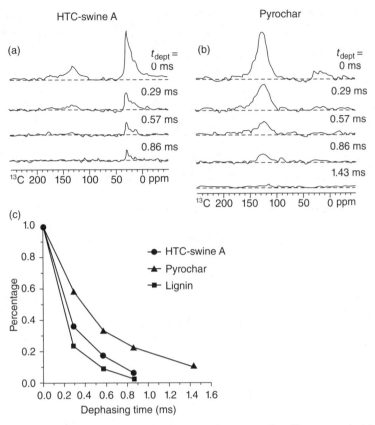

Figure 8.3 *Series of direct polarization/TOSS spectra after $^1H-^{13}C$ recoupled long-range dipolar dephasing of the indicated durations t_{deph} of (a) HTC-swine A and (b) pyrochar. (c) Long-range dipolar dephasing curves for HTC-swine A and pyrochar. The aromatic signals were integrated between 107 and 142 ppm. Circles: HTC-swine A; triangles: pyrochar; squares: lignin. For reference, data for lignin are also shown. The data points have been corrected for regular T_2 relaxation. (Reproduced with permission from [65]. © 2011 American Chemical Society.)*

pyrolysis than HTC as indicated by higher aromatics and low alkyl groups in pyrochar. In pyrochar, aromatic carbons (82.0%) became the predominant components, with the remainder being mostly alkyl hydrocarbons (11.0%). As observed by Ro *et al.* [59], nitrogen from peptides/proteins (NCH and COO/N—C=O groups) was substantially lost during swine manure pyrolysis.

8.2.3   Potential Hydrochar Uses

A variety of applications for both pyro- and hydrochars have been proposed in numerous fields, ranging from innovative materials to renewable energy. This diversity stems partly from the wide range of researchers and companies currently exploring the potential of char

Table 8.6 Quantitative structural information obtained from ^{13}C NMR of raw swine manure and its chars. (Adapted with permission from [65]. © 2011 American Chemical Society.)

	ppm								
	190–220	165–190	165–145	112–145		112–60	60–48		48–0
Sample	Aldehyde ketone	COO/N–C=O	Aromatic C–O	Nonprotonated aromatic/ olefinic C	Protonated aromatic/ olefinic C	O-alkyl C	NCH	O-CH₃	Alkyl
Raw swine manure	1.1	11.0	0.5	3.2	1.7	12.6	5.9	1.2	62.7
HTC-swine A[a]	1.8	5.6	5.2	14.1	7.0	1.8	1.8	0	62.6
Pyrochar	0.9	2.8	6.4	54.9	20.7	2.2	1.1	0	11.0

[a]Hydrochar washed with acetone.

applications as well as from its flexibility. The advantages of hydrochar use over pyrochar are that heterogeneous wet organic residues and waste streams can be processed without preliminary separating and drying, and with the HTC process there is a greater flexibility to design innovative carbon materials to fulfill a specific function. The structure, size, and functionality of the hydrochar can be varied by changing the carbonization time, feedstock type, and concentration, as well as by using additives and stabilizers. For example, the type of functional groups on the hydrochar can be modified by the addition of certain compounds to make the materials more hydrophilic and highly dispersible in water [66, 67].

The potential uses of hydrochar as an innovative material range from bulk applications such as adsorbents, especially for water purification purposes or soil amendments for the improvement of soil characteristics and plant growth, to specialty applications such as catalysts in chemical processing or for energy storage in batteries. Further uses for hydrochar are in energy production as a substitute for fossil fuels in conventional combustion processes or in novel fuel cells and engines. The possibility of using hydrochar as a method for carbon sequestration is also being investigated: the transformation of easily biodegradable waste streams to stable, more recalcitrant hydrochars may offer a means to store carbon, either in soil amendments (i.e., carbon storage by increase of the soil carbon pool size) or in storage vessels.

8.2.3.1   Adsorbents

In order to increase their capacity to adsorb compounds, chars are normally subjected to an activation step, thus becoming "activated carbon." The sorption properties of activated carbons in general are extremely versatile, and can be used in production and environmental protection processes to remove a variety of inorganic and organic constituents from water or gases. Chars are effective in removing heavy metals [68], arsenates [69], organic dyes [70], as well as many other toxic substances [71].

Chars made from carbonizing biomass and animal manures have been found to have high surface area and sorption capacity after activation. Lima *et al.* reported that the poultry- and turkey-based activated pyrochars had greater sorption capacity toward copper than commercial activated carbon [72–74]. Recently, Sun *et al.* [14] found that hydrochars made from hydrothermally carbonizing chicken litter and swine manure showed more effective sorption capacity for endocrine-disrupting chemicals than pyrochars. Hydrochars' logarithmic single-point organic carbon normalized distribution coefficients ($\log K_{OC}$) of bisphenol A and 17α-ethinyl estradiol were higher than that of pyrochars. For nonpolar compounds such as phenanthrene, sorption capacities of both hydrochar and pyrochar were similar, demonstrating that manure-based hydrochar could adsorb a wide spectrum of both polar and nonpolar organic contaminants. Further research on this aspect of hydrochar is currently pursued by many researchers globally and more detailed information about the potential for using hydrochar as an environmental adsorbent will be unraveled in the near future.

8.2.3.2   Energy Source

HTC of waste streams has emerged as a potential alternative strategy to produce a renewable solid fuel source. Pressure to reduce greenhouse gas emissions has prompted the re-evaluation of current approaches and spurred the development of hydrochar production as a source of alternative, renewable energy. Currently, less than 1% of total energy generation in the United States originates from waste sources (including biogenic MSW, landfill gas, sludge waste, agricultural byproducts, and other biomass solids, liquids, and gases [18]), far below its total energy generation capacity. It should be noted that this is not necessarily the case in other countries.

An important advantage of the HTC process is that it produces a homogenized and stabilized char from highly heterogeneous organic waste streams, thus improving the handling, transportation, and storage and increasing its potential as renewable energy source. Ultimately, it is expected that the char may be used as a solid fuel source and be combusted. Muthuraman *et al.* [28] found an additional advantage – the blending of hydrochar produced from MSW and Indian coal resulted in significant reduction in coal ignition temperature.

Many of the experiments evaluating the conversion of waste streams via HTC have focused on evaluating the energy-related properties of the hydrochar (see Table 8.3). Through the hydrothermal reactions, the energy density of the wastes is increased up to 40%. Hydrochar has an energy density equivalent to lignite coals or higher, ranging from 15 to 30 MJ kg^{-1}, and it increases with reaction severity [9, 12, 27, 75]. Berge *et al.* [9] and Ramke *et al.* [12] report that the hydrochar energy content correlates well with carbon content of the solids.

Berge [15] conducted an analysis between the energy potentially generated from the char resulting from the carbonization of paper, food waste, and model mixed MSW and that potentially generated from currently used waste management processes (landfilling and incineration), and reports that there is the potential for waste carbonization to surpass the energy generation from current waste management processes. Note that if using the hydrochar for energy generation, any carbon sequestered will be released.

A further appealing aspect of the HTC process is the ability to influence the size, form, and properties of the char by varying process conditions, opening up new possibilities in energy production. Instead of being restricted to combustion in conventional furnaces, colloidal hydrochar can be used in fuel cells as a fuel [76, 77], or its material properties can be exploited as a solid phase for hydrogen storage [78] or as a catalyst in low-temperature fuel cells to enhance reaction rates [79].

8.3 Soil Application

8.3.1 History of the Idea to Sequester Carbon in Soils Using Chars/Coals

The idea to apply recalcitrant carbonized materials to soils to increase soil fertility was mainly born from research on Amazon dark earth (ADE) soils (also known as "terra preta"), building on the work of Wim Sombroek [80, 81]; it was him and his coworkers' dream to replace current CO$_2$-emitting slash-and-burn agriculture by slash-and-char practices to create terra preta *nova* sites for sustainable forest farming and forest protection [82–84]. In the face of accelerating rainforest destruction, such approaches for sustainable development (even building on ancient traditions) should urgently be pursued, perhaps via carbon-sequestration- and rainforest-protection-targeted development aid, provided by the industrialized nations. As it happened, the time suddenly seemed to be ripe for this certain idea. Thus, the idea to use charcoal, or "biochar" (which is commonly used to denote pyrochar in soil applications) as a means of sequestering carbon in soils, previously taken from the atmosphere via photosynthesis in soils, was also put forward in the early 1990s by Ogawa [85] and discussed in more detail by Okrimori *et al.* [86]. In the best case, stable-carbon application should be associated with beneficial agricultural effects as exemplified by the terra preta soils [81, 87] as well as in Asian cultures where charcoal has been traditionally used in soil substrate production (e.g., charcoal *bokashi* in Japan [88]).

Until a few decades ago, it was unclear if the darker, nutrient-rich, and fertile ADE soils in Brazil were of natural or anthropogenic origin; now it is clear beyond doubt that they developed due to human activities [89–91]. The exact method or purpose for their creation is, and likely will, largely remain unknown, but more and more places with ADE soils and archeological evidence of urban garden settlements have been found at the shores of the rivers of the Amazon basin [91–93], as well as other signs of ancient geoengineering in central America (e.g., [94]).

It is now recognized that these indigenous populations that (purposefully or unintended) created these soils could not have been hunter-gatherers, but rather settled cultures (e.g., employing pottery [89]), and that they must have been many [91, 95, 96]. Thus, the original testimony, given by the Dominican friar Gaspar de Carvajal who accompanied the expeditions of the Spanish Conquistador Francisco de Orellana, seems to be true. For about 450 years, their expedition reports on densely populated river banks in the Amazon basin were discarded as tell-tale or fairy-tale stories.

The recent frequent findings of ADE sites, rich in human waste such as pot shards, fish bone, and signs for human feces [91], provide a late rehabilitation of part of their reports. They indicate that indeed perhaps a million people or more lived there and that what we see as a "native" jungle might instead be 470-year-old secondary forest [93, 96]. However, it still remains a mystery why these cultures vanished, why missionaries following a few decades after de Orellana did not find people there (giving him the "story-teller" reputation). It is assumed that the first Europeans were also the last to see them, because they probably brought deadly infectious diseases to these ancient cultures [96]. If this is true, it is hardly bearable to imagine what tragedies must have taken place on the shores of the rivers in the Amazon basin.

In essence, the long-term stability of old, aged black carbon or biochar is exemplified in terra preta soils that contain considerable amounts of black carbon, most of which is 500–2000 years old, sometimes up to 7000 years (^{14}C dating [89–91, 97]). Considerable biochar stocks of 50 tons of C ha^{-1} have been found down to 1 m depth in such soils, despite the climatic conditions strongly favoring decomposition and despite the lack of additional biochar additions in the last 450 years [82, 90, 91]. Indeed, the oldest carbon pool in ecosystems prone to natural fire is comprised of black carbon/charcoal [98, 99]. Lehmann *et al.* [99] conservatively assessed mean residence times (MRTs) for naturally generated biochars in Australian woodlands to be 1300–2600 years (range: 718–9259 years). The MRT of charcoal buried in sea sediments or anaerobic places may be even higher [100]. Although the term "long-term stability" clearly needs to be explored experimentally to produce more precise numbers, it is obvious that a wider use of such recalcitrant carbon fractions in soils will increase the carbon pool size, thus providing a sink for atmospheric CO_2. However, there is always a long way from a brilliant idea into frequently applied use in reality – a way that today is still paved with many unknowns.

8.3.2 Consideration of Hydrochar Use in Soils

While knowledge on biochar use in soils slowly starts to accumulate, knowledge on hydrochar use in soils is in its infancy compared to biochar [11]. Before we visit the first available results and draw conclusions, the following points need to be taken into consideration. Any hydrochar (as well as biochar) application to soils aimed at achieving soil carbon sequestration can only claim success if beneficial effects on plant yield, soil water availability, soil fertility, or other positive amelioration effects can be shown. In addition, it must be kept in mind that hydrochar (and also biochar) has a high energetic value: both carbon products were designed for use as fossil fuel substitutes, namely for burning in the first place (even historically, where charcoal was produced for ore smelting); hence they will have a monetary value that will be tied to fuel prices. Thus, as long as "soil carbon sequestration" is not paid for, soil amendment will only take place when the created value associated with soil amendment *is greater than that of the "fuel hydrochar."* If this value only arises in the long-term (several years, decades) or starts to be recognized as a public interest (e.g., restoration of degraded soils, remediation of contaminated soils) then soil application will depend on our societies' (political) will to *give* it the same or even a greater monetary value than it would have as a fuel. Today, no prices tied to agricultural use exist: knowledge on beneficial use (where a certain effect can be guaranteed, with a specific char, for a specific soil problem) is still incomplete. In addition, large suppliers of biochar or hydrochar do not exist in most countries and legal regulations on char use in soils also do not yet exist.

To consider the agricultural use of hydrochar in particular with regard to carbon sequestration, several open questions must be explored experimentally: (i) Is hydrochar stable (i.e., recalcitrant and resistant to degradation)? Does it make sense to use hydrochar for carbon sequestration if it is less stable than pyrogenic biochar? (ii) Does it have, or can it at least *theoretically* have, positive effects on plant growth and soil fertility or on nutrient retention in soils (beyond waste nutrient recycling)? Are there problems or toxic effects associated with its soil use? (iii) What are the effects of hydrochar application to soils on the greenhouse gas balance of crop production? This requires knowledge on not only direct changes in the emission or consumption of stable greenhouse gases (N_2O and CH_4) from/into agricultural soils after hydrochar application, but also of changes in the initial stock of the soil organic carbon (SOC) pool. These questions are discussed in subsequent sections.

The most important question, however, cannot be answered experimentally but rather by consensual agreement: *What reference system will we use for comparison to obtain a true value for carbon sequestration via hydrochar application?* Any reference system must be chosen carefully, it must be realistic and conservative, and it must be based on well-established scientific evidence, to ensure that real-world soil carbon pools are increased in the end. Otherwise, a useless but costly carbon sequestration trading business might spring up that sequesters money in some pockets, but not stable carbon in real-world soils.

8.3.3 Stability of Hydrochar in Soils

8.3.3.1 Carbon Sequestration by Chars

The suitability of chars for carbon sequestration in soils will depend on several factors that are not well known to date: (i) the overall carbon balance of the production process, (ii) the chars' long-term stability, (iii) secondary carbon turnover effects, such as an increase in the photosynthetic carbon input (e.g., biomass growth increases), or losses of old SOC by priming, and finally (iv) on the chars' effect on the fluxes of other climate-relevant greenhouse gases such as N_2O or CH_4 [101], which should be included as CO_2-equivalents, calculated based on their 100-year global warming potential [102].

The carbon yield (i.e., the amount of biomass carbon that be converted to char) can vary considerably depending on the carbonization technique and its operation, as detailed by others (e.g., [11]). Commonly in dry pyrolysis, 50% or more of the biomass carbon is converted to bio-oils, tars, H_2 and CO (syngas), or CO_2 [87, 103], while in the HTC process 75–90% of the biomass carbon may remain in the end-product hydrochar [11, 104]. Thus, the "carbon-yield" of hydrochar is usually larger than that of dry pyrolysis. However, this will only be an advantage if hydrochar is as stable as biochar or, at least, not considerably less stable. In addition, the time horizon used for consideration is crucial. So far, no clear, common definition exists: Do we consider the carbon to be "sequestered" (i.e., withdrawn from the atmosphere) when it remains in a given soil or at a given storage location for 50 years, for 100 years, for 1000 years – or "forever" (what ever that means)? If carbon is considered to be "sequestered" in wooden furniture such as a wardrobe we are surely considering a time horizon that is completely different from the situation where carbon is sequestered in charcoal and buried anaerobically in sea sediments [100]. Three orders of magnitude or more may lie between these two forms of sequestration (e.g., MRTs of 10 versus 10 000 years); but the former, "carbon sequestration in woody structures," figures far more prominently in public discussion than the latter.

For char application to soils to be included in carbon-trading schemes, reliable factors are needed that enable the assessment of sequestered CO_2 equivalents with reliable (at least centennial) mean residence times. Biochar or hydrochar properties, soil, climatic, and management conditions may vary widely, but all will influence the chars' recalcitrance. A systematic analytical characterization of biochar/hydrochar properties is urgently demanded, but has not yet been established [105]. Some general deductions can be made from chemical and physical char properties towards its long-term stability, such as aromaticity or the H/C versus O/C atomic ratios of a char (i.e., its place within the van Krevelen diagram) [106, 107].

8.3.3.2 Decomposition, Mineralization, and Fate of Chars in Soils

Although biochar is often considered to be "inert," it will finally be decomposed and mineralized over sufficiently long time scales; otherwise the worlds' carbon stocks would finally end up in biochar [101, 106]. Nevertheless, as discussed above, biochar contributes the longest-living organic carbon pool in soils [108–110].

Hydrochars are comparable to lignite (or brown) coal with less aromatic structures when sorted into a van Krevelen diagram (see Figure 8.4). The process water of hydrochar production contains high fractions of labile dissolved organic carbon (i.e., see BOD and COD values Table 8.3) [9, 11]. Even when the hydrochar slurry is pressed or otherwise dried after production, the solids will contain large labile fractions, either because of some process water dried onto the char or because the char itself contains labile fractions. In a study where hydrochar/soil mixtures were incubated, they produced substantially higher CO_2 emissions than the control soil without amendment or than soil/biochar mixtures [111]. In another study with plants, the HTC-amended soil showed nearly the same CO_2 emissions than the

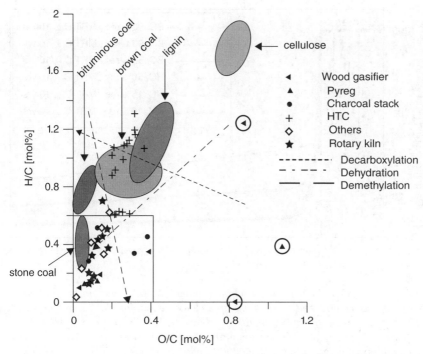

Figure 8.4 *van Krevelen diagram comparing atomic ratios of biochar and hydrochar in comparison to other carbonaceous organic materials such as cellulose, lignin, and brown, bituminous, and stone coal. The square indicates the range of desirable char properties with regard to carbon stability as a soil amendment. The lines for decarboxylation, dehydration, and demethylation are given according to van Krevelen and Schuyer [114]. (Reprinted with permission from [107]. © 2012 Journal of Environmental Quality.)*

feedstock-amended soil (feedstock: *Miscanthus x giganteus* chips), while the CO_2 emissions from *Miscanthus*-biochar-amended soil did not differ from a control without amendment [112]. It is thus reasonable to assume (as long as the experimental evidence is limited) that hydrochar, and here in particular the initial available labile carbon, will decompose quicker than a biochar produced from the same material. However, hydrochar will likely decompose slower than uncarbonized material, as found in a 12-month incubation study using wheat straw, bark-HTC, charcoal and a low-temperature conversion char [113]; in this study, Qayyum *et al.* [113] used three different soils and employed compensation fertilization with mineral nitrogen to obtain the same total nitrogen supply to each soil at the start.

A perception of the overall carbon loss during production versus char stability compared to uncarbonized biogenic residues is provided in Figure 8.5. A two-component decay model was used where a labile carbon fraction with a MRT of 10 years was assumed, as well as a recalcitrant fraction with a MRT of 1000 years. The MRT is the average time that an initial carbon addition is present in soil and is the inverse of the decay rate coefficient k. The decay rate of SOC to CO_2 is frequently modeled by assuming an exponential decay.

$$C_{\text{left}} = C_{\text{lab}} * e^{(-1/k-\text{lab} * t)} + C_{\text{rec}} * e^{(-1/k-\text{lab} * t)} \qquad (8.1)$$

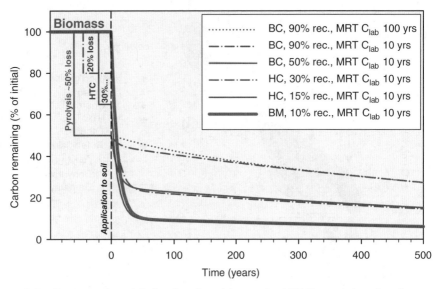

Figure 8.5 *Conceptual model of carbon from biomass (= 100%) remaining after direct application, or carbonization and subsequent application, assuming 50% remaining carbon in biochar after pyrolysis or 65–80% remaining carbon after HTC. BC = biochar; HC = hydrochar; BM = biomass; rec. = recalcitrant; for further explanations see text. (Modified from and reprinted with permission from [101]. © 2009 Taylor & Francis.)*

where C_{left} is the proportion of initial carbon left in soil after a certain time has elapsed, C_{lab} is the labile carbon pool with a MRT of 10 years (only dotted line in Figure 8.5: 100 years), C_{rec} is the recalcitrant carbon pool with a MRT of 1000 years, k is the decay constant of the labile (k-lab) or recalcitrant (k-rec) carbon pool, and t is the time after soil application.

Although the assumptions behind this approach (Eq. 8.1) will likely be correct for biochar (which consists mainly of recalcitrant material) and for biomass (consisting mostly of labile carbon materials) this may not be entirely true for hydrochar where we do not yet have a good perception of the decay behavior of its different fractions. Nevertheless, for the purpose of comparison with biochar or with direct application of uncharred biomass we assumed, as an upper boundary ("best case"), a carbon recovery of 80% and a recalcitrant fraction of 30% (which may be rather high, given the low average black carbon content of 21 hydrochars of around 5% in the study of Schimmelpfennig and Glaser [107]). As a lower boundary ("worst case"), we choose a low recalcitrant fraction only slightly above that of biomass and a larger carbon loss during the production process with only 65% carbon recovery.

Figure 8.5 reveals several points. (i) More carbon remains in soil if a char possesses a sufficiently large fraction of long-term stable, recalcitrant carbon; here biochar with its greater proportion will likely result in the larger amount of carbon sequestered over centennial time scales. (ii) Given that a hydrochar has a notable recalcitrant fraction it may result in the same carbon sequestered after about 50 years have passed due to the trade-off between its larger labile proportion, but higher initial carbon recovery. (iii) A hydrochar with a lower carbon recovery and low proportion of recalcitrant carbon will not be much different from uncarbonized biomass in its long-term carbon sequestration proportions. As already pointed out by Lehmann *et al.* [101], it becomes obvious here that it is the recalcitrant fraction that really matters. Modifying the MRT of the labile pool (e.g., for the most stable biochar in Figure 8.5) did not make much of a difference for the long-term carbon sequestration function.

Also, the incubation time in a given study will matter. While Kuzyakov *et al.* [108] calculated MRTs of about 2000 years from their 3.2-year incubation study in the lab with ^{14}C-labeled biochar, Steinbeiss *et al.* [115] reported MRTs between 4 and 29 years for two ^{13}C-labeled hydrochars made from glucose (without nitrogen) and yeast (5% nitrogen) in a 4-month incubation study. However, studies with durations of less than 100–500 years (which means essentially all experimental studies) will very likely underestimate the "true" MRT. The shorter the study, the stronger the underestimation will be. This is illustrated in Figure 8.6 for hypothetical data, where the decay of a char is again approximated with a double-exponential decay curve. Using only the first 2 years of a 100-year dataset, Lehmann *et al.* [101] calculated the MRT of the recalcitrant fraction to be 57 years. In contrast, using the entire 100-year dataset resulted in a MRT of 2307 years. Overestimation of char degradability may therefore occur when evaluating data from shorter incubation times (e.g., smaller datasets). For the reliable estimation of MRTs, long-term incubation studies and knowledge on the decay kinetics of the various carbonaceous fractions (e.g., labile, recalcitrant) are required for hydrochars. Furthermore, effects on other properties need to be considered as well. For example, with biochar, decomposition of the labile carbon substances on the surfaces of the chars and surface oxidation lead to increases in the cation exchange capacity [116, 117].

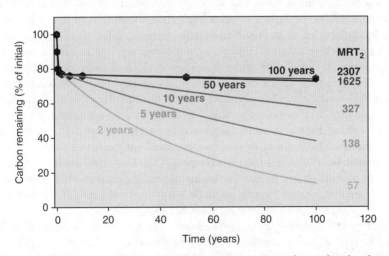

Figure 8.6 *Double-exponential decay model (see Figure 8.5 and text) fitted to hypothetical data of char decay after 0.1, 0.5, 2, 5, 10, 50, or 100 years, assuming data availability for either the first 2, 5, 10, 50, or 100 years. MRT$_2$ = mean residence time of the recalcitrant carbon pool of the char, derived from k-rec (see text) and given in years. (Reprinted with permission from [101]. © 2009 Taylor & Francis.)*

Since it will be hard to carry out studies with durations of at least a decade or more, we propose the following, probably more feasible approach, which may be combined with functional group and structure analyses. Incubations should be performed over 1 year where every "ingredient" for fast biological degradation of any degradable carbon in the hydrochar is provided or imposed upon the char. Mixing with the most fertile soils, incubation at high temperatures and optimum soil moistures, application of, for example, wood-rotting, lignin-degrading fungi (see below), "priming" (comineralization) by application of easily degradable carbon substrates, optimal nutrient supply, planting of mycorrhizal fast-growing plants, and drying/rewetting as well as freezing/thawing cycles; several such treatments should be combined to obtain the highest possible degradation rate in the shortest time feasible. Those combined treatments that produce the highest char carbon loss in the same char may be used later in standardized incubations to characterize various chars for the biological recalcitrance of the labile to moderately stable carbon fraction. Subsequently, maximum decomposition rates should be compared to the chemical characteristics of the chars to obtain a "biological stability" regression with the associated char characteristics (related to feedstock and hydrochar production parameters). Measuring decomposition rates in such a manner on hydrochars may serve as a first proxy to obtain the MRTs of several hydrochar carbon fractions with different stability, characterizing their size and MRT. Chemical characterization of the remaining char carbon fractions before and after rigorous biological attack may provide further insights into the magnitude and structure of its long-term recalcitrant fraction. Even now, without a wealth of data, it can be assumed from the chemical characteristics that hydrochars produced at higher temperatures and pressures will be more stable than weakly carbonized hydrochars made from the same material [107].

However, decomposition of char is not the only pathway of "loss" when one aims at determining the amount of remaining char in a given soil horizon over time. Char may be lost from soils due to surface erosion, run-off, and transport to the subsoil (via rain water as small particles or as dissolved organic carbon). Other potential mechanisms of char intrusion to deeper subsurface layers include bioturbation (e.g., earthworms), kryoturbation (i.e., mixture and movement by frost/thaw cycles), or anthropogenic management [106, 118, 119]. Carbon losses due to erosion or transport to deeper soil layers can be significant and occur over relatively short periods of time. Major *et al.* [118] reported a migration rate of 379 kg carbon ha^{-1} $year^{-1}$ from the biochar application of 116 ton ha^{-1} to the top 10 cm of a grassland soil. The char migrated downward 5–20 cm over a 2-year study in Columbia. Respiratory or dissolution-related carbon losses were reported to be small in comparison (2.2% and 1%, respectively [118]). Due to the above-mentioned characteristics of hydrochar, and its less-particulate nature, the proportions might be reversed: we hypothesize that a larger fraction might be lost by mineralization in the first years, but that the downward migration may be considerably lower.

Char particles may shatter or fractionate into smaller particles via different physical or biological forces, such as freeze/thaw or swelling/shrinking of clay minerals, or in-growth of plant roots and fungal hyphae. Changes in particle size expose additional surface area that may increase char oxidization and/or degradation. In old charcoal-containing soils, biochar particles are very small (e.g., terra preta or chernozems); most of the biochar is included within microaggregates where it seems to be protected from further decomposition [98, 120]. Addition of labile carbon substrates has been reported to increase biochar decomposition; however, the extent was small [101, 106, 108]. White-rot fungi (usually decomposing lignin) or other basidiomycetes are able to slowly decompose lignite, sub-bituminous coal, or biochar via excretion of exoenzymes, such as laccase, mangane peroxidase, or phenol oxidase [100, 121]. A nitrogen-rich hydrochar was preferentially decomposed by fungi [115]. Rillig *et al.* [122] recently reported a considerably stimulated root colonization of arbuscular mycorrhizal fungi in mixtures of up to 20 vol% beet-root hydrochar chips. We observed in incubation studies that various fungi grew quickly on a mixture of brown earth with 8% (weight) hydrochar (sugar beet or bark) until they formed a dense, hydrophobic white layer on the top after 3–4 weeks at 22 °C [111]. It is unknown if fungal decomposition will continue in the field when the labile fraction of the hydrochar has been mineralized. The dense fungal mycelia and fruiting bodies were gone after 150 days of incubation [111]; the same was observed in other experiments with another hydrochar in the presence of plants [112, 123]. In the field it will likely depend on the presence of easily degradable carbon if a hydrochar (in particular, one that is nutrient-rich) will be quickly mineralized. Thus, degradation may be more pronounced over the years in the presence of (mycorrhizal) plants.

8.3.3.3   Carbon Sequestration Potential: Soil Carbon Priming or Buildup?

An intriguing find in terra preta soils is that the SOC stocks are higher than in adjacent Ferralsols, even disregarding the black carbon [90]. In contrast, Wardle *et al.* [124] reported an increased loss of organic matter over 10 years from mesh bags that contained charcoal compared to those without charcoal. Such a finding in a boreal forest soil may be explained by the promotion of nitrification, which has been frequently observed in boreal forest soils when charcoal is added (after forest fire or burnings, e.g., [125, 126]). Thus, addition of biochar or hydrochar to soils or litter layers must be carefully investigated with regard to possible priming effects that endanger the existing old soil carbon pool. However, so far there is not much evidence, at least for biochar, that it will promote old SOC decomposition. An exception may be forest soils where inhibitory phenolic compounds are adsorbed and hence the microbial activity is stimulated (such as the above-mentioned needle-litter dominated boreal forest soils). Moreover, in terra preta soils the increased SOC contents (*in addition* to biochar) do not indicate a long-term SOC loss due to biochar application [90, 98, 120].

Biochar and hydrochar seem to promote fungal growth (e.g., that of arbuscular mycorrhiza) [122, 127]. These fungi produce the protein glomalin which, as a binding agent, significantly promotes soil aggregation [128, 129]. Hence, both chars may in the long-term increase the protection of nonbiochar SOC by increasing soil aggregation through fungal promotion [122] or by formation of organomineral complexes and aggregates [101]. On the other hand, the labile, easily degradable carbon in hydrochar may present a risk if it leads to a priming effect on the old SOC. This aspect needs to be investigated.

Another possible way to cause an indirect positive effect on the soil carbon pool by char application is to increase the net primary productivity. If the net primary productivity increases the root biomass and turnover will increase, the litter carbon input will also rise and result in a SOC buildup in addition to the char carbon input. The annually repeated inputs of organic material (residues, litter, i.e., plant growth) will subsequently be mineralized and transferred into stable humus fractions. Alternatively, the presence of the char particles may provide "condensation nuclei" for the buildup of *a larger proportion* of more stable humus fractions. This would also promote the increase of the total SOC pool besides char-applied carbon. However, both mechanisms may also be working in the opposite direction. With biochar, it is likely that it may have positive effects on the soil carbon buildup besides the applied char carbon (see above [90]). With hydrochar, however, there is simply not enough knowledge available to make predictions.

8.3.4   Influence of Hydrochar on Soil Fertility and Crop Yields

8.3.4.1   Change of Soil Characteristics with Hydrochar

As mentioned before, the use of hydrochar in soils has seldom been investigated so far; the idea of using hydrochar as a soil amendment has its origin in biochar research and in the nanoscale properties of the hydrochar product [13, 11, 130–134].

Biochar application to soils has been shown to have a beneficial effect on soil physico-chemical properties. Biochar presence may (i) enhance the water-holding capacity (WHC) [111, 135–137]), aeration, and hydraulic conductivity of soils [82, 135, 138], (ii) reduce the tensile strength of hard-setting soils [139, 140], (iii) increase the cation-exchange capacity of soils [82, 117, 141], resulting in an improved nutrient retention or higher nutrient use efficiency [142–144], (iv) in combination with increases in pH values [90, 140], and (v) stimulate growth, activity, and the metabolic efficiency of the microbial biomass [145, 146], including (vi) arbuscular mycorrhiza [122, 127, 147] and (vii) nitrogen-fixing rhizobiota [148], and it could also (viii) attract beneficial earthworm activity [140, 149–152].

Hydrochar will affect soil properties based on the same principles. It will very likely reduce the tensile strength, increase the hydraulic conductivity at least in some soils, and enhance the soil WHC due to its microstructure [130–133]. Hydrochars will likely not have the very large, internal surfaces that biochars have when produced at temperatures above 450 °C since hydrochars are produced at lower temperatures [153, 154]. Usually, typical surface areas measured via the Brunauer, Emmett, and Teller (BET) method range between 3 and 30 m^2 g^{-1}. Pyrolyzing the same feedstock at temperatures above 420–430 °C can result in much higher surfaces ranging from 100 to 800 m^2 g^{-1} (or even up to 1700 m^2 g^{-1} with activated charcoal), depending on the maximum pyrolysis temperature, production process, and feedstock [107]. Therefore, water retention curves of hydrochar/soil mixtures may be different to biochar/soil mixtures so that the effect of hydrochar on plant growth (via the "water supply" pathway) will probably resemble that of peat or compost additions to soils.

Measurements of the WHC of hydrochar after production and after two types of post-processing (pressing and drying) showed that drying considerably reduced the WHC, but not pressing the hydrochar out [11]. Some hydrochars become hydrophobic when oven- or completely air-dried; we observed this in particular in planting experiments when dry soil/hydrochar mixtures were wetted for the first time. Theoretically, the presence of a greater number of carboxyl groups in production-fresh hydrochars than in biochars should result in hydrophilic behavior and give hydrochar a larger cation-exchange capacity directly after production [107, 130, 132, 133]. It is unknown to date but likely that hydrochars will undergo ageing processes similar to biochars, where the number of functional groups on the biochar surfaces increases over time [116, 117].

Since the pH of production-fresh hydrochars is rather acidic [9, 107, 131, 155] this may have beneficial effects, in particular in alkaline soil where certain micronutrients are immobilized due to the high pH, while it may have less beneficial effects in acidic soils.

8.3.4.2 Hydrochar, Soil Fertility, and Crop Yield

Biochar can lose some of the initial feedstock nutrients during pyrolysis (e.g., nitrogen via volatilization as NO_x) or the nutrients may become unavailable to plants by inclusion into aromatic stable structures [156, 157]. In contrast, hydrochar, with its lower production temperature, will retain more nutrients in a plant-available form (depending on the feedstock), either in the hydrochar itself or in the aqueous phase. The conservation of nutrients from waste materials and sludge for agricultural use thus may, in the face of problems such as the declining phosphate deposits worldwide, become one of the most interesting features of hydrochar. Sustainable nutrient recycling may in the end govern hydrochar use in agriculture. However, the current experimental evidence is very limited. In the following, we thus give a brief comparison to biochar with theoretical considerations for hydrochar, followed by the available (sparse) evidence.

It is well documented that biochar application can improve crop yields (reviewed in [158]). However, yield increases depend on the combined biochar/soil properties, absolute amounts of biochar and on the concomitant nitrogen application (nitrogen form as well as amount [150, 157, 159, 160]). Analysis of recent literature [11] has shown that biochar application can increase yields (i) in degraded or low-fertility soils, rather than at already-fertile sites [137, 139, 160, 161], (ii) in weathered tropical rather than in temperate soils ([83, 160] versus [149, 157]), (iii) in combination with NPK fertilizers or nutrient-releasing substances (composts, manures) [83, 139, 150, 161], or (iv) when the chars themselves were sources of nutrients (e.g., biochar from poultry litter [140]). It should be noted, however, that nutrient supply, pH, and other soil parameter changes are not always sufficient to fully explain observed biochar effects [157, 160].

In the first hydrochar field study undertaken in Germany [162, 163], labile carbon in the hydrochar was found to initially induce nitrogen deficiency by nitrogen immobilization. The authors conducted a full factorial field study where they applied 10 ton ha^{-1} of two different hydrochars (from beer draff/spent grain and sugar beet pulp, with C/N ratios of 16 and 38, respectively), combined with four levels of nitrogen fertilization (0, 50, 100, and 150 kg N ha^{-1}). The soil was a fertile Haplic/Stagnic Luvisol near Göttingen, Germany. The authors reported that at this hydrochar application level, no significant effect on the soil physicochemical properties such as field capacity, pneumatic conductivity, bulk density, or water-stable aggregation was found. (A concomitantly included compost treatment with 10 ton ha^{-1} also did not have any effect on these parameters.) The results of Gajic and Koch [163] with regard to sugar beet yield confirmed the above-mentioned hypothesis of nitrogen immobilization by the hydrochar: they did not observe germination reduction, but a significant growth reduction, in particular in the young seedlings. This was more pronounced with the hydrochar that had the larger C/N ratio (made from sugar beet pulp). Also, the relative sugar yield per hectare was reduced by hydrochar application. Moreover, the growth limitation was somewhat overcome by higher rates of nitrogen fertilization [163]. The authors reported a quick initial mineralization of both applied hydrochars: in the plots fertilized with 100 kg N ha^{-1} they assessed the CO_2 efflux via closed chambers and acidic traps. They detected a loss of more than 10% of the applied carbon within 2 months; the carbon loss was even significantly larger than that of the applied compost [162]. However, after about 3 months the CO_2 release from the hydrochar-amended plots slowed down to rates not different to the unamended control or the compost treatment

[162]. Thus, their results agree well with the available incubation studies reporting high initial mineralization rates [111, 113, 115]. Those initial high CO_2 efflux rates are the most likely reason for the observed nitrogen deficiency in growing seedlings (see below) due to a burst of microbial growth with the labile fraction in the hydrochar. It should be considered that during composting, a considerable fraction of such a labile initial biomass carbon fraction is also mineralized while in the HTC process, a higher proportion of the initial biomass carbon is retained in the solid HTC product. We therefore propose to use in future studies hydrochars that have been washed free of labile carbon compounds; it is plausible to assume that the strong nitrogen immobilization will be alleviated or absent without the labile carbon fraction.

However, there have been first reports of (phyto)toxic behavior of some hydrochars. Busch *et al.* [151] observed significantly negative effects in the cress germination test for phytotoxic gases (none of the seeds germinated), in the salad germination test (ISO-17126) with increasing amounts of char mixed into fine sand, and also in the barley germination and growth test where a sugar beet hydrochar was mixed in growing volumes with unfertilized standard peat substrate. However, in a subsequent regrowth of the harvested barley seedlings, the hydrochar-amended plants developed significantly larger biomasses than the control plants although there were less germinated plants present for regrowth [151]. In the earthworm avoidance test (ISO-17512), however, the animals significantly avoided the hydrochar-amended vessel side while they actively preferred the biochar-amended side in a concomitantly running test [151]. Rillig *et al.* [134] observed significant growth-reducing effects of hydrochar on *Taraxacum officinalis* plants. These initial problems with some hydrochars (all carbonized at temperatures and pressures around 200 °C and 2 MPa) have also been observed in other biotoxicity tests with different kinds of hydrochar, but not with all hydrochars that were ever tested [164]. In addition, the "nitrogen block" (likely microbial nitrogen immobilization) that is ameliorated by nitrogen fertilization has repeatedly been observed. Gajic and Koch [163] not only observed this in their field study, but also in a greenhouse experiment where the fine-ground hydrochar was more homogenously mixed into the soil. We observed the same [112] where the same feedstock was applied either as uncarbonized feedstock, as hydrochar, or as biochar: the growth of the radish plants in a fertile clayey loam was greatly retarded in the presence of feedstock or hydrochar (Figure 8.7).

In another experiment with radish, five plants were grown in 500-ml pots in a sandy loam. In this study, four biochars (from peanut hull, maize, wood chip sievings, and barbeque charcoal) or two hydrochars (bark and sugar beet) were mixed into the soil at 8% (w/w). The soil and the chars are the same as those described for experiment II in Kammann *et al.* [111]. The pots with the mixtures were brought to equal water holding capacities (60%) and were kept in the greenhouse at 22 °C and 12/12-h day/night cycles. Regular (every 2 days) weighing and watering was performed to keep the soils at optimal water-holding capacities (50–60% of the respective maximum of the mixture). While seeds germinated and grew normally in the control and biochar-amended soils, germination and growth was greatly retarded in the hydrochar-amended treatments; the radish plants remained small and yellow-leafed. Instead, we noticed the growth of several reddish small fruiting bodies of unknown fungi on the pot surfaces (hydrochar treatments only). Three weeks after seedling emergence the pots were fertilized (100 kg N ha^{-1}) with a full-compound fertilizer solution (same as in Kammann *et al.* [137]). From the fertilization date onwards, the young radish

Figure 8.7 *Growth of radish plants in a clay loam soil (SOC 3.5%, pH 6) amended with 5% (w/w; from left to right): nothing (control); feedstock, hydrochar, biochar, or a 1 : 1 hydrochar/biochar mixture. The retarded growth with the pure feedstock, pure hydrochar, and the 1 : 1 soil-mix containing hydrochar is easily visible.*

plants growing on the hydrochar-amended soils became green and showed a completely normal growth, confirming the observations of Gajic and Koch [163]. At harvest, the radish plants reached the same leaf chlorophyll content as the controls (an indication for the nitrogen supply; Figure 8.8b). However, they were not able to completely compensate the initial deficiency-caused delay until the final harvest (Figure 8.8a). The biochar treatments as well as one of the hydrochar treatments all significantly increased the water use efficiency (WUE) over the growth period of 6 weeks compared to the control (Figure 8.8b). It is reasonable to assume that the discrepancy between the biochar- and hydrochar-amended plants in their WHC is due to the initial "nitrogen block" observed with the hydrochar plants, similar to the observations of Gajic and Koch [163]. A higher initial application of nitrogen in the hydrochar treatments might have resulted in the same significantly higher water use efficiency with hydrochar than observed here with four different biochars (Figure 8.8b).

All hydrochar growth experiments described here used hydrochars with relatively high C/N ratios; hydrochars were moderately carbonized (around 200 °C) either in water or steam and did not contain large amounts of plant-available nutrients [162]. It must be urgently identified if the initial negative effects that seem to vanish or were alleviated after some time of char ageing are simply due to a kind of "nitrogen blocking" (which is also frequently observed when carbon-rich organic materials such as harvest residue or immature compost are incorporated into the soil) or if some toxic compounds may be present [151]. So far, we have not been able to identify such substances in analyses for polyaromatic hydrocarbons, heavy metals, polychlorinated biphenyls, or other potentially harmful persistent organic pollutants in any of the hydrochars that caused reduced germination in tests [164]. A more nutrient-rich hydrochar from manure or sludge waste streams might have entirely different effects. Theoretically, the chemical structure (carboxyl groups) and the greater nutrient

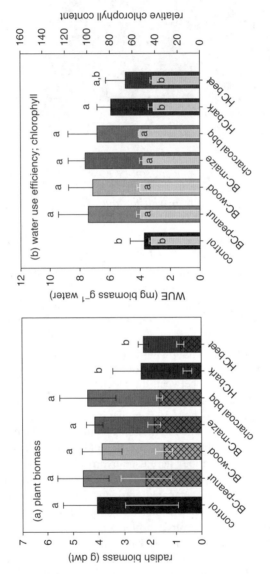

Figure 8.8 (a) Radish biomass yield of hypocotyl (cross-hatched) and leaves (open) in soil/char mixtures. (b) Water use efficiency (mg yield produced g^{-1} water consumed) and relative leaf chlorophyll content at the harvest (method description in [137]). Means ± standard deviation, n = 5 pots per treatment with n = 5 plants per pot. Different letters indicate significant differences between treatments (analysis of variance + Student–Newman–Keuls post-hoc test, p < 0.05). BC = biochar; HC = hydrochar; biochars/hydrochars used here are equal to those used in Experiment II in [111]. Peanut = peanut hull residue; wood = wood chip sievings (dominated by needles); maize = maize silage; bark = bark and wood residue; beet = beet-root chip residue.

(in particular, nitrogen) retention with HTC still make it a highly interesting option for closing nutrient cycles, hygienization of wet wastes, and conversion of waste streams into fertilizers. Hydrochar may not come as a ready-made tool, but demand that some effort is put into its development into a suitable agricultural management strategy. There is simply not enough known on the effects of hydrochar on plant growth to draw any general conclusions so far – more research is urgently required to identify risks, as well as benefits.

8.3.5   Greenhouse Gas Emissions from Char-Amended Soils

Quantification of greenhouse gas emissions (CO_2, N_2O, and CH_4) following char application to soils is crucial because any positive carbon sequestration effect could become negligible if the emissions of other potent greenhouse gases increase after char application. For CO_2 emissions from soils, the possibility of "priming" of old SOC as been discussed earlier. Hence, this section mainly focuses on N_2O and CH_4 fluxes.

8.3.5.1   Sources of N₂O Emissions from Soils

N_2O is a potent greenhouse gas with a global warming potential of 298 [165]. It is predominantly produced during heterotrophic denitrification of NO_3^- to N_2 as a gaseous intermediate and usually to a lesser extent by nitrification of NH_4^+ to NO_3^- as a byproduct (for reviews, see, e.g., [166, 167]). N_2O emissions from agricultural soils are particularly high after application of nitrogen-rich fertilizer (particular in the presence of labile carbon, e.g., manures, slurries), during the conversion of aerobic soils to anaerobicity, during freeze/thaw cycles in spring, and/or in the presence of urine patches [168].

Neglecting N_2O emissions may cause considerable misinterpretation of the real carbon (i.e., CO_2-equivalent) sink capacity of agricultural [169] or seminatural [170] ecosystems. Reductions of N_2O emissions as a result of char application would considerably improve the greenhouse gas balance of biochar-grown agricultural products.

8.3.5.2   Effects of Char Application to Soils on N₂O Emissions

Rondon *et al.* (cited in [154]) reported reduced N_2O emissions after biochar application. Since then, several more reports of reduced N_2O emissions in the presence of biochar have followed [11]. Others have also observed significant reductions of N_2O emissions with biochar, either without plants or in the presence of plants [111]. In addition we observed reductions in a low-SOC as well as high-SOC soil with two different biochars applied at 4% (w/w) in the presence of earthworms, which are known for their high N_2O production [152]. The effects of hydrochar on N_2O formation in soils have to our knowledge not been investigated so far except for our study [111]. In an incubation study of soil mixed with 8% (w/w) of a bark and beet-root hydrochar, we observed significant reductions of the N_2O emissions in unfertilized soil directly after application up to 3 weeks. However, after stirring (to mimic ploughing) and, in particular, after nitrogen fertilization the N_2O emissions rose considerably and exceeded those of the control soil without amendment; the N_2O emissions with 8% (w/w) additions of four different biochars were significantly reduced (see Experiment II in [111]). However, the effect depends on the chosen comparison system. In two other studies where feedstock, hydrochar, and biochar (made from the same feedstock) were applied at equal rates we always observed the highest N_2O emissions from the feedstock mixtures, followed by the hydrochar mixtures while the N_2O emissions from the biochar mixtures were lower (sometimes significantly) compared to the untreated control. The feedstock and hydrochar N_2O emissions, however, were larger than those from the unamended control [112, 123]. It is essential to compare measured N_2O emissions after hydrochar application to a meaningful reference system and to provide information on the chosen reference system.

Specific mechanisms of changes in N_2O fluxes are not currently well-understood and is a rather under studied phenomenon. Libra *et al.* [11] reviewed several potential mechanisms, including: (i) the reduction in anaerobic microsites in soil, (ii) a change in soil pH, (iii) nitrogen immobilization in soil, (iv) stimulation of plant growth, (v) change in nitrogen transformation pathways in soil, and (vi) chemical reduction of N_2O. Many open questions remain and are detailed by Libra *et al.* [11].

8.3.5.3   *Effect of Biochar Soil Application on CH₄ Emission and Uptake Rates*

As organic matter anaerobically decays, CH_4 production will result. CH_4 oxidation by aerobic methanotrophic bacteria (i.e., CH_4 uptake into the soil) is also an ubiquitous process that has been reported to occur under oxic conditions [171]. Char application to soils may increase the CH_4 sink activity, which would be of global significance (see [11] for additional details). The effect that biochar or hydrochar application may impart on CH_4 production or CH_4 oxidation is unknown. Van Zwieten *et al.* [172] suggests that biochar may have a positive influence on CH_4 emissions. They reported that CH_4 production declined to zero in the presence of biochar in a grass stand and in a soybean field [172]. Zhang *et al.* [173] reported that the CH_4 emissions in rice paddies in China increased by 34% and 41% with 40 ton ha^{-1} biochar addition and with/without nitrogen fertilization; however, in this study the N_2O emissions decreased substantially, while rice yields increased by about 10%. Thus, the overall greenhouse gas emission/yield ratio was improved by biochar use in this south-eastern Chinese rice paddy study, despite the increase in CH_4 emissions, which the authors attributed to labile substances that might have been present initially in the biochar [173], but otherwise they did not have an explanation for the increase.

CH_4 uptake in a poor acidic tropical soil increased by 200 mg CH_4 m^{-2} year^{-1} compared to the controls [172], which is quite a large reduction. Priem and Christensen [174] reported that CH_4 uptake rates declined in a savannah that had recently been burned, but it was not clear if noteworthy amounts of black carbon had formed during burning. Spokas *et al.* [175, 176] reported reductions in the CH_4 uptake rates with some biochars and unaltered rates with others. Karhu *et al.* [136], on the other hand, observed significantly increased (more than 100%) CH_4 uptake rates in a northern Europe agricultural soil where the WHC and also aeration were increased by biochar addition. In a greenhouse study with *Lolium perenne*, we found no negative effects (only in tendency positive) of pure biochar addition to an agricultural soil with low native CH_4 uptake rates [111]. In contrast, in a clayey loam high-SOC grassland soil we observed strong stimulations of the endogenous CH_4 uptake rates – not only with biochar, but also with a hydrochar, compared to an unamended control [112, 123]. Changes in the CH_4 fluxes due to char amendment (such as those outlined above) will depend on gas transport in soil. Libra *et al.* [11] provide a detailed discussion on this topic. Briefly, reduced soil compaction and improved soil aeration may stimulate

CH_4 consumption, since O_2 and CH_4 diffusion are regulated by soil water content [177], which has been defined as a key factor.

8.3.6 Best-Practice Considerations for Biochar/Hydrochar Soil Application

Recommendations on how to apply char to agricultural fields are first emerging from the initial field trials. Pyrochars are mostly dry and their application without wetting can result in considerable losses to the air during the transport, spreading, and incorporation into the soil [158]. This can cause dust clouds, which carry the risk of dust explosions, ignited by a spark. Furthermore, fine biochar dust particles in the air may cause health problems [158]. Major and Husk [149] estimated in a commercial biochar field trial in Canada that about 30% of the char was lost by dust erosion during transport and incorporation into the soil. Another concern is that black carbon fine particles and soot have a considerable greenhouse potential [165] so that dust and aerosols must be avoided.

Therefore, appropriate strategies for dry chars would be to mix them with wetter materials such as compost, liquid manures or slurries, or with water. Subsequently the char must be incorporated into the soil by ploughing, disking, or deep-banded application.

However, hydrochar is wet when it leaves the production process, thus no dust losses should take place when it is used directly out of the production process. Gaijć *et al.* reported that they manually spread two different wet hydrochars (from sugar beet and from beer draff) onto the soil surface; the material had to stay there and dry for 10 days, because the wet hydrochar adhered in clods to the tires of the machinery. Our own observations at the start of a field experiment also showed this "clodding" of *Miscanthus* hydrochar [123], only in this case the material had been dried and ground with a 10-mm mesh to a finer particle size. On the other hand, if a wet hydrochar is not dried we frequently observed fungal growth. It may, therefore, be necessary to find a "water content window" (probably 10–15%) where the hydrochar is neither at risk of aerosol formation nor at risk of quick fungal colonization and degradation. So far, the experience with hydrochar storage, handling, and field application is very limited.

Strategies of mixing the biochar (respectively, hydrochar) with a nutrient-rich substance such as green waste compost, slurry, or manure, or of cocomposting or even animal feeding before field application [178] will have the positive side-effect of "loading" the char with nutrients from the start; such techniques may even be comparable to the strategies that the former inhabitants of Amazonia used when creating terra preta [82, 90].

8.4 HTC Technology: Commercial Status and Research Needs

Recently, HTC units for processing thousands of tons biomass residuals per year have been built or are planned (e.g., in the United States, Germany, and Spain), although the research on the details and dependencies of HTC reactions is still very much in its infancy. This of course is not unusual in most fields searching for pragmatic solutions; in the field of waste solids and liquids, processes for treating wet biomass residuals are continually being sought. These wet organic materials can vary widely in chemical composition, volatile and noncombustible fraction, moisture content, particle size, and energy content. In general, multistep process combinations are required to achieve the desired treatment goals, which range from destruction to resource recovery. Currently, many combinations of thermal, chemical, biological, and mechanical processes are already available for treating wet organic wastes. Most often all four types of processes are combined to reach the required level of treatment, using large amounts of energy and equipment. System complexity has grown over the years as treatment limits are lowered and new target compounds are added. Multiple steps for each target compound are often necessary since they can partition between the solid/liquid/gas phases. For example, process chains in municipal wastewater treatment plants usually consist of more than 15 process steps, requiring large expenditures for personnel, capital costs as well as for energy. The chemical conversion of wet organic material to hydrochar through HTC could substantially change the energy balance in many waste treatment plants as well as produce value-added products.

Conceptually, in the process design, pre- and postprocessing steps surround a central reaction process. A HTC process for treating industrial wastes in an urban setting can take advantage of the pre- and postprocessing steps that are part of an existing municipal infrastructure for the collection and disposal of waste streams. For instance, liquids can be discharged to efficient municipal wastewater treatment plants. However, pre-existing infrastructure can be a disadvantage when its inertia to change hinders the introduction of new collection systems for segregated concentrated waste streams (e.g., ecosanitation, biowaste). In a rural setting without such infrastructure, all necessary pre- and postprocessing steps must be accounted for in the process design and operation. Treatment goals for agricultural residues may also differ – resource recovery or pathogen destruction may be the central treatment goals.

8.4.1   Commercial Status

There has been a high level of research activity and commercial development on HTC over the last decade driven in part by the rediscovery of HTC by some researchers for the production of functional carbonaceous materials [5, 6] and the recognition of its potential as a method for carbon sequestration [13]. Its subsequent popularization [179, 180], combined with research funding programs, has resulted in the formation of several spin-off companies in Europe (e.g., in Germany, Spain, and Switzerland). Legislation propagating sustainable resource-efficient materials flow management and renewable sources for energy production set the stage for the recent R&D activity. For example, in Germany, the Closed Substance Cycle and Waste Management Act (KrW-/AbfG 1994) with its tiered management hierarchy (from prevention, reuse, recycling, material, and energy recovery to disposal) paved the way by requiring waste separation with the goal to make full use of substances and materials bound in wastes [181]. In addition, disposal of wastes with an organic content over 5% TOC in landfills was banned in 2005, requiring the development of alternative treatment methods.

A further boost to HTC development was the passage of renewable energy source legislation (e.g., California's Renewable Energy Program [182], Germany's Renewable Energy Sources Act (Erneuerbare-Energien-Gesetz) [183]). The Renewable Energy Sources Act from 2000 and its subsequent amendments, with its guaranteed feed-in tariffs and connection requirement for electricity produced from renewable sources, has been a strong driving force behind the significant growth in renewable energies in the German electricity sector. Electricity produced from hydrochar from certain feedstocks also qualifies for the guaranteed 20-year payment for produced electricity. An additional driving force for the production of hydrochar in Germany is the Renewable Energy Heat Act (Erneuerbare-Energien-Wärmegesetz) from 2008 [184], which stipulates that owners of new buildings must cover part of their heat supply with renewable energies.

Of the over eight German companies active in HTC process development, most are planning systems to produce hydrochar for energy recovery with throughputs ranging from a few thousand to 50 000 metric ton year^{-1} of biomass (http://www.topagrar.com/news/Neue-Energie-News-Erste-Biokohle-ab-2012-am-Markt-100132.html). The first full-scale HTC reactor system in Germany was built in 2010 by a Swiss company (reactor volume of 14.4 m^3, throughput of 8400 metric ton year^{-1}) to convert wet organic waste from beer production to hydrochar and then electricity (http://www.ava-co2.com/web/media/downloads _EN/medienmitteilungen/Press_Release_EN.pdf). In Spain, up to 2000 metric ton year^{-1} of landscape, yard, and forest cuttings are converted to hydrochar (http://www.ingeliahtc. com/English/plantaHTC.html), while in the United States a roughly 100-fold larger HTC unit (capacity 245 000 metric ton year^{-1}) is processing wastewater sludge from five Californian municipalities in the Los Angeles region as a renewable fuel for a cement kiln (http://www.enertech.com/facilities/sitedevelopments/rcrf.html).

Thus, current economic viability is mainly based on renewable energy source regulations and subsidies. Most companies, however, keep an eye on research in the soil application area or actively supply hydrochar to researchers in order to support evaluation of hydrochar potential for increasing soil fertility and/or carbon sequestration. Reactor types vary from batch or continuous-flow mixed tanks to plug flow tubular reactors. Most are modular in design; some are built for mobility in containers.

8.4.2   General Research Needs

Thus, we see that process development as well as the evaluation of the economic and environmental feasibility of a HTC process is very context- and site-specific. Nevertheless, fundamental and systematic investigations still need to be undertaken to determine the general suitability of pyrolysis and char-based concepts for the various environmental applications discussed in this chapter. In particular, investigations on how process conditions affect the physicochemical characteristics of the conversion products and byproducts as well as their ecotoxicity and fates are required. Parallel research on understanding the effects of hydrochar application on soils must continue to determine which properties are responsible for what interactions. With this systematic knowledge, appropriate process combinations can be designed to produce the desired hydrochar and treat any byproducts.

In addition, scenarios for technical implementation need to be evaluated, which include a comprehensive management concept for the feedstock production and collection, treatment of byproducts and recovery of nutrients, and distribution and use of the char. A comparative life cycle analysis should consider the substantial costs and environmental impacts of the current collection and treatment processes in use today for wet organic wastes.

8.4.2.1   *Knowledge Gaps in Process Design and Operation*

Research has just begun to determine the effects of feedstock and process parameters on process energetics, product distribution between phases, and product quality, qualitatively and quantitatively. For some parameters, qualitative trends of their effect on char yield, characteristics, and product distribution are relatively clear (e.g., reaction temperature), while others cannot be generalized yet, owing to the high variation in reported results (e.g., residence time, concentration of feedstock, and the ratio of liquid/gas volumes). Due to the heterogeneity of biomass residuals as feedstock (in composition and form), the complexity of the reaction mechanisms, and the numerous parameters, a broad data basis is required in order to be able to generalize results and develop process models. Some important questions to be investigated on a wide range of systems (in no particular order) are:

- What is the fate of emerging contaminants (personal care products, pharmaceuticals, nanomaterials, and endocrine-disrupting compounds) found in these waste streams? Are they thermally destroyed during carbonization?
- What is the fate of other, potentially toxic compounds (e.g., polychlorinated biphenyls) sometimes found in these waste streams?
- What is the fate of inorganics during HTC? Of particular importance is the fate of chloride.
- How degradable is the process water? Can it be recycled and reused? Can heat be recovered from it?
- Can a catalyst be added to allow for the controlled manipulation of char properties?
- What are the optimal reaction temperatures and times for different waste streams?

The answers to these questions may not only be very dependent on the feedstock and process parameters, but also on the scale of the reaction system since heat and mass transfer may play important roles in the reaction rates. Unfortunately, the cost and safety precautions required for pressurized HTC reaction systems hamper the ability of researchers to experiment at different scales. Most investigations are made at very small scale in batch reactors ranging from 4 ml to 2 l with particulate feedstock. The rates of heat and mass transfer may change substantially in larger-scale reactors with different geometries and mixing patterns, resulting in different product properties. This should be kept in mind when planning experiments and analyzing results.

Most effects still need to be quantified and physical/mathematical models developed for them. Most effects are relevant for environmental as well as energetic applications, indeed, sometimes inversely. For example, higher reaction temperatures produce chars with a higher carbon content and, therefore, with a HHV. This energetic improvement is usually accompanied by a decrease in char yield, resulting in lower carbon efficiencies that reduce the carbon capture potential of the HTC process. Changes in process parameters may also have further consequences downstream – increasing reaction temperature usually results in an increasing amount of colloidal carbon particles and loss of structural features of the original feedstock [185], affecting further processing steps (e.g., solids separation). Moreover, many questions remain on temperature effects, especially with regard to the fate of organic contaminants in the feed such as pharmaceutically active compounds or production of byproducts.

8.4.2.2   *Knowledge Gaps in Char Characterization*

In order to advance our understanding of processes, products, and applications, comprehensive characterization of the chars produced under the various conditions needs to be carried out and reported. This is an essential step in the search to relate char properties to effects in environmental applications, and requires a concerted effort of key players across disciplines, producers, and users to choose the relevant characteristics that need to be measured and develop testing methods for them. There are several important questions that need to be addressed to begin to understand how to most effectively take advantage of the benefits of HTC-derived char:

- Is the process environmentally friendly (e.g., implications associated with liquid-, solid-, and gas-phase parameters, emerging contaminant fate)?
- Can the resulting char be used for environmental remediation, energy generation, and/or soil augmentation? Do harmful contaminants leach from the hydrochar? What is the chemical stability of the hydrochar?
- From a life cycle assessment perspective, when does it make sense to use HTC within a waste management system?

Communication between char producers and users must be developed in order to exploit the ability to influence char properties in the production process and ensure the quality of products. It is important to note that this is an iterative process, particularly at the current stage of development where much is still to be learned about how feedstock and process conditions determine the product chemical composition and efficiencies of conversion. This is especially critical in biochar applications where short-term process developments can substantially change material properties, while long-term research is required to study char–soil–plant interactions and which properties are responsible for the interactions.

References

(1) Cui, X.J., Antonietti, M., and Yu, S.H. (2006) Structural effects of iron oxide nanoparticles and iron ions on the hydrothermal carbonization of starch and rice carbohydrates. *Small*, **2**, 756–759.
(2) Demir-Cakan, R. *et al.* (2009) Carboxylate-rich carbonaceous materials via one-step hydrothermal carbonization of glucose in the presence of acrylic acid. *Chemistry of Materials*, **21**, 484–490.
(3) Fang, Z. *et al.* (2006) CTAB-assisted hydrothermal synthesis of Ag/C nanostructures. *Nanotechnology*, **17**, 3008–3011.
(4) Qian, H.S. *et al.* (2006) Synthesis of uniform Te@carbon-rich composite nanocables with photoluminescence properties and carbonaceous nanofibers by the hydrothermal carbonization of glucose. *Chemistry of Materials*, **18**, 2102–2108.
(5) Yu, S.H. *et al.* (2004) From starch to metal/carbon hybrid nanostructures: hydrothermal metal-catalyzed carbonization. *Advanced Materials*, **16**, 1636–1640.
(6) Wang, Q. *et al.* (2001) Monodispersed hard carbon spherules with uniform nanopores. *Carbon*, **39**, 2211–2214.
(7) Chang, Y.H., Chen, W.C., and Chang, N.B. (1998) Comparative evaluation of RDF and MSW incineration. *Journal of Hazardous Materials*, **58**, 33–45.
(8) Sevilla, M., Fuertes, A.B., and Mokaya, R. (2011) High density hydrogen storage in superactivated carbons from hydrothermally carbonized renewable organic materials. *Energy & Environmental Science*, **4**, 1400–1410.

(9) Berge, N.D. *et al.* (2011) Hydrothermal carbonization of municipal waste streams. *Environmental Science & Technology*, **45**, 5696–5703.

(10) Funke, A. and Ziegler, F. (2010) Hydrothermal carbonization of biomass: a summary and discussion of chemical mechanisms for process engineering. *Biofuels Bioproducts & Biorefining*, **4**, 160–177.

(11) Libra, J.A. *et al.* (2011) Hydrothermal carbonization of biomass residuals: a comparative review of the chemistry, processes and applications of wet and dry pyrolysis. *Biofuels*, **2**, 71–106.

(12) Ramke, H.G. *et al.* (2009) Hydrothermal carbonization of organic waste. Twelfth International Waste Management and Landfill Symposium, Sardinia.

(13) Titirici, M.-M., Thomas, A., and Antonietti, M. (2007) Back in the black: hydrothermal carbonization of plant material as an efficient chemical process to treat the CO_2 problem? *New Journal of Chemistry*, **31**, 787–789.

(14) Sun, K. *et al.* (2011) Sorption of bisphenol A, 17 α-ethnyl estradiol and phenanthrene on thermally and hydrothermally produced biochars. *Bioresource Technology*, **102**, 5757–5763.

(15) Berge, N.D. (2011) Hydrothermal carbonization of municipal solid waste: comparison to current waste management techniques and associated environmental implications. Thirteenth International Waste Management and Landfill Symposium, Cagliari.

(16) Lehmann, J. and Joseph, S. (2009) *Biochar for Environmental Management – Science and Technology*, Earthscan, London.

(17) Schuhmacher, J.P., Huntjens, F.J., and van Krevelen, D.W. (1960) Chemical structure and properties of coal XXVI – studies on artificial coalification. *Fuel*, **39**, 223–234.

(18) US Energy Information Administration (2009) *Net Generation by Energy Source: Total (Electric Power Monthly)*. USEIA, Washington, DC. Available from: http://www.eia.doe.gov.

(19) US Environmental Protection Agency (2012) *Inventory of US Greenhouse Gas Emissions and Sinks: 1990–2012*, USEPA, Washington, DC.

(20) Holm, J.V. *et al.* (1995) Occurrence and distribution of pharmaceutical organic-compounds in the groundwater downgradient of a landfill (Grindsted, Denmark) – response. *Environmental Science & Technology*, **29**, 3074–3074.

(21) Kjeldsen, P. *et al.* (2002) Present and long-term composition of MSW landfill leachate: a review. *Critical Reviews in Environmental Science and Technology*, **32**, 297–336.

(22) Yamamoto, T. *et al.* (2001) Bisphenol A in hazardous waste landfill leachates. *Chemosphere*, **42**, 415–418.

(23) Burkholder, J. *et al.* (2007) Impacts of waste from concentrated animal feeding operations on water quality. *Environmental Health Perspectives*, **115**, 308–312.

(24) Halling-Sorensen, B. *et al.* (1998) Occurrence, fate and effects of pharmaceutical substances in the environment – a review. *Chemosphere*, **36**, 357–394.

(25) Gomez, M.J. *et al.* (2007) Pilot survey monitoring pharmaceuticals and related compounds in a sewage treatment plant located on the Mediterranean coast. *Chemosphere*, **66**, 993–1002.

(26) Jones, O.A.H., Voulvoulis, N., and Lester, J.N. (2005) Human pharmaceuticals in wastewater treatment processes. *Critical Reviews in Environmental Science and Technology*, **35**, 401–427.

(27) Hwang, I.-H. *et al.* (2010) Recovery of solid fuel from municipal solid waste using hydrothermal treatment. Third International Symposium on Energy from Biomass and Waste, Venice.

(28) Muthuraman, M., Namioka, T., and Yoshikawa, K. (2010) A comparison of co-combustion characteristics of coal with wood and hydrothermally treated municipal solid waste. *Bioresource Technology*, **101**, 2477–2482.

(29) Vesilind, P.A., Worrell, W., and Reinhart, D.R. (2002) *Solid Waste Engineering*, Brooks/Cole, Pacific Grove, CA.

(30) Tchobanoglous, G., Theisen, H., and Vigil, S. (1993) *Integrated Solid Waste Management: Engineering Principles and Management Issues*, McGraw Hill, New York.

(31) Barlaz, M.A. (1998) Carbon storage during biodegradation of municipal solid waste components in laboratory-scale landfills. *Global Biogeochemical Cycles*, **12**, 373–380.

(32) Luo, S.Y. *et al.* (2010) Effect of particle size on pyrolysis of single-component municipal solid waste in fixed bed reactor. *International Journal of Hydrogen Energy*, **35**, 93–97.

(33) Caputo, A.C. and Pelagagge, P.M. (2002) RDF production plants: I. Design and costs. *Applied Thermal Engineering*, **22**, 423–437.

(34) Buah, W.K., Cunliffe, A.M., and Williams, P.T. (2007) Characterization of products from the pyrolysis of municipal solid waste. *Process Safety and Environmental Protection*, **85**, 450–457.

(35) US Environmental Protection Agency (2009) *Municipal Solid Waste Generation, Recycling, and Disposal in the United States: Facts and Figures for 2008*, USEPA, Washington, DC.

(36) Bogner, J. *et al.* (2007) Waste management, in *Climate Change 2007: Mitigation. Contribution of Working Group III to the Fourth Assessment Report of the Intergovernmental Panel on Climate Change* (eds B. Metz *et al.*), Cambridge University Press, Cambridge.

(37) Troschinetz, A.M. and Mihelcic, J.R. (2009) Sustainable recycling of municipal solid waste in developing countries. *Waste Management*, **29**, 915–923.

(38) Periathamby, A., Hamid, F.S., and Khidzir, K. (2009) Evolution of solid waste management in Malaysia: impacts and implications of the solid waste bill, 2007. *Journal of Material Cycles and Waste Management*, **11**, 96–103.

(39) US Environmental Protection Agency (2008) Land, in *EPA's Report on the Environment*, USEPA, Washington, DC.

(40) Hockett, D., Lober, D.J., and Pilgrim, K. (1995) Determinants of per-capita municipal solid-waste generation in the southeastern United-States. *Journal of Environmental Management*, **45**, 205–217.

(41) Moqsud, M.A. and Hayashi, S. (2006) An evaluation of solid waste management practice in Japan. *Daffodil International University Journal of Science and Technology*, **1**, 39–44.

(42) European Environment Agency (2011) Diverting waste from landfill: Effectiveness of waste-management policies in the European Union. EEA Report 7/2009. EEA, Copenhagen.

(43) An, D.W. *et al.* (2006) Low-temperature pyrolysis of municipal solid waste: influence of pyrolysis temperature on the characteristics of solid fuel. *International Journal of Energy Research*, **30**, 349–357.

(44) Baggio, P. *et al.* (2008) Energy and environmental analysis of an innovative system based on municipal solid waste (MSW) pyrolysis and combined cycle. *Applied Thermal Engineering*, **28**, 136–144.

(45) Bhuiyan, M.N.A. *et al.* (2010) Pyrolysis kinetics of newspaper and its gasification. *Energy Sources A – Recovery Utilization and Environmental Effects*, **32**, 108–118.

(46) Cheng, H.F. and Hu, Y.N. (2010) Municipal solid waste (MSW) as a renewable source of energy: current and future practices in China. *Bioresource Technology*, **101**, 3816–3824.

(47) Garcia, A.N., Esperanza, M.M., and Font, R. (2003) Comparison between product yields in the pyrolysis and combustion of different refuse. *Journal of Analytical and Applied Pyrolysis*, **68–69**, 577–598.

(48) He, M.Y. *et al.* (2010) Syngas production from pyrolysis of municipal solid waste (MSW) with dolomite as downstream catalysts. *Journal of Analytical and Applied Pyrolysis*, **87**, 181–187.

(49) Xiao, G. *et al.* (2009) Gasification characteristics of MSW and an ANN prediction model. *Waste Management*, **29**, 240–244.

(50) Goto, M. *et al.* (2004) Hydrothermal conversion of municipal organic waste into resources. *Bioresource Technology*, **93**, 279–284.

(51) Lu, L., Namioka, T., and Yoshikawa, K. (2011) Effects of hydrothermal treatment on characteristics and combustion behaviors of municipal solid wastes. *Applied Energy*, **88**, 3659–3664.

(52) Onwudili, J.A. and Williams, P.T. (2007) Hydrothermal catalytic gasification of municipal solid waste. *Energy & Fuels*, **21**, 3676–3683.

(53) Gollehon, N. *et al.* (2001) Confined Animal Production and Manure Nutrients (Agriculture Information Bulletin AIB-771), ERS/USDA, Washington, DC.

(54) McNab, W.W. Jr. *et al.* (2007) Assessing the impact of animal waste lagoon seepage on the geochemistry of an underlying shallow aquifer. *Environmental Science and Technology*, **41**, 753.

(55) Stone, K.C. *et al.* (1998) Impact of swine waste application on ground and stream water quality in an eastern coastal plain watershed. *Transactions of the American Society of Agricultural Engineers*, **41**, 1665–1670.

(56) Szogi, A.A., Vanotti, M.B., and Stansbery, A.E. (2006) Reduction of ammonia emissions from treated anaerobic swine lagoons. *Transactions of the American Society of Agricultural Engineers*, **49**, 217–225.

(57) Zhang, R.H. and Westerman, P.W. (1997) Solid–liquid separation of animal manure for odor control and nutrient management. *Applied Engineering in Agriculture*, **13**, 657–664.

(58) Ro, K.S. *et al.* (2007) Catalytic wet gasification of municipal and animal wastes. *Industrial & Engineering Chemistry Research*, **46**, 8839–8845.

(59) Ro, K.S., Cantrell, K.B., and Hunt, P.G. (2010) High-temperature pyrolysis of blended animal manures for producing renewable energy and value-added biochar. *Industrial & Engineering Chemistry Research*, **49**, 10125–10131.

(60) Sweeten, J.M. *et al.* (2006) Combustion-fuel properties of manure or compost from paved vs. un-paved cattle feedlots. ASABE Meeting, Portland, OR, presentation paper 06-4143.

(61) Chastain, J.P. *et al.* (2001) Removal of solids and major plant nutrients from swine manure using a screw press separator. *Applied Engineering in Agriculture*, **17**, 355–363.

(62) Chastain, J.P., Vanotti, M.B., and Wingfield, M.M. (2001) Effectiveness of liquid–solid separation for treatment of flushed dairy manure: a case study. *Applied Engineering in Agriculture*, **17**, 343–354.

(63) He, B.J. *et al.* (2000) Thermochemical conversion of swine manure: an alternative process for waste treatment and renewable energy production. *Transactions of the American Society of Agricultural Engineers*, **43**, 1827–1833.

(64) Young, L. and Pian, C.C.P. (2003) High-temperature, air-blown gasification of dairy-farm wastes for energy production. *Energy*, **28**, 655–672.

(65) Cao, X. *et al.* (2011) Chemical structures of swine-manure chars produced under different carbonization conditions investigated by advanced solid-state ^{13}C nuclear magnetic resonance (NMR) spectroscopy. *Energy Fuels*, **25**, 388–397.

(66) Baccile, N. *et al.* (2009) Structural characterization of hydrothermal carbon spheres by advanced solid-state MAS ^{13}C NMR investigations. *Journal of Physical Chemistry*, **113**, 9644–9654.

(67) Titirici, M.-M., Antionietti, M., and Baccile, N. (2008) Hydrothermal carbon from biomass: a comparison of the local structure from poly- to monosaccharides and pentoses/hexoses. *Green Chemistry*, **10**, 1204–1212.

(68) Babel, S. and Kurniawan, T.A. (2003) Low-cost adsorbents for heavy metals uptake from contaminated water: a review. *Journal of Hazardous Materials*, **97**, 219–243.

(69) Mohan, D. and Pittman, J. (2007) Arsenic removal from water/wastewater using adsorbents-a critical review. *Journal of Hazardous Materials*, **142**, 1–53.

(70) Crini, G. (2006) Non-conventional low-cost adsorbents for dye removal: a review. *Bioresource Technology*, **97**, 1061–1085.

(71) Radovic, L.R. (2004) *Chemistry and Physics of Carbon*, CRC Press, Boca Raton, FL.

(72) Lima, I.M., Boateng, A.A., and Klasson, K.T. (2009) Pyrolysis of broiler manure: char and product gas characterization. *Industrial & Engineering Chemistry Research*, **48**, 1292–1297.

(73) Lima, I.M., McAloon, A., and Boateng, A.A. (2008) Activated carbon from broiler litter: process description and cost of production. *Biomass and Bioenergy*, **32**, 568–572.

(74) Lima, I.M. and Marshall, W.E. (2005) Granular activated carbons from broiler manure: physical, chemical and adsorptive properties. *Bioresource Technology*, **96**, 699–706.

(75) Hoekman, S.K., Broch, A., and Robbins, C. (2011) Hydrothermal carbonization (HTC) of lignocellulosic biomass. *Energy and Fueld*, **25**, 1802–1810.

(76) Murugan, A.V., Muraliganth, T., and Manthiram, A. (2009) One-pot microwave-hydrothermal synthesis and characterization of carbon-coated LiMPO₄ (M = Mn, Fe and Co) cathodes. *Journal of the Electrochemical Society*, **156**, A79–A83.

(77) Paraknowitsch, J.P., Thomas, A., and Antionietti, M. (2009) Carbon colloids prepared by hydrothermal carbonization as efficient fuel for indirect carbon fuel cells. *Chemistry of Materials*, **21**, 1170–1172.

(78) Strobel, R. *et al.* (2006) Hydrogen storage by carbon materials. *Journal of Power Sources*, **159**, 781–801.

(79) Antolini, E. (2003) Formation, microstructural characteristics and stability of carbon supported platinum catalysts for low temperature fuel cells. *Journal of Material Science*, **38**, 2995–3005.

(80) Sombroek, W.G. (ed.) (1966) *Amazon Soils: A Reconnaissance of the Soils of the Brazilian Amazon Region*, Verslagen van Landbouwkundige Onderzoekingen, Wageningen.

(81) Marris, E. (2006) Putting the carbon back: black is the new green. *Nature*, **442**, 624–626.

(82) Glaser, B., Lehmann, J., and Zech, W. (2002) Ameliorating physical and chemical properties of highly weathered soils in the tropics with charcoal – a review. *Biology and Fertility of Soils*, **35**, 219–230.

(83) Steiner, C. *et al.* (2007) Long term effects of manure, charcoal and mineral fertilization on crop production and fertility on a highly weathered Central Amazonian upland soil. *Plant and Soil*, **291**, 275–290.

(84) Glaser, B. (2007) Prehistorically modified soils of central Amazonia: a model for sustainable agriculture in the twenty-first century. *Philosophical Transactions of the Royal Society B*, **362**, 187–196.

(85) Ogawa, M. (1991) The earth's green saved with charcoal. *Nihon Keizai Shinbun (Newspaper of Japanese Economics, Nikkei)*, 8 January.

(86) Okimori, Y., Ogawa, M., and Takahashi, F. (2003) Potential of CO₂ emission reductions by carbonizing biomass waste from industrial tree plantation in South Sumatra, Indonesia. *Mitigation and Adaptation Strategies for Global Change*, **8**, 261–280.

(87) Lehmann, J. (2007) A handful of carbon. *Nature*, **447**, 143–144.

(88) Ogawa, M. and Okimori, Y. (2010) Pioneering works in biochar research, Japan. *Australian Journal of Soil Research*, **48**, 489–500.

(89) Neves, E.G. *et al.* (2003) Historical and socio-cultural origins of Amazonian Dark Earths, in *Amazonian Dark Earths: Origin, Properties, Management* (eds J. Lehmann *et al.*), Kluwer, Dordrecht, pp. 29–50.

(90) Glaser, B. *et al.* (2001) The 'Terra Preta' phenomenon: a model for sustainable agriculture in the humid tropics. *Naturwissenschaften*, **88**, 37–41.

(91) Glaser, B. and Birk, J.J. (2012) State of the scientific knowledge on properties and genesis of Anthropogenic Dark Earths in Central Amazonia (terra preta de Índio). *Geochimica et Cosmochimica Acta*, **82**, 39–51.

(92) Heckenberger, M.J. *et al.* (2003) Amazonia 1492: pristine forest or cultural parkland? *Science*, **301**, 1710–1714.

(93) Heckenberger, M.J. *et al.* (2008) Pre-Columbian urbanism, anthropogenic landscapes, and the future of the Amazon. *Science*, **321**, 1214–1217.

(94) Renard, D. *et al.* (2012) Ecological engineers ahead of their time: the functioning of pre-Columbian raised-field agriculture and its potential contributions to sustainability today. *Ecological Engineering*, **45**, 30–44.

(95) Stokstad, E. (2003) Amazon archaeology: 'pristine' forest teemed with people. *Science*, **301**, 1645–1646.

(96) Mann, C.C. (2008) The western Amazon's "garden cities". *Science*, **321**, 1151.

(97) Solomon, D. *et al.* (2007) Molecular signature and sources of biochemical recalcitrance of organic C in Amazonian Dark Earths. *Geochimica et Cosmochimica Acta*, **71**, 2285–2298.

(98) Schmidt, M.W.I. *et al.* (1999) Charred organic carbon in German chernozemic soils. *European Journal of Soil Science*, **50**, 351–365.

(99) Lehmann, J. *et al.* (2008) Australian climate – carbon cycle feedback reduced by soil black carbon. *Nature Geoscience*, **1**, 832–835.

(100) Czimczik, C.I. and Masiello, C.A. (2007) Controls on black carbon storage in soils. *Global Biogeochemical Cycles*, **21**, GB3005.

(101) Lehmann, J. *et al.* (2009) Stability of biochar in soil, in *Biochar for Environmental Management – Science and Technology* (eds J. Lehmann and S. Joseph), Earthscan, London, pp. 183–205.

(102) Forster, P. *et al.* (2007) Changes in atmospheric constituents and in radiative forcing, in *Climate Change 2007: The Physical Science Basis. Contribution of Working Group I to the Fourth Assessment Report of the Intergovernmental Panel on Climate Change* (eds S. Solomon *et al.*), Cambridge University Press, Cambridge, pp. 129–234.

(103) Gaunt, J.L. and Lehmann, J. (2008) Energy balance and emissions associated with biochar sequestration and pyrolysis bioenergy production. *Environmental Science & Technology*, **42**, 4152–4158.

(104) Flora, J.R.V. *et al.* (2010) Hydrothermal carbonization of animal wastes for carbon sequestration and energy generation. 239th American Chemical Society National Meeting, San Francisco, CA.

(105) Joseph, S. *et al.* (2009) Developing a biochar classification and test methods, in *Biochar for Environmental Management – Science and Technology* (eds J. Lehmann and S. Joseph), Earthscan, London, pp. 107–126.

(106) Hammes, K. and Schmidt, M.W.I. (2009) Changes of biochar in soil, in *Biochar for Environmental Management – Science and Technology* (eds J. Lehmann and S. Joseph), Earthscan, London, pp. 169–181.

(107) Schimmelpfennig, S. and Glaser, B. (2012) One step forward toward characterization: some important material properties to distinguish biochars. *Journal of Environmental Quality*, **41**, 1001–1013.

(108) Kuzyakov, Y. *et al.* (2009) Black carbon decomposition and incorporation into soil microbial biomass estimated by ^{14}C labeling. *Soil Biology & Biochemistry*, **41**, 210–219.

(109) Nguyen, B. *et al.* (2008) Long-term black carbon dynamics in cultivated soil. *Biogeochemistry*, **89**, 295–308.

(110) Cheng, C.-H. *et al.* (2008) Stability of black carbon in soils across a climatic gradient. *Journal of Geophysical Research*, **113**, G02027.

(111) Kammann, C. *et al.* (2012) Biochar and hydrochar effects on greenhouse gas (CO_2, N_2O, CH_4) fluxes from soils. *Journal of Environmental Quality*, **41**, 1052–1066.

(112) Ha, M.-K. (2011) *Can Biochar Soil Amendments Improve the Greenhouse-Gas Emission to Yield Ratio?* Justus-Liebig University, Gießen.

(113) Qayyum, M.F. *et al.* (2012) Kinetics of carbon mineralization of biochars compared with wheat straw in three soils. *Journal of Environmental Quality*, **41**, 1210–1220.

(114) van Krevelen, D.W. and Schuyer, J. (1957) *Coal Science*, Elsevier, Amsterdam.

(115) Steinbeiss, S., Gleixner, G., and Antonietti, M. (2009) Effect of biochar amendment on soil carbon balance and soil microbial activity. *Soil Biology and Biochemistry*, **41**, 1301–1310.

(116) Cheng, C.-H. and Lehmann, J. (2009) Ageing of black carbon along a temperature gradient. *Chemosphere*, **75**, 1021–1027.

(117) Cheng, C.-H. *et al.* (2006) Oxidation of black carbon by biotic and abiotic processes. *Organic Geochemistry*, **37**, 1477–1488.

(118) Major, J. *et al.* (2010) Fate of soil-applied black carbon: downward migration, leaching and soil respiration. *Global Change Biology*, **16**, 1366–1379.

(119) Nguyen, B.T. and Lehmann, J. (2009) Black carbon decomposition under varying water regimes. *Organic Geochemistry*, **40**, 846–853.

(120) Brodowski, S. *et al.* (2006) Aggregate-occluded black carbon in soil. *European Journal of Soil Science*, **57**, 539–546.

(121) Wengel, M. *et al.* (2006) Degradation of organic matter from black shales and charcoal by the wood-rotting fungus *Schizophyllum commune* and release of DOC and heavy metals in the aqueous phase. *Science of the Total Environment*, **367**, 383–393.

(122) Rillig, M.C. *et al.* (2010) Material derived from hydrothermal carbonization: Effects on plant growth and arbuscular mycorrhiza. *Applied Soil Ecology*, **45**, 238–242.

(123) Schimmelpfennig, S. and Kammann, C. (2011) GHG-fluxes from a biochar field trial accompanied by an incubation study-preliminary results: could biochar possibly be one solution to anthropogenic caused climate change? European Biochar Symposium, Halle.

(124) Wardle, D.A., Nilsson, M.-C., and Zackrisson, O. (2008) Fire-derived charcoal causes loss of forest humus. *Science*, **320**, 629.

(125) DeLuca, T.H. *et al.* (2006) Wildfire-produced charcoal directly influences nitrogen cycling in ponderosa pine forests. *Soil Science Society of America Journal*, **70**, 448–453.

(126) Ball, P.N. *et al.* (2010) Wildfire and charcoal enhance nitrification and ammonium-oxidizing bacterial abundance in dry montane forest soils. *Journal of Environmental Quality*, **39**, 1243–1253.

(127) Warnock, D. *et al.* (2007) Mycorrhizal responses to biochar in soil – concepts and mechanisms. *Plant and Soil*, **300**, 9–20.

(128) Rillig, M. *et al.* (2007) Role of proteins in soil carbon and nitrogen storage: controls on persistence. *Biogeochemistry*, **85**, 25–44.

(129) Wright, S.F. and Upadhyaya, A. (1998) A survey of soils for aggregate stability and glomalin, a glycoprotein produced by hyphae of arbuscular mycorrhizal fungi. *Plant and Soil*, **198**, 97–107.

(130) Fuertes, A.B. *et al.* (2010) Chemical and structural properties of carbonaceous products obtained by pyrolysis and hydrothermal carbonisation of corn stover. *Australian Journal of Soil Research*, **48**, 618–626.

(131) Titirici, M.-M. and Antonietti, M. (2010) Chemistry and materials options of sustainable carbon materials made by hydrothermal carbonization. *Chemical Society Reviews*, **39**, 103–116.

(132) Mumme, J. *et al.* (2011) Hydrothermal carbonization of anaerobically digested maize silage. *Bioresource Technology*, **102**, 9255–9260.

(133) Sevilla, M., Macia-Agullo, J.A. and Fuertes, A.B. (2011) Hydrothermal carbonization of biomass as a route for the sequestration of CO_2: chemical and structural properties of the carbonized products. *Biomass & Bioenergy*, **35**, 3152–3159.

(134) Rillig, M.C. *et al.* (2010) Material derived from hydrothermal carbonization: effects on plant growth and arbuscular mycorrhiza. *Applied Soil Ecology*, **45**, 238–242.

(135) Laird, D.A. *et al.* (2010) Impact of biochar amendments on the quality of a typical Midwestern agricultural soil. *Geoderma*, **158**, 443–449.

(136) Karhu, K. *et al.* (2011) Biochar addition to agricultural soil increased CH_4 uptake and water holding capacity – results from a short-term pilot field study. *Agriculture Ecosystems & Environment*, **140**, 309–313.

(137) Kammann, C. *et al.* (2011) Influence of biochar on drought tolerance of *Chenopodium quinoa* Willd and on soil–plant relations. *Plant and Soil*, **345**, 195–210.

(138) Oguntunde, P.G. *et al.* (2008) Effects of charcoal production on soil physical properties in Ghana. *Journal of Plant Nutrition and Soil Science/Zeitschrift fur Pflanzenernahrung und Bodenkunde*, **171**, 591–596.

(139) Chan, K.Y. *et al.* (2007) Agronomic values of greenwaste biochar as a soil amendment. *Australian Journal of Soil Research*, **45**, 629–634.

(140) Chan, K.Y. *et al.* (2008) Using poultry litter biochars as soil amendments. *Australian Journal of Soil Research*, **46**, 437–444.

(141) Liang, B. *et al.* (2006) Black carbon increases cation exchange capacity in soils. *Soil Science Society of America Journal*, **70**, 1719–1730.

(142) Singh, B.P. *et al.* (2010) Influence of biochars on nitrous oxide emission and nitrogen leaching from two contrasting soils. *Journal of Environmental Quality*, **39**, 1224–1235.

(143) Ding, Y. *et al.* (2010) Evaluation of biochar effects on nitrogen retention and leaching in multi-layered soil columns. *Water, Air & Soil Pollution*, **213**, 47–55.

(144) Steiner, C. *et al.* (2008) Nitrogen retention and plant uptake on a highly weathered central Amazonian Ferralsol amended with compost and charcoal. *Journal of Plant Nutrition and Soil Science*, **171**, 893–899.

(145) Kolb, S.E., Fermanich, K.J., and Dornbush, M.E. (2009) Effect of charcoal quantity on microbial biomass and activity in temperate soils. *Soil Science Society of America Journal*, **73**, 1173–1181.

(146) Steiner, C., Teixeira, M., and Zech, W. (2007) Soil respiration curves as soil fertility indicators in perennial central Amazonian plantations treated with charcoal and mineral or organic fertilisers. *Tropical Science*, **47**, 218–230.

(147) Thies, J.E. and Rillig, M.C. (2009) Characteristics of biochar: biological properties, in *Biochar for Environmental Management: Science and Technology* (eds J. Lehmann and S. Joseph), Earthscan, London, pp. 85–105.

(148) Rondon, M. *et al.* (2007) Biological nitrogen fixation by common beans (*Phaseolus vulgaris* L.) increases with bio-char additions. *Biology and Fertility of Soils*, **43**, 699–708.

(149) Major, J. and Husk, B. (2010) *Commercial Scale Agricultural Biochar Field Trial in Québec, Canada, Over Two Years: Effects of Biochar on Soil Fertility, Biology, Crop Productivity and Quality*, BlueLeaf, Drummondville.

(150) van Zwieten, L. *et al.* (2010) Effects of biochar from slow pyrolysis of papermill waste on agronomic performance and soil fertility. *Plant and Soil*, **327**, 235–246.

(151) Busch, D. *et al.* (2012) Simple biotoxicity tests for evaluation of carbonaceous soil additives: Establishment and reproducibility of four test procedures. *Journal of Environmental Quality*, **41**, 1023–1032.

(152) Augustenborg, C.A. *et al.* (2012) Biochar and earthworm effects on soil nitrous oxide and carbon dioxide emissions. *Journal of Environmental Quality*, **41**, 1203–1209.

(153) Downie, A., Crosky, A., and Munroe, P. (2009) Physical properties of biochar, in *Biochar for Environmental Management – Science and Technology* (eds J. Lehmann and S. Joseph), Earthscan, London, pp. 13–32.

(154) Lehmann, J. (2007) Bio-energy in the black. *Frontiers in Ecology and the Environment*, **5**, 381–387.

(155) Funke, A. and Ziegler, F. (2010) Hydrothermal carbonization of biomass: a summary and discussion of chemical mechanisms for process engineering. *Biofuels, Bioproducts and Biorefining*, **4**, 166–177.

(156) Cao, X. and Harris, W. (2010) Properties of dairy-manure-derived biochar pertinent to its potential use in remediation. *Bioresource Technology*, **101**, 5222–5228.

(157) Gaskin, J.W. *et al.* (2010) Effect of peanut hull and pine chip biochar on soil nutrients, corn nutrient status, and yield. *Agronomy Journal*, **102**, 623–633.

(158) Blackwell, P., Riethmuller, G., and Collins, M. (2009) Biochar application to soil, in *Biochar for Environmental Management: Science and Technology* (eds J. Lehmann and S. Joseph), Earthscan, London, pp. 207–226.

(159) van Zwieten, L. *et al.* (2010) A glasshouse study on the interaction of low mineral ash biochar with nitrogen in a sandy soil. *Australian Journal of Soil Research*, **48**, 569–576.

(160) Kimetu, J. *et al.* (2008) Reversibility of soil productivity decline with organic matter of differing quality along a degradation gradient. *Ecosystems*, **11**, 726–739.

(161) Oguntunde, P.G. *et al.* (2004) Effects of charcoal production on maize yield, chemical properties and texture of soil. *Biology and Fertility of Soils*, **39**, 295–299.

(162) Gajić, A., Koch, H.-J., and Märländer, B. (2011) HTC-Biokohle als Bodenverbesserer – Erste Ergebnisse aus einem Feldversuch mit Zuckerrüben [HTC-biochar as a soil conditioner – first results from a field trial with sugar beet]. Institut für Zuckerrübenforschung, Göttingen.

(163) Gajić, A. and Koch, H.-J. (2011) Sugar beet (*Beta vulgaris* L.) growth reduction caused by hydrochar is related to nitrogen supply. *Journal of Environmental Quality*, **41**, 1067–1075.

(164) Kammann, C. *et al.* (2011) Chancen und Risiken von Biokohle – Forschungsstand dern der Justus-Liebig Universität Gießen [Chances and risks of biochar – state of research at the Justus-Liebig-University Gießen]. ANS-Symposium "Biochar – climate savior or bluff package?", Botanical Garden, Berlin.

(165) IPCC (2007) *Climate Change 2007: The Physical Science Basis. Contribution of Working Group I to the Fourth Assessment Report of the Intergovernmental Panel on Climate Change* (eds S. Solomon *et al.*), Cambridge University Press, Cambridge.

(166) Firestone, M.K. *et al.* (1989) Microbiological basis of NO and N_2O production and consumption in soil, in *Exchange of Trace Gases between Terrestrial Ecosystems and the Atmosphere* (ed. S. Bernhard), John Wiley & Sons, Ltd, Chichester, pp. 7–21.

(167) Granli, T. and Bøckmann, O.C. (1994) Nitrous oxide from agriculture. *Norwegian Journal of Agricultural Sciences*, Suppl. 12, 1–128.

(168) Clough, T.J. *et al.* (2004) Lime and soil moisture effects on nitrous oxide emissions from a urine patch. *Soil Science Society of America Journal*, **68**, 1600–1609.

(169) Crutzen, P.J. *et al.*(2007) N_2O release from agro-biofuel production negates global warming reduction by replacing fossil fuels. *Atmospheric Chemistry and Physics*, **8**, 389–395.

(170) Kammann, C. *et al.* (2008) Elevated CO_2 stimulates N_2O emissions in permanent grassland. *Soil Biology and Biochemistry*, **40**, 2194–2205.

(171) Conrad, R. (2007) Microbial ecology of methanogens and methanotrophs, in *Advances in Agronomy* (ed. D. Sparks), Elsevier, Amsterdam, pp. 1–63.

(172) van Zwieten, L. *et al.* (2009) Biochar and emissions of non-CO_2 greenhouse gases from soil, in *Biochar for Environmental Management – Science and Technology* (eds J. Lehmann and S. Joseph), Earthscan, London, pp. 227–249.

(173) Zhang, A. *et al.* (2010) Effect of biochar amendment on yield and methane and nitrous oxide emissions from a rice paddy from Tai Lake plain, China. *Agriculture, Ecosystems & Environment*, **139**, 469–475.

(174) Priem, A. and Christensen, S. (1999) Methane uptake by a selection of soils in Ghana with different land use. *Journal of Geophysical Research*, **104**, 23617–23622.

(175) Spokas, K.A. *et al.* (2009) Impacts of woodchip biochar additions on greenhouse gas production and sorption/degradation of two herbicides in a Minnesota soil. *Chemosphere*, **77**, 574–581.

(176) Spokas, K.A. and Reicosky, D.C. (2009) Impact of sixteen different biochars on soil greenhouse gas production. *Annals of Environmental Science*, **3**, 179–193.

(177) Castro, M.S. *et al.* (1994) Soil moisture as a predictor of methane uptake by temperate forest soils. *Canadian Journal of Forest Research*, **24**, 1805–1810.

(178) McHenry, M.P. (2010) Carbon-based stock feed additives: a research methodology that explores ecologically delivered C biosequestration, alongside live weights, feed use efficiency, soil nutrient retention, and perennial fodder plantations. *Journal of the Science of Food and Agriculture*, **90**, 183–187.

(179) Röthlein, B. (2006) Zauberkohle aus dem Dampfkochtopf [Magic Coal from a pressure cooker]. *MaxPlanckForschung*, **2**, 20–25.

(180) Wust, C. (2006) Kohle aus dem Kochtopf [Coal from a pot]. *Der Spiegel*, **30**, 24 July. Available from: http://www.spiegel.de/spiegel/0,1518,428493,00.html.

(181) Federal Ministry for the Environment Nature Conservation and Nuclear Safety (2011) Closed-loop waste management: recovering wastes – conserving resources [Brochure]. Available from: http://www.bmu.de/files/pdfs/allgemein/application/pdf/broschuere_kreislaufwirtschaft_en_bf.pdf.

(182) California Energy Commission (2011) California Renewable Energy Overview and Programs. CEE, Sacramento, CA. Available from: http://www.energy.ca.gov/renewables/.

(183) German Renewable Energy Sources Act (Erneuerbare-Energien-Gesetz) (2010) Act on Granting Priority to Renewable Energy Sources of 29 March 2000, last amended by the Act as of 8/11/10. *Federal Law Gazette*, I, 1170.

(184) German Renewable Energy Heat Act (Erneuerbare-Energien-Wärmegesetz)) (2011) Act on the Promotion of Renewable Energies in the Heat Sector of 2008, last amended by article 2 as of 4/12/11. *Federal Law Gazette*, I, 619.

(185) Titirici, M.M. *et al.* (2007) A direct synthesis of mesoporous carbons with bicontinuous pore morphology from crude plant material by hydrothermal carbonization. *Chemistry of Materials*, **19**, 4205–4212.

9

Scale-Up in Hydrothermal Carbonization

Andrea Kruse[1,2], Daniela Baris[1], Nicole Tröger[1], and Peter Wieczorek[3]

[1]*Karlsruhe Institute of Technology (KIT), Institute for Catalysis Research and Technology, Germany*

[2]*University Hohenheim, Institute of Agricultural Engineering, Conversion Technology and Life Cycle Assessment of Renewable Resources, Germany*

[3]*Artec Biotechnologie GmbH, Germany*

9.1 Introduction

Chapter 7 showed that hydrothermal carbonization (HTC)-derived materials have gained a reputation for being useful in a large variety of applications. Among these, soil enrichment (see Chapter 8) and biocoal as CO_2-negative fuel for decentralized energy production require large amounts of HTC-derived materials to be produced. An increase in size of production is a common way to decrease the production costs. If the throughput of a plant is increased by an increase in the volume of the reactor, the pumps, and so on, the investment costs do not increase by the same factor as the volume or capacity. The relationship between size increase, here the ratio of capacities, and investment cost enhancement is given by:

$$I_2 = I_1 \left(\frac{Cap._2}{Cap._1} \right)^{0.67} \tag{9.1}$$

where I_1 is the investment cost of the plant with capacity $Cap._1$ and I_2 is the investment cost of the plant with capacity $Cap._2$ [1]. If the capacity of a plant is, for example, increased by a factor of 10, the investment costs only increase by a factor of 4.7. This effect is visualized

---

Sustainable Carbon Materials from Hydrothermal Processes, First Edition. Edited by Maria-Magdalena Titirici.
© 2013 John Wiley & Sons, Ltd. Published 2013 by John Wiley & Sons, Ltd.

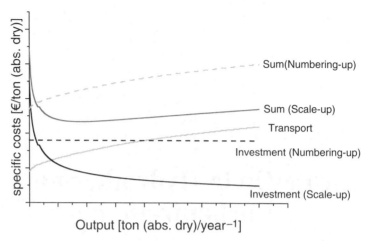

Figure 9.1   *Schematic view of the dependence of specific production cost as a function of production size.*

in Figure 9.1 where the relationship between relative costs and throughput are shown. Here, only the investment is shown, because the consumables increase more or less linearly with throughput. The decrease of the relative costs with throughput leads usually to very large plants (e.g., in the chemical industry). In the case of biomass, especially biomass with a high water content, transport results in significant costs. Large distances are necessary to supply large plants, which limits the size of a plant. This leads to a slight minimum of cost at a certain throughput (Figure 9.1). The other possibility to increase the production is to build more than one plant at the same place. Such a pure "numbering-up" is very unlikely, but in this case the specific cost would be constant, as shown in Figure 9.1

The increase of the throughput by increase of volume, usually called scale-up, is connected with a certain risk. The principle of similarity and scaling rules predict that processes in a large system behave like processes in a small reactor if they are similar. Similar means that the dimensionless numbers like the Reynolds number, Grasshof number, and so on, are identical [2, 3]. In reality, similarity is a goal that often cannot be reached. Also, in many cases, the process is not known well enough to identify all the significant dimensionless ratios that are the basis of the similarity.

A method to avoid the risk of scaling-up is to apply numbering-up (i.e., not the size of a reactor, but the number of reactors is increased). This is the way the company AVA-CO_2 (www.ava-co2.com) wants to increase production (Figure 9.2). Here, the production is increased by adding modules. As illustrated in Figure 9.1, the lower cost decrease with throughput is acceptable if the reactor is not too small and additional positive aspects are considered. These positive aspects are not only the avoided risks of scaling-up, but also avoiding the time and money to do the R&D work for new reactors and other equipment. The company is early in the market, which might be more important than a slightly lower price. In addition, it is not a "pure" numbering-up that is planned, because the infrastructure would be used for all reactors. In fact, the relative costs will be reduced with increasing output and do not stay constant, as shown in Figure 9.1. If transportation is not necessary,

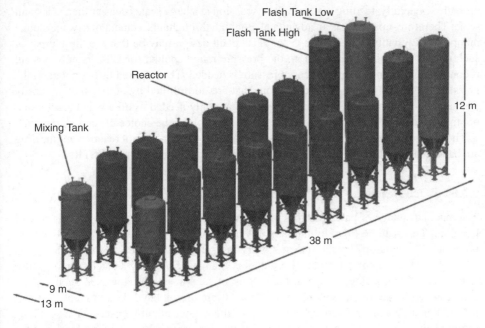

Figure 9.2 *Illustration of the AVA-CO2 concept with two modules, each with six reactors. (Printed with kind permission of AVA-CO2. © 2013 AVA-CO2 Schweiz AG (Source: Web-page AVA-CO2.).)*

because the HTC plant is built near the source of feedstock (e.g., near a food company), the size will be determined by the available flow of biomass for conversion. As shown in Figure 9.1, in this case the plant also should not be too small and the biomass should be available (nearly) the whole year round.

9.2 Basic Aspects of Process Development and Upscaling

Every technical plant consist of a pretreatment process (e.g., milling of the biomass), the reactor, and the separation processes (e.g., separation of HTC coal from water). In lab-scale experiments. the focus is on the reaction; in technical processes, the pretreatment and separation investment and productions cost are usually higher than the costs connected with the reactor.

Concerning the pretreatment, the most important question is what type of biomass has to be processed. Biomass with fibers like grass needs other milling equipment than hard wood without fibers. Biomass from food production or other treated residues might be pumpable without milling, which should also be the case for sewage sludge. The discussion section of this chapter (Section 9.3) shows that there are no physical reasons to reduce the particle size for high-water-content biomass, especially with regard to the relatively high costs. The particle size needed is determined by the feeding system. Here, we need to distinguish between feeding systems to fill a batch reactor and pumps to feed against an increased pressure. Generally, feeding of a particle suspension at a large scale is easier than at a small scale. Therefore, upscaling is an advantage. For high throughputs, cement or sewage sludge pumps work reliably. The challenge for HTC plant design may be that the throughput of such pumps is too high. In addition, the pressure range applied for HTC is not constant if continuous feeding against process pressure is needed. The required throughput and the pressure range leads to working conditions that are not fulfilled by pumps that are easily available on the market. If the feeding systems are only needed to fill a batch reactor and not to work against pressure, a company is more flexible in the choice of the feeding unit.

The second aspect is the reactor choice. The basic reactor concepts of reaction engineering are batch reactors, tubular reactors and continuous stirred tank reactors (CSTRs).

9.2.1   Batch Reactors

The main advantage of batch reactors in view of their application for HTC is their ease of handling. They can be filled by humans or various feeding systems. Therefore, there are no principle limitations for the maximum dry mass content. The reaction time can easily be varied if, for example, the type of biomass is changed. Emptying a batch reactor is also relative easy, by opening a valve at the bottom of such a reactor. However, such a valve can become blocked and therefore large diameters are useful, but difficult to realize. Here, an intelligent design with active feeding out of the reactor might be useful. This opening of the reactor and the connected heat and pressure changes will reduce the lifetime of the reactor. Such frequent heat and pressure changes do not occur at continuous operation, as described below. One important disadvantage of batch reactors is the time needed to fill and to empty such a reactor. These time periods should be as short as possible, because these are periods where money is spent, but no coal is produced. The reaction time of HTC is some hours and therefore this is not so important as in other technical processes with a lower reaction time, but still has to be considered. The best solution to solve this problem is a multibatch concept, as applied by AVA-CO_2. Here, at the same time one reactor is filled, one reactor is emptied and carbonization occurs in the other reactors (Figure 9.2). Wall thickness increases with increasing volume, which reduces heat transfer rates if external heating is used. This would also lead to the choice of a multibatch process rather than one single, large autoclave. In addition, a number of autoclaves are safer than a single, large autoclave because it is very unlikely that all reactors would fail at the same time and the energy content is proportional to the volume (at the same pressure). On the other hand, as discussed in Section 9.1, some small reactors are more expensive than one large reactor. The most important challenge for use of a batch reactor is heat management. Heating from the outside is difficult because a batch reactor has a more disadvantageous surface/volume ratio than a tube. This ratio becomes smaller (i.e., worse) if the volume is increased. A good solution is to heat the reactor from the inside by steam injection and a well-designed injection system is necessary to distribute the steam inside the reactor. Heat management means also the recovery of energy. The only way to recover heat in a batch process is to use the product flow to heat the incoming flow. This is not easy to manage; for example, a common tube-in-tube heat exchanger would be plugged by two different suspensions flowing through it and such a unit is usually only applied for continuous working plants. In the AVA-CO2 concept, the flash tanks have the function to store heat until it is needed to heat-up a reactor that has been just filled with biomass.

9.2.2   Tubular Reactors

If a tubular reactor is applied, then feeding of the suspension against pressure is necessary. The low flow rates as consequence of the high reaction times are challenging. This means that pumps for suspension with low flow rates are difficult to find and that the tubes are easily plugged, especially if the flow direction changes to make the plant as compact as possible. Such challenges and solutions are described in Section 9.5, which highlights some practical applications. The pumpability of biomass slurry differs significantly. In some cases, slurry of more than 10% dry mass content (or in other cases more than 30%) is not pumpable at all, therefore the maximum dry mass content for continuously operating HTC plants is limited. A significant advantage is that no time is necessary for filling or emptying the reactor. In addition, the heat transfer is faster because the surface/volume ratio is higher. In one example, oil flowing around the tubes is used to transport heat from the end to the beginning of the reactor (see Section 9.5). The low reaction heat of the slightly exothermic reaction can also be used. Feeding of a suspension into the reactor is difficult at low flow rates, but feeding out is the more challenging process. Fortunately, HTC coal is (e.g., in the case of food production residue conversion) relative uniform and small in particle size. In addition, it is more brittle than biomass. Thus, HTC coal slurry is easier to handle than biomass slurry. Such a feeding-out is also described in Section 9.5.

9.2.3   CSTRs

As for the tubular reactor, feeding into the reactor has to be made possible; as for a batch reactor, the maximum size is limited because of safety concerns. A significant advantage of a CSTR is the uniform heat distribution inside the reactor. The most important disadvantage of a single CSTR is the residence time distribution. This means that highly converted particles with a high carbon content leave the reactor together with low and nearly unconverted biomass. An interesting concept has been introduced by the company TerraNova Energy (www.terranova-energy.com) [4]. Here, a heat exchanger heats up the biomass with the heat of the product flow. Oil is used for this heat exchange, which reduces significantly the risk of plugging compared with direct heat transfer from one suspension to another. This enables efficient heat recovery [4]. In addition, the heat exchanger tubes act as tubular reactors, upstream und downstream in the CSTR, and this avoids any early unconverted biomass leaving the reactor (Figure 9.3). The focus of TerraNova is on sewage sludge conversion.

9.2.4   Product Handling

As a result of the loss of chemical-bonded oxygen during carbonization, HTC coal is relatively hydrophobic in contrast to biomass. This means that mechanical dewatering leads to a relatively large reduction of the water content. This is particularly important if the HTC coal is burned for energy purposes. In special cases, like sewage sludge, the lower cost of dewatering alone would make HTC economically interesting compared with direct burning of sewage sludge. The type of mechanical dewatering applied depends on the throughput, the manpower costs (location), and the customer's needs. Filter presses, for example, produce plats of pressed HTC coal, which might be difficult to use if the HTC coal need to be spread out on a field for soil improvement. A thermal drying is necessary

Figure 9.3   *Heat exchanger and CSTR of TerraNova Energy. (Reproduced with permission from [4]. © 2011 WILEY-VCH Verlag GmbH & Co. KGaA, Weinheim.)*

to get really dry coal. This avoids molding, but increases the risk of burning or explosion during drying or transport.

9.3   Risks of Scaling-Up

We would like to provide one example to illustrate the challenge of scale-up. If a chemical reaction of a solid occurs, then usually the solid is milled for small-scale lab equipment so as to be able to be fed (e.g., pumped) in to the reactor. However, for a large-scale reactor, feeding of solid is much easier and milling is not necessary. In addition, milling leads to additional costs, which would therefore be avoided. In a small particle, the temperature difference due to heat transfer limitations is smaller than in a large particle. If in a large-scale plant the particles are larger, the temperature in the middle of the particle is lower than in the small-scale equipment. This may have the consequence that the complete reaction takes longer or unwanted reactions occur in the center of the particle. This is a typical challenge of dry biomass conversions like pyrolysis. Usually, if these effects occur, the reaction process shows a significant dependence on the heating rate, because fast heating-up leads to higher temperature differences. In other cases, the particle size simply has no significant influence, such as when there are no heat transfer limitations due to high heat transfer coefficients or other processes are of major influence.

HTC is a process involving a solid, in this case biomass, that has to be heated up in water. The biomass used for HTC usually has a high water content and, therefore, the heat transfer inside the particle should be rather good. It can be assumed that the challenge of heat transfer limitation inside the particle is of minor importance. This could be proven by experimental investigations.

There is also another heat transfer that has to be considered. If a reactor is heated from the outside, a volume increase leads to a worse surface/volume ratio and it is difficult to get the same heating rate inside a large reactor compared to a smaller reactor. An addition, the wall thickness is increased with the external diameter – this also leads to a reduced heat transfer of larger batch reactors compared with smaller reactors and batch reactors compared with tubular reactors:

$$s = \frac{D_a \cdot p}{23\frac{K}{S} - p} \tag{9.2}$$

where s is the necessary wall thickness, D_a is the external diameter, p is the pressure inside, K is the strength value of the material used, and S is the safety factor (e.g., 1.5).

Large batch reactors should be heated from the inside. Therefore, for example, AVA-CO2 heats by the injection of steam, not by a heated reactor surface. The company Artec (www.artec-biotechnologie.com), as another example, uses tubes that are easier to heat from the outside than a large reactor. Anyway, also in this case it is difficult to reach "similar" conditions in such as a way that the dimensionless numbers are identical. On the other hand, HTC also involves different steps like hydrolysis of carbohydrates, water elimination, and polymerization. These reactions likely have different temperature dependencies. Therefore, the heating-up process or how long the mixtures stay at what temperature is important. However, during cooling-down, compounds still solved in the water may show further reaction or be adsorbed on the HTC coal. To determine if such an effect may play a role, experiments were conducted in small reactors that were heated and cooled in different ways. Such small reactors are used because the difference between the internal and external temperature is small.

9.4 Lab-Scale Experiments

9.4.1 Experimental

Small autoclaves have been used to test if the heating rate and the cooling-down procedure have a significant influence on HTC [5–7]. These autoclaves are filled with the feedstock and heated up in a gas chromatography (GC) oven. The GC oven might be cold and is heated-up at a certain heating rate. If the GC oven is already hot, the heating rate inside the autoclave is up to around 25 K min^{-1}. After the reaction time, the autoclave is cooled down. Various different methods have been tested.

After cooling-down, the autoclaves are opened in a closed containment. The amount of gas and its composition is measured. The mixture inside the autoclave is filtered and the solid is dried. The amount and the elementary composition (Elementar Vario EL 3) of the solid product are determined. The liquid is analyzed for furfurals (LiChroCART 250:4 LiChrospher 100 RP-18 column, a water/acrylnitrile eluent mixture, ultraviolet detector) and selected carbonyl compounds (Rezex ROA-Organic Acid H$^+$, 10^{-4} mol l^{-1} H$_2$SO$_4$ as eluent, ultraviolet detector (210 nm) + refractive index detector) by high-performance liquid chromatography. This aqueous phase is also extracted by ethyl acetate and analyzed for selected phenols by GC (Rxi^{TM}-5ms column, flame ionization detector). The selection of compounds is based on earlier studies in which key compounds for different reaction pathways were identified [8–10].

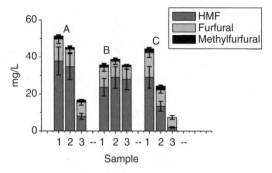

Figure 9.4   *Concentration of HMF, furfural, and methylfurfural after carbonization with various dry matter contents of digestate and various heating rates to 230 °C (4 h reaction time).*

9.4.2   Results and Discussion

For the digestate, the different heating and cooling procedures have no significant influence on the solid composition. In addition, scanning electron microscopy (SEM) images show no visible difference. With regard to the proposed use or clean-up, the composition of the aqueous phase is of special interest. Therefore, the focus here is on key compounds in the process water. Figure 9.4 shows the concentration of hydroxymethylfurfural (HMF), furfural, and methylfurfural at three different dry mass contents and heating procedures. The furfurals are of special interest because, on the one hand, they may polymerize to HTC coal [11, 12] and, on the other hand, they have a significant impact on plant germination (manuscript in preparation). Usually, the concentration of HMF decreases with dry matter content, because for a polymerization a high-order reaction rate can be assumed. This means that the reaction rate is significant higher at increased concentration. A higher reaction rate of polymerization should result in a lower concentration of the monomer. This is observed here, except for the very fast heating-up, where the reaction seems to be less complete. The differences are too small to be measureable (e.g., as solid yield). In addition, Figure 9.5 shows that the polymerization of HMF is affected by the heating rate. Here, HMF is the feedstock, and a high heating rate leads to less sharply formed and less spherical particles. Other compounds like selected acids and an aldehyde, as shown in Figure 9.6, have less significance. In addition, the cooling procedure was also tested. Cooling with water is the normal procedure to stop the reaction. Faster cooling with ice or a slower decrease of temperature shows no significant influence on the concentration of furfurals (Figure 9.7). The influence on phenols is small or not significant (results are not shown).

9.5   Praxis Report

A discussion of the similarity or influence of heating-up might be of more theoretical interest; however, in practical applications other problems like finding the right pump or heating system are dominant. Therefore, a praxis report from Artec applying a tubular reactor for HTC is provided.

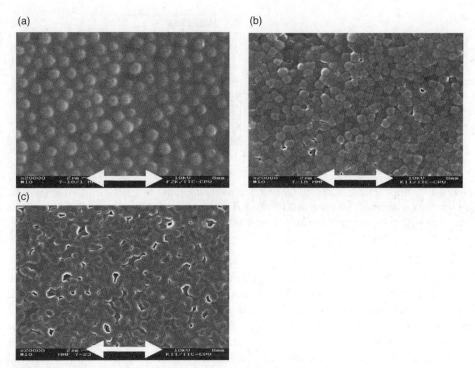

Figure 9.5 *SEM pictures of carbonized HMF (10% dry matter, 50 bar, 2 h at 200 °C) and different heating rates: (a) 1, (b) 2, and (c) around 25 K min⁻¹. The white arrows indicate a length of 2 μm.*

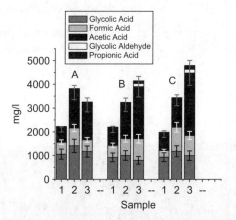

Figure 9.6 *Concentration of gylcolic, formic, acetic, and propionic acids as well as glycolic aldehyde after carbonization with various dry matter contents of digestate and various heating rates to 230 °C (4 h reaction time).*

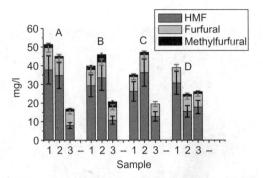

Figure 9.7   *Concentration of HMF, furfural, and methylfurfural after carbonization with various dry matter contents of digestate at 230 °C (4 h reaction time). Different cooling procedures were investigated. Time until the autoclave reaches a temperature of around 30 °C is given in brackets.*

The basic idea was that in a tubular reactor the heating-up, the formation of heat by the reaction, and the cooling-down occur at the same time. The whole process was divided into zones and therefore heat formed in one part of the reactor could be transported to the heating-up zone. Unfortunately, the heat formation was lower than expected, but the systems works reliably by the support of a heater. In addition, Artec states that a tubular reactor is more efficient than a batch reactor because high throughput with low material needs are possible (see also Section 9.2).

The first plant, called MOLE I, is a continuous reactor with a volume of 180 l, and can be used up to 220 °C and 2.4 MPa (Figure 9.8). Reactions times up to 8 h were tested. This plant has run more than 3000 h of operation. Different kind of biomass, including sewage sludge, saw dust, straw and pomace, were successfully converted. The challenges for continuous plants for HTC and how they are solved can now be described:

- Feeding of biomass suspension along a long tube and under pressure was a hurdle to overcome. The biomass is filled in via a funnel. The most important part is a pressure

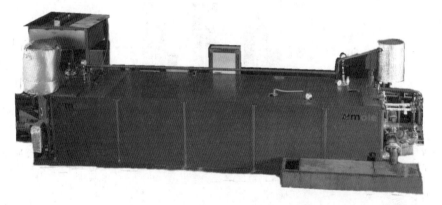

Figure 9.8   *MOLE I: Artec's first plant.*

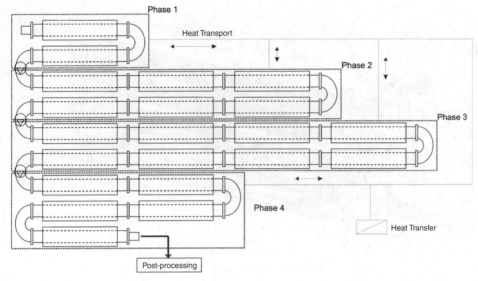

Figure 9.9 *Tube arrangement of MOLE I.*

lock, consisting of two valves, in combination with a piston pump, which the biomass has to pass before entering the reactor. If the first valve is open, biomass is sucked in by the pump. Then this first valve is closed and the second is opened. At this time the pump presses the feedstock into the reactor. At the end of the reactor another piston pump with a two-valve pressure lock, working in the opposite direction, maintains the pressure while the product slurry leaves the reactor.

- When MOLE I was built, it was difficult for Artec to obtain the necessary equipment on the market. For the pressure gate, first slides and then later ball valves are used, which work without problems.
- Space requirements for such a long tube are a challenge. The reactor is calculated for reaction times of up to 8 h, which means it is a tube of 9 m length with an internal diameter of 150 mm. The compact design includes a construction as shown in Figure 9.9. This construction limits the dry mass contents to values between 10 and 13%. At higher dry mass content, the tube is plugged.
- Heat management has to carefully planned. The reactor consists of a double-tube construction and a thermostable oil flows around the reactor tubes. As shown in Figure 9.9, the reactor is divided in four zones corresponding to the four phases of the HTC process. Oil transports the heat formed in zone 2 (and 3) to zone 1. Additional heat for this first oil cycle is supplied by a heater (household oil heating boiler). The cooling in zone 4 is realized by a second oil cycle.

After optimization of MOLE I, a downscaling to a continuous lab-scale plant with a volume of 20 l was the first project conducted. This also has the pressure lock system, but because of the small scale no heat management was installed. Zones 1–3 are heated electrically; zone 4 is simply not insulated and is cooled by the surrounding air. The tube

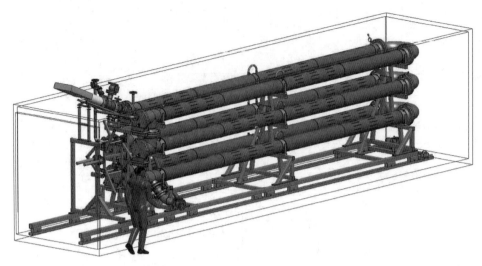

Figure 9.10   *Tube arrangement in Art.coal 3000k.*

has a diameter 120 mm and is 3 m long, which is a different diameter/length ratio than in MOLE I.

Plant Artcoal 3000k has a size that is calculated for economical technical application and is under development by Artec. The volume of 3000 l is calculated for use in small villages or large farms and without the need of biomass transport. The compact design is even more challenging than for MOLE I. The solution is illustrated in Figure 9.10 (diameter 250 mm and tube length 40 m). Tubes with a U-form, not a S-form like in MOLE I, are installed in three layers. The biomass falls from the higher to the lower U-tube as shown in Figure 9.10. In this case active feeding along the tube is necessary, which is assumed to allow higher dry mass contents. The input of biomass will be in a similar way as in MOLE I, with some small variations: the pressure lock valves have an open diameter of 150 mm, which is less expensive than having valves with the same diameter as the tube. In addition, another pump will be needed.

Artec's focus is to limit the investment costs, which thus determines the choice of equipment.

9.6   Conclusions

Different technical solutions for an industrial application of HTC have been discussed. All the plants produce coal with sufficient yield and properties. The choice of technical realization is determined, which was the focus and most important aspect for the companies planning the process. There might be some differences in the process water composition for different technical concepts or scales, but in summary HTC is very easy to handle in view of scale-up.

References

(1) Vogel, G.H. (2005) Process evaluation, in *Process Development: From the Initial Idea to the Chemical Production Plant*, Wiley-VCH GmbH, Weinheim, pp. 329–363.

(2) Zlokarnik, M. (2005) Dimensionsanalyse, in *Scale-Up: Modellübertragung in der Verfahrenstechnik*, Wiley-VCH Verlag GmbH, Weinheim, pp. 3–15.

(3) Zlokarnik, M. (2005) Maßstabsinvarianz des pi-Raumes – Grundlage der Modellübertragung, in *Scale-Up: Modellübertragung in der Verfahrenstechnik*, Wiley-VCH Verlag GmbH, Weinheim, pp. 25–30.

(4) Buttmann, M. (2011) Klimafreundliche Kohle durch Hydrothermale Karbonisierung von Biomasse [Climate friendly coal from hydrothermal carbonization of biomass]. *Chemie Ingenieur Technik*, **83** 1890–1896.

(5) Kruse, A., Badoux, F., Grandl, R., and Wüst, D. (2012) Hydrothermale Karbonisierung: 2. Kinetik der Biertreber-Umwandlung [Hydrothermal carbonization: 2. Kinetics of Draff Conversion]. *Chemie Ingenieur Technik*, **84**, 509–512.

(6) Dinjus, E., Kruse, A., and Tröger, N. (2011) Hydrothermale Karbonisierung: 1. Einfluss des Lignins in Lignocellulosen [Hydrothermal carbonization: 1. Influence of lignin in lignocelluloses]. *Chemie Ingenieur Technik*, **83**, 1734–1741.

(7) Dinjus, E., Kruse, A., and Tröger, N. (2011) Hydrothermal carbonization: 1. Influence of lignin in lignocelluloses. *Chemical Engineering & Technology*, **34**, 2037–2043.

(8) Kruse, A. (2008) Supercritical water gasification. *Biofuels, Bioproducts and Biorefining*, **2**, 415–437.

(9) Sinag, A., Kruse, A., and Rathert, J. (2004) Influence of the heating rate and the type of catalyst on the formation of selected intermediates and on the generation of gases during hydropyrolysis of glucose with supercritical water in a batch reactor. *Industrial & Engineering Chemistry Research*, **43**, 502–508.

(10) Kruse, A., Schwarzkopf, V., and Sinag, A. (2003) Key compounds of the hydropyrolysis of glucose in supercritical water in the presence of K2CO3. *Industrial & Engineering Chemistry Research*, **42**, 3516–3521.

(11) Libra, J.A., Ro, K.S., Kammann, C. *et al.* (2010) Hydrothermal carbonization of biomass residuals: a comparative review of the chemistry, processes and applications of wet and dry pyrolysis. *Biofuels*, **2**, 71–106.

(12) Titirici, M.M., Antonietti, M., and Baccile, N. (2008) Hydrothermal carbon from biomass: a comparison of the local structure from poly- to monosaccharides and pentoses/hexoses. *Green Chemistry*, **10**, 1204–1212.

Index

Sustainable Carbon Materials from Hydrothermal Processes, First Edition. Edited by Maria-Magdalena Titirici.
© 2013 John Wiley & Sons, Ltd. Published 2013 by John Wiley & Sons, Ltd.